LONDON MATHEMATICAL

Managing Editor: C. M. SERIES, Mathematics Institute, University of Warwick, Coventry
CV4 7AL, United Kingdom

London Mathematical Society Student Texts 39

Set Theory for the Working Mathematician

Krzysztof Ciesielski

West Virginia University

CAMBRIDGE
UNIVERSITY PRESS

PUBLISHED BY THE PRESS SYNDICATE OF THE UNIVERSITY OF CAMBRIDGE
The Pitt Building, Trumpington Street, Cambridge CB2 1RP, United Kingdom

CAMBRIDGE UNIVERSITY PRESS
The Edinburgh Building, Cambridge CB2 2RU, United Kingdom
40 West 20th Street, New York, NY 10011-4211, USA
10 Stamford Road, Oakleigh, Melbourne 3166, Australia

First published 1997

Printed in the United States of America

Typeset in Computer Modern

Library of Congress Cataloging-in-Publication Data

Ciesielski, Krzysztof, 1957–

Set theory for the working mathematician / Krzysztof Ciesielski.

p. cm. – (London Mathematical Society student texts : 39)

Includes bibliographical references and index.

ISBN 0-521-59441-3 (hardback). – ISBN 0-521-59465-0 (paperback)

1. Set theory.   I. Title. II. Series.
QA248.C475   1997
511.3′22 – dc21   97-10533
                                                        CIP

A catalog record for this book is available from the British Library.

ISBN 0-521-59441-3 hardback
ISBN 0-521-59465-0 paperback

To Monika, Agata, and Victoria
and to the memory of my Mother, who taught me mathematics

Contents

Part III The power of recursive definitions   77

Part IV When induction is too short   127

Preface

The course presented in this text concentrates on the typical methods of modern set theory: transfinite induction, Zorn's lemma, the continuum hypothesis, Martin's axiom, the diamond principle $\Diamond$, and elements of forcing. The choice of the topics and the way in which they are presented is subordinate to one purpose – to get the tools that are most useful in applications, especially in abstract geometry, analysis, topology, and algebra. In particular, most of the methods presented in this course are accompanied by many applications in abstract geometry, real analysis, and, in a few cases, topology and algebra. Thus the text is dedicated to all readers that would like to apply set-theoretic methods outside set theory.

The course is presented as a textbook that is appropriate for either a lower-level graduate course or an advanced undergraduate course. However, the potential readership should also include mathematicians whose expertise lies outside set theory but who would like to learn more about modern set-theoretic techniques that might be applicable in their field.

The reader of this text is assumed to have a good understanding of abstract proving techniques, and of the basic geometric and topological structure of the n-dimensional Euclidean space $\mathbb{R}^n$. In particular, a comfort in dealing with the continuous functions from $\mathbb{R}^n$ into $\mathbb{R}$ is assumed. A basic set-theoretic knowledge is also required. This includes a good understanding of the basic set operations (union, intersection, Cartesian product of arbitrary families of sets, and difference of two sets), abstract functions (the operations of taking images and preimages of sets with respect to functions), and elements of the theory of cardinal numbers (finite, countable, and uncountable sets.) Most of this knowledge is included in any course in analysis, topology, or algebra. These prerequisites are also discussed briefly in Part I of the text.

The book is organized as follows. Part I introduces the reader to axiomatic set theory and uses it to develop basic set-theoretic concepts. In particular, Chapter 1 contains the necessary background in logic, discusses the most fundamental axioms of ZFC, and uses them to define basic set-theoretic operations. In Chapter 2 the notions of relation, function, and Cartesian product are defined within the framework of ZFC theory. Related notions are also introduced and their fundamental properties are discussed. Chapter 3 describes the set-theoretic interpretation of the sets of natural numbers, integers, and real numbers. Most of the facts presented in Part I are left without proof.

Part II deals with the fundamental concepts of "classical set theory." The ordinal and cardinal numbers are introduced and their arithmetic is developed. A theorem on definition by recursion is proved and used to prove Zorn's lemma. Section 4.4 contains some standard applications of Zorn's lemma in analysis, topology, and algebra.

Part III is designed to familiarize the reader with proofs by transfinite induction. In particular, Section 6.1 illustrates a typical transfinite-induction construction and the diagonalization argument by describing several different constructions of subsets of $\mathbb{R}^n$ with strange geometric properties. The two remaining sections of Chapter 6 introduce the basic elements of descriptive set theory and discuss Borel and Lebesgue-measurable sets and sets with the Baire property. Chapter 7 is designed to help the reader master the recursive-definitions technique. Most of the examples presented there concern real functions, and in many cases consist of the newest research results in this area.

Part IV is designed to introduce the tools of "modern set theory": Martin's axiom, the diamond principle $\diamond$, and the forcing method. The overall idea behind their presentation is to introduce them as natural refinements of the method of transfinite induction. Thus, based on the solid foundation built in Part III, the forcing notions and forcing arguments presented there are obtained as "transformed" transfinite-induction arguments. In particular, the more standard axiomatic approach to these methods is described in Chapter 8, where Martin's axiom and the diamond principle are introduced and discussed. Chapter 9 is the most advanced part of this text and describes the forcing method. Section 9.1 consists of some additional prerequisites, mainly logical, necessary to follow the other sections. In Section 9.2 the main theoretical basis for the forcing theory is introduced while proving the consistency of ZFC and the negation of CH. In Section 9.3 is constructed a generic model for ZFC+$\diamond$ (thus, also for ZFC+CH). Section 9.4 discusses the product lemma and uses it to deduce a few more properties of the Cohen model, that is, the model from Section 9.2. The book finishes with Section 9.5 in which is proved the simultaneous consistency of Martin's axiom and the negation of the continuum hypothesis. This proof, done by iterated forcing, shows that even in the world of the "sophisticated recursion method" of forcing the "classical" recursion technique is still a fundamental method of set theory – the desired model is obtained by constructing forcing extensions by transfinite induction.

It is also worthwhile to point out here that readers with different backgrounds will certainly be interested in different parts of this text. Most advanced graduate students as well as mathematical researchers using this book will almost certainly just skim Part I. The same may be also true for some of these readers for at least some portion of Part II. Part III and the first chapter of Part IV should be considered as the core of this text and are written for the widest group of readers. Finally, the last chapter (concerning forcing) is the most difficult and logic oriented, and will probably be of interest only to the most dedicated readers. It certainly can be excluded from any undergraduate course based on this text.

Part I

Basics of set theory

Chapter 1

Axiomatic set theory

1.1 Why axiomatic set theory?

Essentially all mathematical theories deal with sets in one way or another. In most cases, however, the use of set theory is limited to its basics: elementary operations on sets, fundamental facts about functions, and, in some cases, rudimentary elements of cardinal arithmetic. This basic part of set theory is very intuitive and can be developed using only our "good" intuition for what sets are. The theory of sets developed in that way is called "naive" set theory, as opposed to "axiomatic" set theory, where all properties of sets are deduced from a fixed set of axioms.

Clearly the "naive" approach is very appealing. It allows us to prove a lot of facts on sets in a quick and convincing way. Also, this was the way the first mathematicians studied sets, including Georg Cantor, a "father of set theory." However, modern set theory departed from the "paradise" of this simple-minded approach, replacing it with "axiomatic set theory," the highly structured form of set theory. What was the reason for such a replacement?

Intuitively, a set is any collection of all elements that satisfy a certain given property. Thus, the following **axiom schema of comprehension,** due to Frege (1893), seems to be very intuitive.

> If φ is a property, then there exists a set $Y = \{X : \varphi(X)\}$ of all elements having property φ.

This principle, however, is false! It follows from the following theorem of Russell (1903) known as Russell's antinomy or Russell's paradox.

3

Russell's paradox   There is no set $S = \{X : X \notin X\}$.

The axiom schema of comprehension fails for the formula $\varphi(X)$ defined as "$X \notin X$." To see it, notice that if S had been a set we would have had for every Y

$$Y \in S \Leftrightarrow Y \notin Y.$$

Substituting S for Y we obtain

$$S \in S \Leftrightarrow S \notin S,$$

which evidently is impossible.

This paradox, and other similar to it, convinced mathematicians that we cannot rely on our intuition when dealing with abstract objects such as arbitrary sets. To avoid this trouble, "naive" set theory has been replaced with axiomatic set theory.

The task of finding one "universal" axiomatic system for set theory that agrees with our intuition and is free of paradoxes was not easy, and was not without some disagreement. Some of the disagreement still exists today. However, after almost a century of discussions, the set of ten axioms/schemas, known as the Zermelo–Fraenkel axioms (abbreviated as ZFC, where C stands for the axiom of choice), has been chosen as the most natural. These axioms will be introduced and explained in the next chapters. The full list of these axioms with some comments is also included in Appendix A.

It should be pointed out here that the ZFC axioms are far from "perfect." It could be expected that a "perfect" set of axioms should be *complete*, that is, that for any statement φ expressed in the language of set theory (which is described in the next section) either φ or its negation is a consequence of the axioms. Also, a "good" set of axioms should certainly be *consistent*, that is, should not lead to a contradiction. Unfortunately, we cannot prove either of these properties for the ZFC axioms. More precisely, we do *believe* that the ZFC axioms are consistent. However, if this belief is correct, we can't prove it using the ZFC axioms alone. Does it mean that we should search for a better system of set-theory axioms that would be without such a deficiency? Unfortunately, there is no use in searching for it, since no "reasonable" set of axioms of set theory can prove its own consistency. This follows from the following celebrated theorem of Gödel.

Theorem 1.1.1 (Gödel's second incompleteness theorem) *Let T be a set of axioms expressed in a formal language $\mathcal{L}$ (such as the language of set theory described in Section 1.2) and assume that T has the following "reasonable" properties.*

(1)   *T is consistent.*

(2)   *There is an effective algorithm that decides for an arbitrary sentence of the language $\mathcal{L}$ whether it is in T or not.*

(3)   *T is complicated enough to encode simple arithmetic of the natural numbers.*

Then there is a sentence φ of the language $\mathcal{L}$ that encodes the statement "T is consistent." However, φ is not a consequence of the axioms T.

In other words, Theorem 1.1.1 shows us that for whatever "reasonable" systems of axioms of set theory we choose, we will always have to rely on our intuition for its consistency. Thus, the ZFC axioms are as good (or bad) in this aspect as any other "reasonable" system of axioms.

So what about the completeness of the ZFC axioms? Can we prove at least that much? The answer is again negative and once again it is a common property for all "reasonable" systems of axioms, as follows from another theorem of Gödel.

Theorem 1.1.2 (Gödel's first incompleteness theorem) *Let T be a set of axioms expressed in a formal language $\mathcal{L}$ (such as the language of set theory described in Section 1.2) and assume that T has the following "reasonable" properties.*

(1)   *T is consistent.*

(2)   *There is an effective algorithm that decides for an arbitrary sentence of the language $\mathcal{L}$ whether it is in T or not.*

(3)   *T is complicated enough to encode simple arithmetic of the natural numbers.*

Then there is a sentence φ of the language $\mathcal{L}$ such that neither φ nor its negation $\neg\varphi$ can be deduced from the axioms T.

A sentence φ as in Theorem 1.1.2 is said to be *independent* of the axioms T. It is not difficult to prove that a sentence φ is independent of the consistent set of axioms T if and only if both $T \cup \{\varphi\}$ and $T \cup \{\neg\varphi\}$ are consistent too. Part of this course will be devoted to studying the sentences of set theory that are independent of the ZFC axioms.

The preceding discussion shows that there is no way to find a good complete set of axioms for set theory. On the other hand, we can find a set of axioms that reach far enough to allow encoding of all set-theoretic operations and all classical mathematical structures. Indeed, the ZFC axioms do satisfy this requirement, and the rest of Part I will be devoted to describing such encodings of all structures of interest.

1.2   The language and the basic axioms

Any mathematical theory must begin with undefined concepts. In the case
of set theory these concepts are the notion of a "set" and the relation "is
an element of" between the sets. In particular, we write "$x \in y$" for "x is
an element of y."

The relation $\in$ is primitive for set theory, that is, we do not define it.
All other objects, including the notion of a set, are described by the axioms.
In particular, all objects considered in formal set theory are sets. (Thus,
the word "set" is superfluous.)

In order to talk about any formal set theory it is necessary to specify
first the language that we will use and in which we will express the axioms.
The correct expressions in our language are called *formulas*. The basic formulas are "$x \in y$" and "$z = t$," where x, y, z, and t (or some other variable
symbols) stand for the sets. We can combine these expressions using the
basic logical connectors of negation $\neg$, conjunction $\&$, disjunction $\vee$, implication $\rightarrow$, and equivalence $\leftrightarrow$. Thus, for example, $\neg\varphi$ means "not φ" and
$\varphi\rightarrow\psi$ stands for "φ implies ψ." In addition, we will use two quantifiers:
existential $\exists$ and universal $\forall$. Thus, an expression $\forall x\varphi$ is interpreted as
"for all x formula φ holds." In addition, the parentheses "(" and ")" are
used, when appropriate.

Formally, the formulas can be created only as just described. However,
for convenience, we will also often use some shortcuts. For example, an ex-
pression $\exists x \in A\varphi(x)$ will be used as an abbreviation for $\exists x(x \in A \& \varphi(x))$,
and we will write $\forall x \in A\varphi(x)$ to abbreviate the formula $\forall x(x \in A\rightarrow\varphi(x))$.
Also we will use the shortcuts $x \neq y$, $x \notin y$, $x \subset y$, and $x \not\subset y$, where, for
example, $x \subset y$ stands for $\forall z(z \in x\rightarrow z \in y)$.

Finally, only variables, the relations $=$ and $\in$, and logical symbols already mentioned are allowed in formal formulas. However, we will often
use some other constants. For example, we will write $x = \emptyset$ (x is an empty
set) in place of $\neg\exists y(y \in x)$.

We will discuss ZFC axioms throughout the next few sections as they
are needed. Also, in most cases, we won't write in the main text the
formulas representing the axioms. However, the full list of ZFC axioms
together with the formulas can be found in Appendix A.

Let us start with the two most basic axioms.

Set existence axiom There exists a set: $\exists x(x = x)$.

Extensionality axiom If x and y have the same elements, then x is equal
to y.

The set existence axiom follows from the others. However, it is the
most basic of all the axioms, since it ensures that set theory is not a trivial

theory. The extensionality axiom tells us that the sets can be distinguish only by their elements.

Comprehension scheme (or schema of separation) For every formula $\varphi(s,t)$ with free variables s and t, set x, and parameter p there exists a set $y = \{u \in x \colon \varphi(u,p)\}$ that contains all those $u \in x$ that have the property φ.

Notice that the comprehension scheme is, in fact, the scheme for infinitely many axioms, one for each formula φ. It is a weaker version of Frege's axiom schema of comprehension. However, the contradiction of Russell's paradox can be avoided, since the elements of the new set $y = \{u \in x \colon \varphi(u,p)\}$ are chosen from a fixed *set* x, rather than from an undefined object such as "the class of all sets."

From the set existence axiom and the comprehension scheme used with the formula "$u \neq u$," we can conclude the following stronger version of the set existence axiom.

Empty set axiom There exists the *empty set* $\emptyset$.

To see the implication, simply define $\emptyset = \{y \in x \colon y \neq y\}$, where x is a set from the set existence axiom. Notice that by the extensionality axiom the empty set is unique.

An interesting consequence of the comprehension scheme axiom is the following theorem.

Theorem 1.2.1 *There is no set of all sets.*

Proof If there were a set S of all sets then the following set

$$Z = \{X \in S \colon X \notin X\}$$

would exist by the comprehension scheme axiom. However, with S being the set of all sets, we would have that $Z = \{X \colon X \notin X\}$, the set from Russell's paradox. This contradiction shows that the set S of all sets cannot exist. $\qquad\square$

By the previous theorem all sets do not form a set. However, we sometimes like to talk about this object. In such a case we will talk about a *class* of sets or the set-theoretic *universe*. We will talk about classes only on an intuitive level. It is worth mentioning, however, that the theory of classes can also be formalized similarly to the theory of sets. This, however, is far beyond the scope of this course. Let us mention only that there are other proper classes of sets (i.e., classes that are not sets) that are strictly smaller than the class of all sets.

The comprehension scheme axiom is a conditional existence axiom, that is, it describes how to obtain a set (subset) from another set. Other basic conditional existence axioms are listed here.

Pairing axiom For any a and b there exists a set x that contains a and b.

Union axiom For every family $\mathcal{F}$ there exists a set U containing the *union* $\bigcup \mathcal{F}$ of all elements of $\mathcal{F}$.

Power set axiom For every set X there exists a set P containing the set $\mathcal{P}(X)$ (the *power set*) of all subsets of X.

In particular, the pairing axiom states that for any a and b there exists a set x such that $\{a, b\} \subset x$. Although it does not state directly that there exists a set $\{a, b\}$, the existence of this set can easily be concluded from the existence of x and the comprehension scheme axiom:

$$\{a, b\} = \{u \in x : u = a \lor u = b\}.$$

Similarly, we can conclude from the union and power set axioms that for every sets $\mathcal{F}$ and X there exist the union of $\mathcal{F}$

$$\bigcup \mathcal{F} = \{x : \exists F \in \mathcal{F} \, (x \in F)\} = \{x \in U : \exists F \in \mathcal{F} \, (x \in F)\}$$

and the power set of X

$$\mathcal{P}(X) = \{z : z \subset X\} = \{z \in P : z \subset X\}.$$

It is also easy to see that these sets are defined uniquely. Notice also that the existence of a set $\{a, b\}$ implies the existence of a singleton set $\{a\}$, since $\{a\} = \{a, a\}$.

The other basic operations on sets can be defined as follows: the union of two sets x and y by

$$x \cup y = \bigcup\{x, y\};$$

the difference of sets x and y by

$$x \setminus y = \{z \in x : z \notin y\};$$

the arbitrary intersections of a family $\mathcal{F}$ by

$$\bigcap \mathcal{F} = \left\{z \in \bigcup \mathcal{F} : \forall F \in \mathcal{F} \, (z \in F)\right\};$$

and the intersections of sets x and y by

$$x \cap y = \bigcap\{x, y\}.$$

The existence of sets $x \backslash y$ and $\bigcap \mathcal{F}$ follows from the axiom of comprehension.

We will also sometimes use the operation of *symmetric difference* of two sets, defined by

$$x \triangle y = (x \setminus y) \cup (y \setminus x).$$

Its basic properties are listed in the next theorem. Its proof is left as an exercise.

Theorem 1.2.2 *For every x, y, and z*

(a) $x \triangle y = y \triangle x$,

(b) $x \triangle y = (x \cup y) \setminus (x \cap y)$,

(c) $(x \triangle y) \triangle z = x \triangle (y \triangle z)$.

We will define an ordered pair $\langle a, b \rangle$ for arbitrary a and b by

$$\langle a, b \rangle = \{\{a\}, \{a, b\}\}. \tag{1.1}$$

It is difficult to claim that this definition is natural. However, it is commonly accepted in modern set theory, and the next theorem justifies it by showing that it maintains the intuitive properties we usually associate with the ordered pair.

Theorem 1.2.3 *For arbitrary a, b, c, and d*
$\langle a, b \rangle = \langle c, d \rangle$ *if and only if $a = c$ and $b = d$.*

Proof Implication $\Leftarrow$ is obvious.

To see the other implication, assume that $\langle a, b \rangle = \langle c, d \rangle$. This means that $\{\{a\}, \{a, b\}\} = \{\{c\}, \{c, d\}\}$. In particular, by the axiom of extensionality, $\{a\}$ is equal to either $\{c\}$ or $\{c, d\}$.

If $\{a\} = \{c\}$ then $a = c$. If $\{a\} = \{c, d\}$, then c must belong to $\{a\}$ and we also conclude that $a = c$. In any case, $a = c$ and we can deduce that $\{\{a\}, \{a, b\}\} = \{\{a\}, \{a, d\}\}$. We wish to show that this implies $b = d$.

But $\{a, b\}$ belongs to $\{\{a\}, \{a, d\}\}$. Thus we have two cases.

Case 1: $\{a, b\} = \{a, d\}$. Then $b = a$ or $b = d$. If $b = d$ we are done. If $b = a$ then $\{a, b\} = \{a\}$ and so $\{a, d\} = \{a\}$. But d belongs then to $\{a\}$ and so $d = a$. Since we had also $a = b$ we conclude $b = d$.

Case 2: $\{a, b\} = \{a\}$. Then b belongs to $\{a\}$ and so $b = a$. Hence we conclude that $\{a, d\} = \{a\}$, and as in case 1 we can conclude that $b = d$.
$\square$

Now we can define an ordered triple $\langle a, b, c \rangle$ by identifying it with $\langle \langle a, b \rangle, c \rangle$ and, in general, an ordered n-tuple by

$$\langle a_1, a_2, \dots, a_{n-1}, a_n \rangle = \langle \langle a_1, a_2, \dots, a_{n-1} \rangle, a_n \rangle.$$

The agreement of this definition with our intuition is given by the following theorem, presented without proof.

Theorem 1.2.4 $\langle a_1, a_2, \dots, a_{n-1}, a_n \rangle = \langle a_1', a_2', \dots, a_{n-1}', a_n' \rangle$ *if and only if* $a_i = a_i'$ *for all* $i = 1, 2, \dots, n$.

Next we will define a *Cartesian product* $X \times Y$ as the set of all ordered pairs $\langle x, y \rangle$ such that $x \in X$ and $y \in Y$. To make this definition formal, we have to use the comprehension axiom. For this, notice that for every $x \in X$ and $y \in Y$ we have

$$\langle x, y \rangle = \{\{x\}, \{x, y\}\} \in \mathcal{P}(\mathcal{P}(X \cup Y)).$$

Hence, we can define

$$X \times Y = \{z \in \mathcal{P}(\mathcal{P}(X \cup Y)) \colon \exists x \in X \, \exists y \in Y \, (z = \langle x, y \rangle)\}. \qquad (1.2)$$

The basic properties of the Cartesian product and its relation to other set-theoretic operations are described in the exercises.

The last axiom we would like to discuss in this section is the infinity axiom. It states that there exists at least one infinite set. This is the only axiom that implies the existence of an infinite object. Without it, the family $\mathcal{F}$ of all finite subsets of the set of natural numbers would be a good "model" of set theory, that is, $\mathcal{F}$ satisfies all the axioms of set theory except the infinity axiom.

To make the statements of the infinity axiom more readable we introduce the following abbreviation. We say that y is a successor of x and write $y = S(x)$ if $y = x \cup \{x\}$, that is, when

$$\forall z [z \in y \leftrightarrow (z \in x \lor z = x)].$$

Infinity axiom (Zermelo 1908) There exists an infinite set (of some special form):

$$\exists x \, [\forall z (z = \emptyset {\rightarrow} z \in x) \ \& \ \forall y \in x \forall z (z = S(y) {\rightarrow} z \in x)].$$

Notice that the infinity axiom obviously implies the set existence axiom.

EXERCISES

1 Prove that if $F \in \mathcal{F}$ then $\bigcap \mathcal{F} \subset F \subset \bigcup \mathcal{F}$.

2 Show that for every family $\mathcal{F}$ and every set A

(a)   if $A \subset F$ for every $F \in \mathcal{F}$ then $A \subset \bigcap \mathcal{F}$, and

(b)   if $F \subset A$ for every $F \in \mathcal{F}$ then $\bigcup \mathcal{F} \subset A$.

3 Prove that if $\mathcal{F} \cap \mathcal{G} \neq \emptyset$ then $\bigcap \mathcal{F} \cap \bigcap \mathcal{G} \subset \bigcap (\mathcal{F} \cap \mathcal{G})$. Give examples showing that the inclusion cannot be replaced by equality and that the assumption $\mathcal{F} \cap \mathcal{G} \neq \emptyset$ is essential.

4 Prove Theorem 1.2.2.

5 Show that $\langle \langle a, b \rangle, c \rangle = \langle \langle a', b' \rangle, c' \rangle$ if and only if $\langle a, \langle b, c \rangle \rangle = \langle a', \langle b', c' \rangle \rangle$ if and only if $a = a'$, $b = b'$, and $c = c'$. Conclude that we could define an ordered triple $\langle a, b, c \rangle$ as $\langle a, \langle b, c \rangle \rangle$ instead of $\langle \langle a, b \rangle, c \rangle$.

6 Prove that $X \times Y = \emptyset$ if and only if $X = \emptyset$ or $Y = \emptyset$.

7 Show that for arbitrary sets X, Y, and Z the following holds.

(a) $(X \cup Y) \times Z = (X \times Z) \cup (Y \times Z)$.

(b) $(X \cap Y) \times Z = (X \times Z) \cap (Y \times Z)$.

(c) $(X \setminus Y) \times Z = (X \times Z) \setminus (Y \times Z)$.

8 Prove that if $X \times Z \subset Y \times T$ and $X \times Z \neq \emptyset$ then $X \subset Y$ and $Z \subset T$. Give an example showing that the assumption $X \times Z \neq \emptyset$ is essential.

Chapter 2

Relations, functions, and Cartesian product

2.1 Relations and the axiom of choice

A subset R of a Cartesian product $X \times Y$ is called a *(binary) relation*.

For a relation R of a Cartesian product $X \times Y$, we usually write aRb instead of $\langle a, b \rangle \in R$ and read: *a is in the relation R to b*, or *the relation R between a and b holds*.

The *domain* $\mathrm{dom}(R)$ of a relation R is defined as the set of all x such that $\langle x, y \rangle \in R$ for some $y \in Y$, that is,

$$\mathrm{dom}(R) = \{ x \in X \colon \exists y \in Y \; (\langle x, y \rangle \in R) \};$$

the *range* $\mathrm{range}(R)$ of a relation R is defined as the set of all y such that $\langle x, y \rangle \in R$ for some $x \in X$, that is,

$$\mathrm{range}(R) = \{ y \in Y \colon \exists x \in X \; (\langle x, y \rangle \in R) \}.$$

The set $Z = \mathrm{dom}(R) \cup \mathrm{range}(R)$ for relation R is called a *field* of R. Notice that $R \subset Z \times Z$. In this case we often say that R is defined on Z.

Examples 1. The relation $R_<$ on the set of real numbers $\mathbb{R}$ defined as $\langle x, y \rangle \in R_<$ if and only if $x < y$ is usually denoted by $<$.[1] Notice that

---

[1] In the examples we will often use notions that you supposedly know from other courses, even if we have not yet defined them within the framework of set theory (such as the set of real numbers $\mathbb{R}$ in this example). This will be used only to help you develop the right intuition. We will try to avoid this kind of situation in the main stream of the course.

$R_<$ is the subset of the plane consisting of those points that are above the identity line $y = x$. The domain and the range of this relation are equal to $\mathbb{R}$.

2. Consider the relation R_{div} on the set of natural numbers $\mathbb{N}$ defined as $\langle x, y \rangle \in R_{\mathrm{div}}$ if and only if x and y are different natural numbers greater than 1 and x divides y. Then $\mathrm{dom}(R_{\mathrm{div}}) = \{2, 3, 4, \ldots\}$ and $\mathrm{range}(R_{\mathrm{div}})$ is the set of all composite natural numbers.

Let $R \subset X \times Y$ and $S \subset Y \times Z$. The relation

$$\{\langle y, x \rangle \in Y \times X : xRy\}$$

is called the *inverse* of R and is denoted by R^{-1}. The relation

$$\{\langle x, z \rangle \in X \times Z : \exists y \in Y \ (xRy \ \& \ ySz)\}$$

is called the *composition* of R and S and is denoted by $S \circ R$.

Note that $(R^{-1})^{-1} = R$, $\mathrm{dom}(R^{-1}) = \mathrm{range}(R)$, $\mathrm{range}(R^{-1}) = \mathrm{dom}(R)$, $\mathrm{dom}(S \circ R) \subset \mathrm{dom}(R)$, and $\mathrm{range}(S \circ R) \subset \mathrm{range}(S)$. Moreover, $(S \circ R)^{-1} = R^{-1} \circ S^{-1}$.

Examples 1. If $\leq$ and $\geq$ are defined on $\mathbb{R}$ in the natural way then $(\leq)^{-1} = \geq$.
2. $(\geq) \circ (\leq)$ is equal to the relation $\mathbb{R} \times \mathbb{R}$.
3. $(\leq) \circ (\leq)$ is equal to $\leq$.

Let R be a binary relation on $X \times X$. We say that R is

reflexive if xRx for every $x \in X$;

symmetric if xRy implies yRx for every $x, y \in X$; and

transitive if xRy and yRz imply xRz for every $x, y, z \in X$.

Examples 1. The relations $<$ and $>$ on $\mathbb{R}$ are transitive, but they are neither reflexive nor symmetric.
2. The relations $\leq$ and $\geq$ on $\mathbb{R}$ are transitive and reflexive, but not symmetric.
3. The relation $\neq$ on $\mathbb{R}$ is symmetric, but is neither reflexive nor transitive.
4. The relation $\emptyset \subset \mathbb{R} \times \mathbb{R}$ is symmetric and transitive, but is not reflexive.
5. The relation R on $\mathbb{R}$ defined by xRy if and only if $y = x^2$ has neither of the three properties.
6. The relation $=$ on $\mathbb{R}$ is reflexive, symmetric, and transitive.

A binary relation $R \subset X \times X$ is an *equivalence relation* on X if it is reflexive, symmetric, and transitive. Equivalence relations are usually denoted by symbols such as $\sim$, $\approx$, or $\equiv$.

Examples 1. The relation $=$ on any set X is an equivalence relation.
2. If L is the family of all straight lines on the plane and R is the relation on L of being parallel, then R is an equivalence relation.
3. Let C be the family of all Cauchy sequences $\langle a_n \rangle = \langle a_1, a_2, \ldots \rangle$ of rational numbers. Define relation $\sim$ on C by $\langle a_n \rangle \sim \langle b_n \rangle$ if and only if $\lim_{n \to \infty}(a_n - b_n) = 0$. Then $\sim$ is an equivalence relation.
4. A nonempty family $\mathcal{I} \subset \mathcal{P}(X)$ is said to be an *ideal* on X if for every $A, B \in \mathcal{P}(X)$

$$A \subset B \, \& \, B \in \mathcal{I} \to A \in \mathcal{I} \quad \text{and} \quad A, B \in \mathcal{I} \to A \cup B \in \mathcal{I}.$$

For every ideal $\mathcal{I}$ on a set $\mathcal{P}(X)$ the relation

$$ArB \quad \text{if and only if} \quad A \triangle B \in \mathcal{I}$$

is an equivalence relation.

Let X be a set. A family $\mathcal{F}$ of nonempty subsets of X is said to be a *partition* of X if $\bigcup \mathcal{F} = X$ and sets belonging to $\mathcal{F}$ are pairwise disjoint, that is, for every $A, B \in \mathcal{F}$ either $A = B$ or $A \cap B = \emptyset$.

For an equivalence relation E on X and $x \in X$ the set

$$[x] = \{y \in X : xEy\}$$

is called the *equivalence class* of x (with respect to E). The family

$$\{[x] \in \mathcal{P}(X) : x \in X\}$$

of all equivalence classes for E is denoted by X/E and is called the *quotient class of X with respect to E*.

For a partition $\mathcal{F}$ of a set X let us define a relation $R_{\mathcal{F}}$ on X by

$$xR_{\mathcal{F}}y \Leftrightarrow \exists F \in \mathcal{F} \, (x \in F \, \& \, y \in F).$$

Theorem 2.1.1

(A) *If E is an equivalence relation on X then the family X/E of all equivalence classes forms a partition of X, that is, for every $x, y \in X$ either $[x] = [y]$ or $[x] \cap [y] = \emptyset$.*

(B) *If $\mathcal{F}$ is a partition of X then $R_{\mathcal{F}}$ is an equivalence relation on X. Moreover, $X/R_{\mathcal{F}} = \mathcal{F}$ and $R_{X/E} = E$.*

The proof is left as an exercise.

Theorem 2.1.1 shows that there is a one-to-one correspondence between the class of all partitions of X and the family of all equivalence relations on X. In general, equivalence classes generalize the notion of equality. Elements within one equivalence class are identified by the relation.

An element x of an equivalence class C is called a *representative* of C. A *set of representatives* of an equivalence relation E is a set that contains exactly one element in common with each equivalence class. Notice that, by Theorem 2.1.1 and the axiom of choice, to be stated shortly, a set of representatives exists for every equivalence relation.

Axiom of choice For every family $\mathcal{F}$ of nonempty disjoint sets there exists a *selector*, that is, a set S that intersects every $F \in \mathcal{F}$ in precisely one point.

The axiom of choice (usually abbreviated as AC) has the conditional existence character of the pairing, union, and power set axioms. However, it has also a very different character, since a selector, which exists by this axiom, does not have to be unique. This nonconstructive character of the axiom of choice was, in the past, the reason that some mathematicians (including Borel and Lebesgue) did not like to accept it. However, the discussion on the validity of the axiom of choice has been for the most part resolved today, in favor of accepting it.

The axiom of choice will be one of the most important tools in this course.

EXERCISES

1 Let $R \subset X \times Y$ and $S \subset Y \times Z$ be the relations. Prove that $\operatorname{dom}(R^{-1}) = \operatorname{range}(R)$, $\operatorname{dom}(S \circ R) \subset \operatorname{dom}(R)$, and $\operatorname{range}(S \circ R) \subset \operatorname{range}(S)$.

2 Show the formulas $(R \cup S)^{-1} = R^{-1} \cup S^{-1}$, $(R \cap S)^{-1} = R^{-1} \cap S^{-1}$, and $(S \circ R)^{-1} = R^{-1} \circ S^{-1}$.

3 Prove the formulas

$$(R \cup S) \circ T = (R \circ T) \cup (S \circ T), \quad T \circ (R \cup S) = (T \circ R) \cup (T \circ S),$$

$$(R \cap S) \circ T \subset (R \circ T) \cap (S \circ T), \quad T \circ (R \cap S) \subset (T \circ R) \cap (T \circ S).$$

Find examples of relations R, S, and T that show that the inclusions in the display cannot be replaced with equations.

4 Find examples of relations on $\mathbb{R}$ that are

(a)   reflexive, but neither symmetric nor transitive;

(b)   reflexive and symmetric but not transitive.

5 Prove Theorem 2.1.1.

2.2 Functions and the replacement scheme axiom

A relation $R \subset X \times Y$ is called a *function* if

$$(\forall x, y_1, y_2)(xRy_1 \ \& \ xRy_2 \rightarrow y_1 = y_2).$$

Functions are usually denoted by the letters $f, g, h, \ldots$. The domain $\text{dom}(f)$ and the range $\text{range}(f)$ of a function f are defined as for relations. For a function f if $\text{dom}(f) = X$ and $\text{range}(f) \subset Y$ then f is called a *function* (or *map* or *transformation*) from X into Y and it is denoted by $f \colon X \to Y$. If, moreover, $\text{range}(f) = Y$ then f is said to be a function (map or transformation) from X *onto* Y, or a *surjective function*. The set of all functions from X into Y is denoted by Y^X.

If $f \in Y^X$ and $x \in X$ then there exists precisely one $y \in Y$ such that xfy. The element y is called the *value* of f at x and is denoted by $f(x)$. Thus, the formula $y = f(x)$ has the same meaning as xfy.

Notice that for $f, g \in Y^X$

$$f = g \Leftrightarrow \forall x \in X \ (f(x) = g(x)).$$

A function $f \colon X \to Y$ is a *one-to-one* (or *injective*) *function* if

$$f(x) = f(y) \to x = y$$

for all $x, y \in X$. A function $f \colon X \to Y$ is a *bijection* if it is one-to-one and onto Y.

For $f \colon X \to Y$, $A \subset X$, and $B \subset Y$ we define

$$f[A] = \{f(x) \colon x \in X\} = \{y \in Y \colon \exists x \in X \ (y = f(x))\}$$

and

$$f^{-1}(B) = \{x \in X \colon f(x) \in B\}.$$

We use square brackets in $f[A]$ rather than regular parentheses to avoid a double meaning for the symbol $f(A)$ when A is at the same time an element of X and its subset. A similar double meaning may happen when the symbol $f^{-1}(B)$ is used. However, in this case it will be always clear from the context which meaning of the symbol we have in mind. The sets $f[A]$ and $f^{-1}(B)$ are called the *image* of A and the *preimage* of B with respect to f, respectively.

Theorem 2.2.1 *If $f \in Y^X$ then f^{-1} is a function if and only if f is one-to-one. Moreover, if f^{-1} is a function then $f^{-1} \in X^Z$, where $Z =$ range(f), and f^{-1} is one-to-one.*

Theorem 2.2.2 *If $f \in Y^X$ and $g \in Z^Y$ then $g \circ f$ is also a function and $g \circ f \in Z^X$. Moreover, $(g \circ f)(x) = g(f(x))$ for every $x \in X$.*

Theorem 2.2.3 *Let $f \in Y^X$ and $g \in Z^Y$. If f and g are one-to-one, then so is $g \circ f$. If f and g are onto, then so is $g \circ f$.*

The proofs are left as exercises.

For the proof of the next theorem we need one more axiom scheme.

Replacement scheme axiom (Fraenkel 1922; Skolem 1922) For every formula $\varphi(s, t, U, w)$ with free variables s, t, U, and w, set A, and parameter p if $\varphi(s, t, A, p)$ defines a function F on A by $F(x) = y \Leftrightarrow \varphi(x, y, A, p)$, then there exists a set Y containing the range $F[A]$ of the function F, where $F[A] = \{F(x) \colon x \in A\}$.

As with the comprehension scheme axiom, the replacement scheme axiom is in fact a scheme for infinitely many axioms, one for each formula φ. In conjunction with the comprehension scheme axiom, the replacement scheme axiom implies that for a function defined by formula $F(x) = y \Leftrightarrow \varphi(x, y, A, p)$ on a set A, the range of F exists, since it can be defined:

$$F[A] = \{y \in Y : \exists x \in A \, (y = F(x))\}.$$

Sometimes the replacement scheme axiom is formulated in a stronger version, which states that the set Y existing by the axiom is *equal* to the range of F. It is worth noticing that such a stronger version of the replacement scheme axiom implies the comprehension scheme axiom.

Now we are ready for the next theorem.

Theorem 2.2.4 *If $\mathcal{F}$ is a family of nonempty sets then there is a function $f \colon \mathcal{F} \to \bigcup \mathcal{F}$ such that $f(X) \in X$ for every $X \in \mathcal{F}$.*

Proof Take a formula $\varphi(X, Y)$ that can be abbreviated to

$$Y = \{X\} \times X.$$

Notice that if $\varphi(X, Y)$ and $\varphi(X, Y')$ hold then $Y = \{X\} \times X = Y'$. In particular, φ satisfies the assumptions of the replacement scheme axiom. So there exists a family

$$\mathcal{G} = \{Y : \exists X \in \mathcal{F} \, (\varphi(X, Y))\} = \{\{X\} \times X : X \in \mathcal{F}\}.$$

Notice that elements of $\mathcal{G}$ are nonempty and that they are pairwise disjoint, since

$$(\{X\} \times X) \cap (\{X'\} \times X') = (\{X\} \cap \{X'\}) \times (X \cap X').$$

Thus, by the axiom of choice, there is a selector f for $\mathcal{G}$. Notice that f is the desired function.   □

The function from Theorem 2.2.4 is called a *choice function* for the family $\mathcal{F}$. In fact, the statement that for every family of nonempty sets there exists a choice function is equivalent to the axiom of choice. The statement follows from the axiom of choice by Theorem 2.2.4. To see that the axiom of choice follows from it define a selector S for a family $\mathcal{F}$ of nonempty pairwise-disjoint sets as $S = \text{range}(f)$ when f is a choice function for $\mathcal{F}$.

For a function $f \in Y^X$ and $A \subset X$ the relation $f \cap (A \times Y)$ is a function that belongs to Y^A. It is called the *restriction* of f to the set A and is denoted by $f|_A$.

A function g is said to be an *extension* of a function f if $f \subset g$. We also say that f is a *restriction* of g.

Theorem 2.2.5 *If f and g are functions then g extends f if and only if* $\text{dom}(f) \subset \text{dom}(g)$ *and* $f(x) = g(x)$ *for every* $x \in \text{dom}(f)$.

The proof is left as an exercise.

Functions for which the domain is the set $\mathbb{N}$ of natural numbers are called *(infinite) sequences*. A *(finite) sequence* is any function whose domain is any natural number $n \, (= \{0, 1, \dots, n-1\})$. If a is any sequence, then we usually write a_m in place of $a(m)$. We will sometimes represent such a sequence by the symbol $\langle a_m \rangle$ or $\{a_m\}$, indicating the domain of the sequence if necessary.

EXERCISES

1 Prove Theorem 2.2.1.

2 Prove Theorem 2.2.2.

3 Prove Theorem 2.2.3.

4 Prove Theorem 2.2.5.

2.3 Generalized union, intersection, and Cartesian product

Let F be a function from a nonempty set T into a family of sets $\mathcal{G}$. We will usually write F_t instead of $F(t)$ and $\{F_t : t \in T\}$ or $\{F_t\}_{t \in T}$ for range(F). The family $\mathcal{F} = \{F_t\}_{t \in T}$ is called an *indexed family*, with T being the *index set*.

The following notation will be used for $\mathcal{F} = \{F_t\}_{t \in T}$:

$$\bigcup_{t \in T} F_t = \bigcup \mathcal{F}, \quad \bigcap_{t \in T} F_t = \bigcap \mathcal{F}.$$

In the text these sets will also appear as $\bigcup_{t \in T} F_t$ and $\bigcap_{t \in T} F_t$, respectively.

When the set T is fixed, we will sometimes abbreviate this notation and write $\{F_t\}_t$ or $\{F_t\}$ in place of $\{F_t\}_{t \in T}$, and $\bigcup_t F_t$ and $\bigcap_t F_t$ for $\bigcup_{t \in T} F_t$ and $\bigcap_{t \in T} F_t$. When the index set of the index family $\mathcal{F}$ is a Cartesian product $S \times T$ then we usually denote its elements by F_{st} instead of $F_{\langle s,t \rangle}$ and we say that $\mathcal{F}$ is a *doubly indexed family*. We denote it as $\{F_{st} : s \in S,\ t \in T\}$ or simply $\{F_{st}\}$. In such a case we will write $\bigcup_{s \in S, t \in T} F_{st}$ or $\bigcup_{s,t} F_{st}$ for $\bigcup_{\langle s,t \rangle \in S \times T} F_{st}$ and $\bigcap_{s \in S, t \in T} F_{st}$ or $\bigcap_{s,t} F_{st}$ for $\bigcap_{\langle s,t \rangle \in S \times T} F_{st}$. If $S = T$ we will also write $\bigcup_{s,t \in T} F_{st}$ and $\bigcap_{s,t \in T} F_{st}$.

The following properties are easy to verify:

$$x \in \bigcup_{t \in T} F_t \Leftrightarrow \exists t \in T\ (x \in F_t) \qquad \text{and} \qquad x \in \bigcap_{t \in T} F_t \Leftrightarrow \forall t \in T\ (x \in F_t);$$

$$\bigcup_{t \in T} F_t = F_p = \bigcap_{t \in T} F_t \qquad \text{for } T = \{p\};$$

$$\bigcup_{t \in T} F_t = F_p \cup F_q \quad \text{and} \quad \bigcap_{t \in T} F_t = F_p \cap F_q \qquad \text{for } T = \{p, q\}.$$

It is also easy to see that for any formula $\varphi(t, x)$ and sets T, X

$$\bigcup_{t \in T} \{x \in X : \varphi(t, x)\} = \{x \in X : \exists t \in T\ (\varphi(t, x))\}$$

$$\bigcap_{t \in T} \{x \in X : \varphi(t, x)\} = \{x \in X : \forall t \in T\ (\varphi(t, x))\}.$$

Other properties of these operations are listed in the exercises.

The behavior of generalized union and intersection under the action of image and preimage of a function is described in the next theorem.

Theorem 2.3.1 *If $f \in Y^X$, $\{F_t\}_{t \in T}$ is an indexed family of subsets of X, $\{G_t\}_{t \in T}$ is an indexed family of subsets of Y, $A, B \subset X$, and $C, D \subset Y$ then*

(a) $f\left[\bigcup_{t \in T} F_t\right] = \bigcup_{t \in T} f[F_t];$

(b) $f\left[\bigcap_{t \in T} F_t\right] \subset \bigcap_{t \in T} f[F_t];$

(c) $f[A] \setminus f[B] \subset f[A \setminus B];$

(d) $f^{-1}\left(\bigcup_{t \in T} G_t\right) = \bigcup_{t \in T} f^{-1}(G_t);$

(e) $f^{-1}\left(\bigcap_{t \in T} G_t\right) = \bigcap_{t \in T} f^{-1}(G_t);$

(f) $f^{-1}(A) \setminus f^{-1}(B) = f^{-1}(A \setminus B).$

The proof is left as an exercise.

For an indexed family $\{F_t \colon t \in T\}$ we define its *Cartesian product* by

$$\prod_{t \in T} F_t = \{h \in Z^T : \forall t \in T \ (h(t) \in F_t)\},$$

where $Z = \bigcup_{t \in T} F_t$. If all the sets F_t are identical, $F_t = Y$, then $\prod_{t \in T} F_t = Y^T$. The set Y^T is called a *Cartesian power* of the set Y.

For $t \in T$ the function $p_t \colon \prod_{t \in T} F_t \to F_t$ defined by $p_t(x) = x(t)$ is called the *projection* of $\prod_{t \in T} F_t$ onto F_t.

Remark For T a two-element set $\{a, b\}$ the Cartesian products $\prod_{t \in T} F_t$ and $F_a \times F_b$ are different. (The first one is a set of functions on T, the second one is a set of ordered pairs.) However, there is a natural identification of every element $\{\langle a, x \rangle, \langle b, y \rangle\}$ from $\prod_{t \in T} F_t$ with $\langle x, y \rangle \in F_a \times F_b$. Therefore we will usually identify these products.

Theorem 2.3.2 *A product $\prod_{t \in T} F_t$ of nonempty sets F_t is nonempty.*

Proof The choice function f for the family $\{F_t \colon t \in T\}$, which exists by Theorem 2.2.4, is an element of $\prod_{t \in T} F_t$. ☐

Remark Theorem 2.3.2 easily implies Theorem 2.2.4. Thus its statement is equivalent to the axiom of choice. However, we do not need the axiom of choice to prove Theorem 2.3.2 if either T is finite or $\prod_{t \in T} F_t = Y^T$.

EXERCISES

1 Prove that for every indexed families $\{F_t\}_{t\in T}$ and $\{G_t\}_{t\in T}$

(a) $\quad \bigcap_{t\in T}(F_t \cap G_t) = \bigcap_{t\in T} F_t \cap \bigcap_{t\in T} G_t,$

(b) $\quad \bigcup_{t\in T}(F_t \cup G_t) = \bigcup_{t\in T} F_t \cup \bigcup_{t\in T} G_t,$

(c) $\quad \bigcap_{t\in T} F_t \cup \bigcap_{t\in T} G_t = \bigcap_{s,t\in T}(F_s \cup G_t) \subset \bigcap_{t\in T}(F_t \cup G_t),$ and

(d) $\quad \bigcup_{t\in T} F_t \cap \bigcup_{t\in T} G_t = \bigcup_{s,t\in T}(F_s \cap G_t) \supset \bigcup_{t\in T}(F_t \cap G_t).$

Give examples showing that the inclusions cannot be replaced by equalities.

2 Show that for every indexed family $\{F_t\}_{t\in T}$ and every set A

(a) $\quad A \setminus \bigcap_{t\in T} F_t = \bigcup_{t\in T}(A \setminus F_t)$ and $A \setminus \bigcup_{t\in T} F_t = \bigcap_{t\in T}(A \setminus F_t),$

(b) $\quad \bigcap_{t\in T}(A \cup F_t) = A \cup \bigcap_{t\in T} F_t$ and $\bigcup_{t\in T}(A \cap F_t) = A \cap \bigcup_{t\in T} F_t.$

3 Prove that for every indexed families $\{F_t\}_{t\in T}$ and $\{G_t\}_{t\in T}$

(a) $\quad (\bigcup_{t\in T} F_t) \times (\bigcup_{t\in T} G_t) = \bigcup_{s,t\in T}(F_s \times G_t),$

(b) $\quad (\bigcap_{t\in T} F_t) \times (\bigcap_{t\in T} G_t) = \bigcap_{s,t\in T}(F_s \times G_t).$

4 Show that $\bigcup_{s\in S}\bigcap_{t\in T} F_{st} \subset \bigcap_{t\in T}\bigcup_{s\in S} F_{st}$ for every doubly indexed family $\{F_{st} : s \in S,\ t \in T\}.$

5 Prove Theorem 2.3.1(a), (b), and (c). Show, by giving examples, that the inclusions in parts (b) and (c) cannot be replaced by equality.

6 Prove Theorem 2.3.1(d), (e), and (f).

7 For $r, s \in \mathbb{R}$ let $A_r = [r, r+1]$ and $B_{rs} = [r, s)$. Calculate $\bigcup_{s\leq 0}\bigcap_{r\geq s} A_r$, $\bigcap_{s\leq 0}\bigcup_{r\geq s} A_r$, $\bigcup_{r\leq 0}\bigcap_{s>r} B_{rs}$, and $\bigcap_{r\leq 0}\bigcup_{s>r} B_{rs}.$

2.4   Partial- and linear-order relations

A binary relation R on X is *antisymmetric* if

$$xRy \ \& \ yRx \rightarrow x = y$$

for every $x, y \in X$.

    A relation R on X is a *(partial-)order relation* if it is reflexive, transitive, and antisymmetric. Order relations are usually denoted by the symbols $\leq$,

$\preceq$, or $\ll$. If $\leq$ is a partial-order relation then the ordered pair $\langle X, \leq \rangle$ is called a *partially ordered set* (abbreviated also as *poset*).

Examples 1. The relations $\leq$ and $\geq$ on $\mathbb{R}$ are order relations.
2. For any set X the relation $\subset$ is an order relation on $\mathcal{P}(X)$.
3. The relation $|$ on the set $\{2, 3, 4, \dots\}$ defined by

$$n | m \quad \text{if and only if} \quad n \text{ divides } m$$

is an order relation.

An element $m \in X$ of an ordered set $\langle X, \leq \rangle$ is *minimal* if for every $x \in X$ the condition $x \leq m$ implies $x = m$. Similarly, an element $M \in X$ is *maximal* if for every $x \in X$ the condition $x \geq M$ implies $x = M$.

An element $m \in X$ is the *smallest element* (*least element* or *first element*) in X if $m \leq x$ for every $x \in X$, and $M \in X$ is the *greatest element* (*largest element* or *last element*) in X if $x \leq M$ for every $x \in X$.

Theorem 2.4.1 *Let $\langle X, \leq \rangle$ be a partially ordered set.*

(a) X *can have at most one greatest and one smallest element.*

(b) *The smallest element of X, if it exists, is the only minimal element of X.*

(c) *The greatest element of X, if it exists, is the only maximal element of X.*

Proof (a) If a and b are the smallest elements of X then $a \leq b$ and $b \leq a$. Hence $a = b$. The argument for the greatest element is the same.

(b) If a is the smallest element of X then it is minimal, since the condition $x \leq a$ combined with $a \leq x$, which is true for every x, implies $x = a$. Moreover, if m is minimal, then $a \leq m$, since a is the smallest element, and, by minimality of m, $m = a$.

(c) The argument is the same as in (b). $\square$

Examples 1. $\langle \mathbb{R}, \leq \rangle$ has neither minimal nor maximal elements.
2. $\langle [0, 1], \leq \rangle$ has 0 as the least element and 1 as the last element.
3. $\emptyset$ is the smallest element of $\langle \mathcal{P}(X), \subset \rangle$. X is the greatest element of $\langle \mathcal{P}(X), \subset \rangle$.
4. Let $\langle \{2, 3, 4, \dots\}, | \rangle$ be defined as before. It does not have any maximal element. A number m is minimal in this order if and only if m is a prime number. (Thus $\langle \{2, 3, 4, \dots\}, | \rangle$ has infinitely many minimal elements!)

A relation R on X is *connected* if

$$xRy \lor yRx$$

for every $x, y \in X$. An order relation $\leq$ is called a *linear-order relation* if it is connected. In this case we also say that $\langle X, \leq \rangle$ (or just X) is *linearly ordered*.

Examples 1. $\langle \mathbb{R}, \leq \rangle$ is linearly ordered.
2. $\langle \mathcal{P}(X), \subset \rangle$ is linearly ordered if and only if X has at most one element.
3. $\langle \{2, 3, 4, \ldots\}, | \rangle$ is not linearly ordered, since neither $2|3$ nor $3|2$.
4. The relation $\leq$ on $\mathbb{N}$ is a linear-order relation.

Theorem 2.4.2 *If $\langle X, \leq \rangle$ is linearly ordered then every minimal element in X is the smallest element and every maximal element in X is the greatest element. In particular, linearly ordered sets can have at most one maximal element and at most one minimal element.*

The proof is left as an exercise.

Notice that if $\leq$ is an order relation on a set X and $Y \subset X$ then the relation $\leq \cap (Y \times Y)$ is an order relation on Y. It is called the *restriction* of $\leq$ to Y. We often write $\langle Y, \leq \rangle$ in place of $\langle Y, \leq \cap (Y \times Y) \rangle$.

Notice also that a subset of a partially ordered set is partially ordered and a subset of a linearly ordered set is linearly ordered.

In general, for any partial order denoted by $\leq$ we will write $\geq$ for $(\leq)^{-1}$ and define relations $<$ and $>$ by

$$x < y \Leftrightarrow x \leq y \ \& \ x \neq y$$

and

$$x > y \Leftrightarrow x \geq y \ \& \ x \neq y.$$

EXERCISES

1 Prove that the restriction $\leq \cap (Y \times Y)$ of an order relation $\leq$ on X is an order relation, provided $Y \subset X$. Show that $\langle Y, \leq \rangle$ is linearly ordered if $\langle X, \leq \rangle$ is linearly ordered. Give an example such that $\langle Y, \leq \rangle$ is linearly ordered, while $\langle X, \leq \rangle$ is not.

2 Find all minimal, maximal, greatest, and smallest elements of $\langle \mathcal{F}, \subset \rangle$, where $\mathcal{F} = \{X \subset \mathbb{N} \colon X \text{ is finite}\}$.

3 Prove Theorem 2.4.2.

4 A binary relation $R \subset X \times X$ is said to be a *preorder relation* if it is transitive and reflexive.

Let $\preceq \subset X \times X$ be a preorder relation.

(a) Show that the relation $\sim$ on X defined by

$$x \sim y \Leftrightarrow x \preceq y \ \& \ y \preceq x$$

is an equivalence relation on X.

(b) Define the relation $\leq$ on the family $X/\sim$ of all equivalence classes for $\sim$ by

$$[x] \leq [y] \Leftrightarrow x \preceq y.$$

Show that the relation $\leq$ is well defined and that it is a partial-order relation.

Chapter 3

Natural numbers, integers, and real numbers

From the results of Section 1.2 it is clear that sets such as $\emptyset$, $\{\emptyset\}$, $\{\emptyset, \{\emptyset\}\}$, $\{\{\emptyset\}\}$, and so forth exist. Using the axiom of infinity we can also conclude that we can build similar infinite sets. But how do we construct complicated sets, such as the sets of natural and real numbers, if the only tools we have to build them are the empty set $\emptyset$ and "braces" $\{\cdot\}$? We will solve this problem by identifying the aforementioned objects with some specific sets.

Briefly, we will identify the number 0 with the empty set $\emptyset$, and the other natural numbers will be constructed by induction, identifying n with $\{0, \dots, n-1\}$. The existence of the set $\mathbb{N}$ of all natural numbers is guaranteed by the infinity axiom. The real numbers from the interval $[0, 1]$ will be identified with the set of functions $\{0, 1\}^{\mathbb{N}}$, where an infinite sequence $a \colon \mathbb{N} \to \{0, 1\}$ is identified with the binary expansion of a number, that is, with $\sum_{n \in \mathbb{N}} a(n)/2^{n+1}$. The details of these constructions are described in the rest of this chapter.

3.1 Natural numbers

In this section we will find a set that represents the set $\mathbb{N}$ of natural numbers in our set-theoretic universe. For this, we need to find for each natural number n a set that represents it. Moreover, we will have to show that the class of all such defined natural numbers forms a set.

When picking the natural numbers, we will have to pick also the ordering relation $<$ between them. Essentially, the only relation that we have available for this purpose is the relation $\in$. Thus, we will choose the natural numbers to satisfy the following principle:

$$m < n \text{ if and only if } m \in n.$$

Also, since natural numbers are going to be distinguished by the relation $\in$ of being an element, it seems to be natural to have the following intuitive principle:

Each natural number n should have n elements.

These two principles give us no choice about our definition.

By the second principle we have $0 = \emptyset$.

Now suppose that we have already defined n and want to define $n + 1$. Since $n < n+1$ we have $n \in n+1$. Also, for every $m \in n$ we have $m < n$, so $m < n+1$ and $m \in n+1$. In particular, $n \subset n+1$ and $n \cup \{n\} \subset n+1$. But n has n elements, so $n \cup \{n\}$ has $n + 1$ elements, since $n \notin n$, as $n \not< n$. Therefore $n \cup \{n\} \subset n+1$ and both sets have $n+1$ elements. Thus, $n + 1 = n \cup \{n\}$.

By the foregoing discussion we see that $0 = \emptyset$, $1 = \{0\} = \{\emptyset\}$, $2 = \{0,1\} = \{\emptyset, \{\emptyset\}\}$, $3 = \{0,1,2\} = \{\emptyset, \{\emptyset\}, \{\emptyset, \{\emptyset\}\}\}$, and so forth.

Now, showing that the class of all such numbers forms a set is another problem. We will use for this the axiom of infinity. We will also make sure that the following three principles are satisfied, where $S(n) = n + 1$ is known as a *successor* of n and P stands for some property of natural numbers.

P1 $0 \neq S(n)$ for every $n \in \mathbb{N}$.

P2 If $S(n) = S(m)$ then $n = m$ for every $n, m \in \mathbb{N}$.

P3 If 0 has property P and for every $n \in \mathbb{N}$

$S(n)$ has property P provided n has property P

then n has property P for every $n \in \mathbb{N}$.

Principles P1–P3, known as the Peano axioms of arithmetic, are the most commonly accepted axioms for the natural numbers. Axiom P3 is the principle of mathematical induction.

Notice that the definition of the successor operator $S(x) = x \cup \{x\}$ from the axiom of infinity coincides with the definition given in this section, since $S(n) = n + 1 = n \cup \{n\}$.

Now we are ready to construct the set $\mathbb{N}$.

Theorem 3.1.1 *There exists exactly one set* $\mathbb{N}$ *such that*

(1) $\emptyset \in \mathbb{N}$,

(2) $x \in \mathbb{N} \to S(x) \in \mathbb{N}$ *for every* x, *and*

(3)   *if* K *is any set that satisfies (1) and (2) then* $\mathbb{N} \subset K$.

Proof By the axiom of infinity there exists at least one set X satisfying (1) and (2). Let

$$\mathcal{F} = \{Y \in \mathcal{P}(X) \colon \emptyset \in Y \;\&\; \forall x \in Y \; (S(x) \in Y)\}$$

and put $\mathbb{N} = \bigcap \mathcal{F}$. It is easy to see that the intersection of any nonempty family of sets satisfying (1) and (2) still satisfies (1) and (2). To see that (3) holds let K satisfy (1) and (2). Then $X \cap K \in \mathcal{F}$ and $\mathbb{N} = \bigcap \mathcal{F} \subset X \cap K \subset K$. Theorem 3.1.1 is proved. □

The set $\mathbb{N}$ from Theorem 3.1.1 will be called the *set of natural numbers*. Natural numbers will usually be denoted by the letters $m, n, p, \ldots$. The operation $S(n)$ is the counterpart to $(n + 1)$. Thus for a natural number n we will henceforth write $n + 1$ instead of $S(n)$.

The set of natural numbers will also be denoted by the symbol ω.

Notice that condition (3) of Theorem 3.1.1 can be rephrased as follows:

$$[0 \in K \;\&\; \forall x \; (x \in K \to S(x) \in K)] \text{ implies } \mathbb{N} \subset K. \qquad (3.1)$$

If we additionally know that $K \subset \mathbb{N}$ then (3.1) can be expressed as

$$[0 \in K \;\&\; \forall n \; (n \in K \to n + 1 \in K)] \text{ implies } \mathbb{N} = K. \qquad (3.2)$$

Property (3.2) is called the *induction property* of the set of natural numbers and the proofs that use it are cited as *proofs by induction*.

It is easy to see that the induction property (3.2) implies axiom P3 if we put $K = \{n \in \mathbb{N} \colon n \text{ has property } P\}$. Axiom P2 follows from Theorem 3.1.2(f). To see that $\mathbb{N}$ satisfies P1 just notice that $0 = \emptyset$ and $S(n) = n \cup \{n\}$ is clearly nonempty.

Essentially all the known properties of the natural numbers can be deduced from these definitions and the ZFC axioms.[1] However, we will restrict ourselves only to those properties that are more connected with our representation of the natural numbers and to those that will be of more use in what follows.

---

[1] In fact, the Peano axioms are strictly weaker than the ZFC axioms. More precisely, everything that can be proved in the Peano arithmetic can be proved in ZFC. However, there are properties of natural numbers that can be deduced from the ZFC axioms but cannot be proved on the basis of the Peano axioms alone.

Theorem 3.1.2 *For every natural numbers m, n, and p*

(a) $m \subset m+1$ *and* $m \in m+1$;

(b)   *if* $x \in m$ *then* $x \in \mathbb{N}$;

(c)   *if* $m \in n$ *then* $m+1 \subset n$;

(d)   *if* $m \in n$ *and* $n \in p$ *then* $m \in p$;

(e) $m \notin m$;

(f)   *if* $n+1 = m+1$ *then* $n = m$;

(g) $m \subset n$ *if and only if* $m = n$ *or* $m \in n$;

(h) $m \subset n$ *or* $n \subset m$.

Proof (a) This follows immediately from the definition $m+1 = m \cup \{m\}$.
    (b) The proof follows by induction. Let

$$K = \{n \in \mathbb{N} : \forall x \, (x \in n \rightarrow x \in \mathbb{N})\}.$$

Notice that $0 = \emptyset \in K$, since there is no $x \in 0$. Now let $n \in K$ and $x \in n+1$. Since $n+1 = n \cup \{n\}$ we have two cases: If $x \in n$ then $x \in \mathbb{N}$, since $n \in K$; if $x \in \{n\}$ then $x = n$ and also $x \in \mathbb{N}$. Thus $n \in K$ implies $n+1 \in K$. Hence, by (3.2), $K = \mathbb{N}$ and (b) is proved.
    (c) The proof is left as an exercise.
    (d) Since $m \in n$ and, by (a), $n \subset n+1$ we have $m \in n+1$. But, by (c), $n \in p$ implies $n+1 \subset p$. So $m \in p$.
    (e) This can be easily proved by induction. It is left as an exercise.
    (f) This is left as an exercise.
    (g) $\Leftarrow$: If $m = n$ then $m \subset n$. If $m \in n$ then, by (a) and (c), $m \subset m+1 \subset n$.
    $\Rightarrow$: Fix $m \in \mathbb{N}$ and consider the set

$$K = \{n \in \mathbb{N} : m \subset n \rightarrow [m \in n \; \vee \; m = n]\}.$$

It is enough to show that $K = \mathbb{N}$.
    First notice that $0 \in K$ since $m \subset 0 = \emptyset$ implies $m = \emptyset = 0$.
    Now assume that $n \in K$. By induction it is enough to show that $n+1 \in K$. So assume that $m \subset n+1 = n \cup \{n\}$. If $m \not\subset n$ then $n \in m$ and, by (c), $n+1 \subset m \subset n+1$, that is, $m = n+1$. If $m \subset n$ then, since $n \in K$, either $m \in n$ or $m = n$. But in both cases $m \in n \cup \{n\} = n+1$; so $n+1 \in K$.
    (h) Define the set

$$K = \{m \in \mathbb{N} : \forall n \in \mathbb{N} \, (n \notin m \rightarrow m \subset n)\}.$$

First we will show that $K = \mathbb{N}$.

First notice that $0 \in K$, since $0 = \emptyset \subset n$ for every n.

Now assume that $m \in K$. We will show that this implies $m + 1 \in K$. So let $n \in \mathbb{N}$ be such that $n \not\subset m + 1 = m \cup \{m\}$. Then $n \neq m$ and $n \not\subset m$. But $m \in K$ and $n \not\subset m$ imply that $m \subset n$. This last condition and $n \neq m$ imply in turn, by (g), that $m \in n$ and, by (c), $m + 1 \subset n$. Thus we conclude that $n \in K$ implies $n + 1 \in K$ and, by induction, that $K = \mathbb{N}$.

Now, to finish the proof, let $n, m \in \mathbb{N}$. If $n \subset m$ then the conclusion is correct. So assume that $n \not\subset m$. Then, by (g), $n \not\subset m$. Since $m \in \mathbb{N} = K$ we conclude $m \subset n$. □

From Theorem 3.1.2 it follows immediately that every natural number $n = \{0, 1, 2, \ldots, n - 1\}$. However, we cannot express it in the formal language of set theory, since the symbol "..." does not belong to its language. We can only express it using infinitely many sentences, one for each natural number. On the other hand, this informal expression is more intuitive for most mathematicians than the formal definition. We will often use it to describe the set of natural numbers less than a fixed number. In particular, we will very often write 2 instead of $\{0, 1\}$.

Let us also define for natural numbers m and n

$$m < n \text{ if and only if } m \in n$$

and

$$m \leq n \text{ if and only if } m \subset n.$$

It is not difficult to see that these definitions imply the properties of $<$ and $\leq$ that we usually associate with them.

We will finish this section by defining on the set $\mathbb{N}$ the arithmetic operations of sum, product, and exponentiation. The definitions are by induction and go as follows:

$$n + 0 = n; \quad n + (m + 1) = (n + m) + 1, \text{ where } k + 1 \text{ stands for } S(k).$$

$$n \, 0 = 0; \quad n \, (m + 1) = (n \, m) + n.$$

$$n^0 = 1; \quad n^{m+1} = n^m \, n.$$

It can be proved, by induction, that these operations satisfy all known properties of the standard operations.

EXERCISES

1 Prove that if $\mathcal{F}$ is a nonempty family such that every $F \in \mathcal{F}$ satisfies (1) and (2) from Theorem 3.1.1 then $\bigcap \mathcal{F}$ also satisfies these conditions.

2 Prove by induction Theorem 3.1.2(c), (e), and (f).

3.2   Integers and rational numbers

Intuitively, we will define integers as differences $n - m$ of two natural numbers. The formalization of this intuition has, however, two problems. First, we do not have an operation $-$ defined as of yet. Thus we will talk about ordered pairs $\langle n, m \rangle$ of natural numbers, thinking of them as $n - m$. However, representing the number $n - m$ as $\langle n, m \rangle$ is not unique: $1 = 2 - 1 = 3 - 2$. So, instead, we will use equivalence classes of pairs of natural numbers, where the equivalence relation E on $\mathbb{N} \times \mathbb{N}$ is defined by the formula:

$$\langle n, m \rangle E \langle n', m' \rangle \Leftrightarrow n + m' = n' + m. \tag{3.3}$$

Now the *set of integers* $\mathbb{Z}$ is defined as the quotient class $(\mathbb{N} \times \mathbb{N})/E$.

We define a relation $\leq$ on $\mathbb{Z}$ by

$$[\langle n, m \rangle] \leq [\langle n', m' \rangle] \Leftrightarrow n + m' \leq n' + m. \tag{3.4}$$

The arithmetic operations on $\mathbb{Z}$ are defined as follows:

$$[\langle n, m \rangle] + [\langle n', m' \rangle] = [\langle n + n', m + m' \rangle].$$
$$[\langle n, m \rangle] - [\langle n', m' \rangle] = [\langle n, m \rangle] + [\langle m', n' \rangle].$$
$$[\langle n, m \rangle] [\langle n', m' \rangle] = [\langle n\,n' + m\,m', n\,m' + m\,n' \rangle].$$

We will leave as an exercise the proof that E is an equivalence relation, that the preceding operations are well defined (i.e., they do not depend on the choice of representatives of equivalence classes), and that they have the properties we know from algebra.

Following the same path, we define the set $\mathbb{Q}$ of *rational numbers* as ordered pairs $\langle a, b \rangle$ of integers representing of a/b. More precisely, we define an equivalence relation E' on $\mathbb{Z} \times \mathbb{Z}$ by

$$\langle a, b \rangle E' \langle a', b' \rangle \;\Leftrightarrow\; [a\,b' = a'\,b \;\&\; b \neq 0 \neq b'] \;\text{ or }\; [b = b' = 0].$$

Then

$$\mathbb{Q} = \{[\langle a, b \rangle] : a, b \in \mathbb{Z} \;\&\; b \neq 0\},$$

where $[\langle a, b \rangle]$ is an equivalence class with respect to E'. We define a relation $\leq$ and arithmetic operations on $\mathbb{Z}$ as follows:

$$[\langle a, b \rangle] + [\langle a', b' \rangle] = [\langle a\,b' + a'\,b, b\,b' \rangle];$$
$$[\langle a, b \rangle] [\langle a', b' \rangle] = [\langle a\,a', b\,b' \rangle];$$
$$[\langle a, b \rangle] \leq [\langle a', b' \rangle] \Leftrightarrow b \geq 0 \;\&\; b' \geq 0 \;\&\; a\,b' \leq a'\,b.$$

As before, we will leave as an exercise the proof that E' is an equivalence relation, that the preceding definitions are well defined, and that they have the desired properties.

EXERCISES

1 Prove that the relation E defined by (3.3) is an equivalence relation.

2 Show that $\leq$ on $\mathbb{Z}$ defined by (3.4) is well defined and is a linear-order relation.

3.3 Real numbers

In the previous section we constructed integers from the natural numbers and rational numbers from the integers. Both constructions were primarily algebraic. This was possible since the rational numbers are the algebraic completion of the set of natural numbers. The same approach, however, cannot be used in a construction of the real numbers, since there are real numbers of no simple algebraic relation to rational numbers.

The crucial geometric property that is often used to described real numbers is that the real line "does not have holes." This intuition was used by Dedekind to define the set $\mathbb{R}$ of *real numbers* as follows.

A subset A of $\mathbb{Q}$ is said to be a Dedekind cut if A does not contain a largest element and is a *proper initial segment* of $\mathbb{Q}$, that is, $\emptyset \neq A \neq \mathbb{Q}$ and A contains every $p \leq q$ provided $q \in A$. Now, $\mathbb{R}$ can be defined as the set of all Dedekind cuts, intuitively identifying A with a real number $\sup A$. Then for Dedekind cuts A and B we can define

$$A + B = \{p + q : p \in A \ \& \ q \in B\}$$

and

$$A \leq B \Leftrightarrow A \subset B.$$

Similarly, we can define the product $A\,B$ of two numbers (though the definition must be done more carefully, since the product of two negative numbers is positive).

Another approach for the definition of real numbers is to use their numerical expansions. This approach is a little more messy, but it better fits the methods of this course. So we will include it here too.

The construction of real numbers that follows will be done in two steps. First, consider the set $2^{\mathbb{N}}$ of all sequences $s \colon \mathbb{N} \to \{0, 1\}$. It is usually called the Cantor set. (The classical Cantor "ternary" set can be obtained from this set by identifying a sequence $a \in 2^{\mathbb{N}}$ with $\sum_{n \in \mathbb{N}} 2a_n / 3^{n+1}$. See Section 6.2 for more details.) We would like to identify a sequence $a \in 2^{\mathbb{N}}$ with the real number $\sum_{n \in \mathbb{N}} a_n / 2^{n+1}$. However, this identification function is not one-to-one. To correct it, we will define $[0, 1]$ as the quotient class

$2^{\mathbb{N}}/E$, where aEb if and only if $a = b$ or there exists an $n \in \mathbb{N}$ such that for every $k \in \mathbb{N}$

$$[k < n \rightarrow a_k = b_k] \,\&\, [a_n = 1 \,\&\, b_n = 0] \,\&\, [k > n \rightarrow (a_k = 0 \,\&\, b_k = 1)]. \quad (3.5)$$

We will leave as an exercise the proof that E is indeed an equivalence relation.

The linear-order relation $\leq$ on $[0, 1]$ can be defined by

$$[a] \leq [b] \Leftrightarrow [a] = [b] \,\vee\, \exists n \in \mathbb{N} \, [a_n < b_n \,\&\, \forall k \in n \, (a_k = b_k)]. \quad (3.6)$$

Again, we will not prove the correctness of this definition.

Numbers 0 and 1 in $[0, 1]$ are defined as equivalence classes of functions constantly equal to 0 and 1, respectively. Then we define

$$[0, 1) = [0, 1] \setminus \{1\}, \quad (0, 1] = [0, 1] \setminus \{0\}, \quad (0, 1) = [0, 1] \setminus \{0, 1\}.$$

The set $\mathbb{R}$ of real numbers is defined as $\mathbb{Z} \times [0, 1)$, where intuitively we identify a pair $\langle k, r \rangle$ with $k + r$. In particular, we can define $\leq$ on $\mathbb{R}$ by

$$\langle k, r \rangle \leq \langle l, s \rangle \Leftrightarrow k < l \,\vee\, (k = l \,\&\, r \leq s). \quad (3.7)$$

In the remainder of this section we will recall some geometric and topological properties of the n-dimensional Euclidean space $\mathbb{R}^n$ ($n \in \mathbb{N}$). In particular, the *distance* between two points $p = \langle p_1, \ldots, p_n \rangle$ and $q = \langle q_1, \ldots, q_n \rangle$ of $\mathbb{R}^n$ is given by the formula

$$d(p, q) = \sqrt{\sum_{i=1}^{n} (p_i - q_i)^2}.$$

The *open ball* in $\mathbb{R}^n$ centered at $p \in \mathbb{R}^n$ and with radius $\varepsilon > 0$ is defined as

$$B(p, \varepsilon) = \{q \in \mathbb{R}^n : d(p, q) < \varepsilon\}.$$

A subset U of $\mathbb{R}^n$ is *open* if U is a union of some family of open balls in $\mathbb{R}^n$. The family τ of all open subsets of $\mathbb{R}^n$ is called the (natural) *topology* on $\mathbb{R}^n$ and is closed under finite intersections and arbitrary unions.

A set $F \subset \mathbb{R}^n$ is *closed* if its complement $\mathbb{R}^n \setminus F$ is open. Notice that finite unions of closed sets and arbitrary intersections of closed sets are also closed.

For a subset S of $\mathbb{R}^n$ its *interior* $\mathrm{int}(S)$ is defined as the largest open subset of S, that is,

$$\mathrm{int}(S) = \bigcup \{U \subset \mathbb{R}^n : U \text{ is open in } \mathbb{R}^n\}.$$

The *closure* of S is the smallest closed set containing S, that is,

$$\mathrm{cl}(S) = \bigcap \{F \supset S \colon F \text{ is closed in } \mathbb{R}^n\}.$$

A subset D of $\mathbb{R}^n$ is *dense* in $\mathbb{R}^n$ if $\mathrm{cl}(D) = \mathbb{R}^n$ or, equivalently, when $D \cap U \neq \emptyset$ for every nonempty open set $U \subset \mathbb{R}^n$. In particular, $\mathbb{Q}^n$ is a dense subset of $\mathbb{R}^n$. A subset N of $\mathbb{R}^n$ is *nowhere dense* if $\mathrm{int}(\mathrm{cl}(N)) = \emptyset$.

Now let us recall a few more specific properties of $\mathbb{R}^n$. A function $f \colon \mathbb{R}^n \to \mathbb{R}^m$ is *continuous* if $f^{-1}(U) = \{x \in \mathbb{R}^n \colon f(x) \in U\}$ is open in $\mathbb{R}^n$ for every open set $U \subset \mathbb{R}^m$. It is easy to see that for every dense subset D of $\mathbb{R}^n$ and continuous functions $f \colon \mathbb{R}^n \to \mathbb{R}^m$ and $g \colon \mathbb{R}^n \to \mathbb{R}^m$ if $f(d) = g(d)$ for every $d \in D$ then $f = g$.

A subset B of $\mathbb{R}^n$ is *bounded* if $B \subset B(p, \varepsilon)$ for some $p \in \mathbb{R}^n$ and $\varepsilon > 0$. Closed, bounded subsets of $\mathbb{R}^n$ are called *compact*. One of the most important properties of compact subsets of $\mathbb{R}^n$ is given in the next theorem.

Theorem 3.3.1 *If $K_0 \supset K_1 \supset K_2 \supset \cdots$ is a decreasing sequence of compact nonempty subsets of $\mathbb{R}^n$ then their intersection $\bigcap_{i=0}^{\infty} K_i$ is nonempty.*

A subset C of $\mathbb{R}^n$ is *connected* if there do not exist two disjoint open sets $U, V \subset \mathbb{R}^n$ such that $U \cap C \neq \emptyset$, $V \cap C \neq \emptyset$, and $C \subset U \cup V$. Recall that any interval in $\mathbb{R}$ as well as any $\mathbb{R}^n$ are connected.

A sequence $\{p_k\}_{k=0}^{\infty}$ of points in $\mathbb{R}^n$ is a *Cauchy sequence* if for every $\varepsilon > 0$ there is a number N such that $d(p_i, p_j) < \varepsilon$ for every $i, j > N$. Every Cauchy sequence in $\mathbb{R}^n$ has a limit point. This fact serves as a basis for the proof of the Baire category theorem, which follows.

Theorem 3.3.2 (Baire category theorem) *If $N_0, N_1, N_2, \ldots$ is a sequence of nowhere-dense subsets of $\mathbb{R}^n$ then its union $\bigcup_{i=0}^{\infty} N_i$ has an empty interior.*

EXERCISES

1 Define the product of two real numbers using Dedekind's definition of $\mathbb{R}$. Show that your product has the distributive property, that is, that $a(b + c) = ab + ac$ for every $a, b, c \in \mathbb{R}$.

2 Prove that the relation E defined by (3.5) is an equivalence relation.

3 Show that $\leq$ on $[0, 1]$ defined by (3.6) is well defined and is a linear-order relation.

4 Prove that $\leq$ on $\mathbb{R}$ defined by (3.7) is a linear-order relation.

Part II

Fundamental tools of set theory

Chapter 4

Well orderings and transfinite induction

4.1 Well-ordered sets and the axiom of foundation

A binary relation R on a set X is said to be *well founded* if every nonempty subset Y of X has an R-minimal element, that is, if

$$\forall Y \subset X \; [Y \neq \emptyset \to \exists m \in Y \; \neg \exists y \in Y \; (yRm)].$$

Examples 1. For every finite linearly ordered set $\langle X, \leq \rangle$ the relation $<$ is well founded.
2. The relation $<$ on the set $\mathbb{N}$ of all natural numbers is well founded. This is known as the well-ordering principle for the natural numbers. It can be easily deduced from the principle of mathematical induction P3. (See also (3.2).) It is also a special case of the next example.
3. The relation $\in$ is well founded on every nonempty set X. This, however, does not follow from the axioms we have studied so far. For this we need the following axiom known as the *axiom of foundation* or *regularity*. This is the last of the axioms of ZFC.

Foundation axiom (Skolem 1922; von Neumann 1925) Every nonempty set has an $\in$-minimal element:

$$\forall x \, [\exists y (y \in x) \to \exists y [y \in x \; \& \; \neg \exists z (z \in x \; \& \; z \in y)]] \, .$$

This axiom is not essential for this course. However, it is good to know some of its basic consequences.

Theorem 4.1.1 $x \notin x$ *for every* x.

Proof To obtain a contradiction assume that there exists a t such that $t \in t$. Put $x = \{t\}$. We will see that the axiom of foundation fails for x.

Evidently, there exists a $y \in x$, since $t \in x$. Moreover, if $y \in x$ then $y = t$. But then, there exists a z, $z = t$, such that $z = t \in x$ and $z = t \in t = y$, contradicting the axiom of foundation. $\square$

Theorem 4.1.2 *There is no infinite $\in$-decreasing sequence, that is, there is no sequence $\langle x_n : n \in \mathbb{N} \rangle$ such that $x_{n+1} \in x_n$ for all $n \in \mathbb{N}$.*

Proof Otherwise the set $\{x_n : n \in \mathbb{N}\}$ would have no $\in$-minimal element.
$\square$

A binary relation $\leq$ on X is a *well-ordering relation* if $\langle X, \leq \rangle$ is linearly ordered and the relation $<$ is well founded. In other words, a linearly ordered set $\langle X, \leq \rangle$ is *well ordered* if every nonempty subset A of X has a smallest element $a \in A$, usually denoted by $\min A$.

It is pretty easy to see that a linearly ordered set $\langle X, \leq \rangle$ is well ordered if and only if it does not contain an infinite strictly decreasing sequence, that is, a sequence $\langle x_n \in X : n \in \mathbb{N} \rangle$ such that $x_{n+1} < x_n$ for every $n \in \mathbb{N}$. However, a formal proof of this fact requires a recursive definition technique, which is still not available to us. Thus its proof will be postponed until Section 4.3 (see Theorem 4.3.2).

Notice also that a subset of a well-ordered set is well ordered.

Examples 1. Every finite linearly ordered set is well ordered.
2. The set $\mathbb{N}$ of all natural numbers is well ordered by the usual order $\leq$.
3. The set $S = \{1 - 1/n : n \in \mathbb{N}, \ n > 0\} \cup \{1\}$ is a well-ordered subset of $\langle \mathbb{R}, \leq \rangle$ since $S_0 = \{1 - 1/n : n \in \mathbb{N}, \ n > 0\}$ is ordered the same way as $\mathbb{N}$ and 1 is greater than any number from S_0.
4. The interval $[0, 1]$ with the usual order is linearly ordered and has the smallest element. However, it is not well ordered, since its subset $(0, 1)$ does not have a smallest element (0 is not an element of $(0, 1)$).
5. Any family of sets linearly ordered by the relation

$$x \varepsilon y \Leftrightarrow x \in y \ \lor \ x = y$$

is well ordered by ε. This follows from the axiom of foundation.

The partially ordered sets $\langle X, \leq \rangle$ and $\langle Y, \preceq \rangle$ are *(order) isomorphic* (or have the same *order type*) if there is a bijection $f \colon X \to Y$ such that

$$a \leq b \ \text{ if and only if } \ f(a) \preceq f(b)$$

for every $a, b \in X$. The function f is called an *order isomorphism* (or just *isomorphism*) between X and Y.

The next two propositions list some basic properties of order isomorphisms. Their proofs are left as exercises.

Proposition 4.1.3 *Let $\langle X, \leq \rangle$, $\langle Y, \preceq \rangle$, and $\langle Z, \ll \rangle$ be partially ordered sets.*

(1) *If $f\colon X \to Y$ is an order isomorphism between X and Y then the inverse function $f^{-1}\colon Y \to X$ is an order isomorphism between Y and X.*

(2) *If $f\colon X \to Y$ is an order isomorphism between X and Y and $g\colon Y \to Z$ is an order isomorphism between Y and Z then $g \circ f\colon X \to Z$ is an order isomorphism between X and Z.*

Proposition 4.1.4 *Let $f\colon X \to Y$ be an order isomorphism between partially ordered sets $\langle X, \leq \rangle$ and $\langle Y, \preceq \rangle$. Then*

(1) *X is linearly ordered (well ordered) if and only if Y is linearly ordered (well ordered);*

(2) *$x \in X$ is the smallest (the greatest, a minimal, a maximal) element of X if and only if $f(x)$ is the smallest (the greatest, a minimal, a maximal) element of Y.*

Examples 1. The intervals $[0, 1]$ and $(0, 1)$ are not order isomorphic since $[0, 1]$ has a smallest element and $(0, 1)$ does not have one. The set $\langle \mathcal{P}(\mathbb{R}), \subset \rangle$ is isomorphic to neither of these sets, since it is not linearly ordered.
2. The sets $\mathbb{N}$ of natural numbers and P of even natural numbers have the same order type. The isomorphism is established by the function $f(n) = 2n$. The order type of this set is denoted by ω and represents an infinite strictly increasing sequence. (In Section 4.2 we will make the notion of "order type" of well-ordered sets more formal.)
3. The order type of the set $S = \{1 - 1/n\colon n \in \mathbb{N},\ n > 0\} \cup \{1\}$ is denoted by $\omega + 1$ and it is not the same as ω, since S has a greatest element and $\mathbb{N}$ does not.
4. The set $\{-n\colon n \in \mathbb{N}\}$ ordered by $\leq$ is isomorphic to $\langle \mathbb{N}, (\leq)^{-1} \rangle = \langle \mathbb{N}, \geq \rangle$ by the isomorphism $f(n) = -n$. The order type of this set is denoted by $\omega^\star$. The sets with order type $\omega^\star$ are infinite strictly decreasing sequences.

A subset S of a partially ordered set $\langle X, \leq \rangle$ is an *initial segment* of X if for every $x, y \in X$
$$x \leq y \ \& \ y \in S \Rightarrow x \in S.$$

An initial segment S of X is *proper* if $S \neq X$. It is easy to see that if $f \colon X \to Y$ is an order isomorphism between $\langle X, \leq \rangle$ and $\langle Y, \preceq \rangle$ then $S \subset X$ is a (proper) initial segment of X if and only if $f[S]$ is a (proper) initial segment of Y.

Note that for every element x_0 of a partially ordered set $\langle X, \leq \rangle$ the set

$$O(x_0) = \{x \in X \colon x < x_0\}$$

is a proper initial segment of X. However, not every proper initial segment of a partially (even linearly) ordered set is of the form $O(x_0)$, as is shown by the subset $(-\infty, 0] = \{x \in \mathbb{R} \colon x \leq 0\}$ of $\mathbb{R}$. On the other hand, such an example cannot be found in a well-ordered set, as is asserted by the next theorem.

Theorem 4.1.5 *Every proper initial segment S of a well-ordered set W is of the form $O(\xi)$ for some $\xi \in W$.*

Proof Since S is a proper subset of W the set $W \setminus S$ is nonempty. Let $\xi = \min(W \setminus S)$. Then $S = O(\xi)$. $\square$

Theorem 4.1.6 (Principle of transfinite induction) *If a set A is well-ordered, $B \subset A$, and for every $x \in A$ the set B satisfies the condition*

$$O(x) \subset B \Rightarrow x \in B, \qquad\qquad (4.1)$$

then $B = A$.

Proof It is enough to show that $A \setminus B = \emptyset$. To arrive at a contradiction assume that $A \setminus B \neq \emptyset$ and let $x = \min(A \setminus B)$. Then, by minimality of x, $O(x) \subset B$. Hence, by (4.1), $x \in B$. This contradicts $x \in A \setminus B$. $\square$

If we apply Theorem 4.1.6 to $A = \mathbb{N}$ then we obtain the following version of the induction schema for the set of natural numbers.

Corollary 4.1.7 *If $B \subset \mathbb{N}$ and for every $n \in \mathbb{N}$ the set B satisfies the condition*

$$n \subset B \Rightarrow n \in B, \qquad\qquad (4.2)$$

then $B = \mathbb{N}$.

The statement of Corollary 4.1.7 also follows immediately from the induction property (3.2). The converse implication is also easy to prove and can be concluded directly, without the use of Theorem 4.1.6.

Proofs using Theorem 4.1.6 are called *proofs by transfinite induction.*
Such proofs are the basic techniques of set theory and will be used as a main
tool in this course. The proof of the next theorem is a classic example of
the use of this technique. To formulate it we need the following definition.

Let $\langle A, \leq \rangle$ and $\langle B, \leq \rangle$ be partially ordered sets. A function $f\colon A \to B$
is an *increasing* function if $x \leq y$ implies $f(x) \leq f(y)$ for every $x, y \in A$;
f is *strictly increasing* if $x < y$ implies $f(x) < f(y)$. Similarly we define
decreasing and *strictly decreasing* functions.

Notice that if $f\colon A \to B$ is strictly increasing then it establishes an
order isomorphism between A and $f[A]$.

Theorem 4.1.8 *If $f\colon A \to A$ is a strictly increasing function defined on
a well-ordered set A then $x \leq f(x)$ for all $x \in A$.*

Proof Let $B = \{x \in A\colon x \leq f(x)\}$. We will show that $B = A$. Let $x \in A$
be such that $O(x) \subset B$. By Theorem 4.1.6 it is enough to show that $x \in B$.
So take $y \in O(x)$. Then $y \leq f(y)$. Moreover, $f(y) < f(x)$, since $y < x$.
Hence $y < f(x)$ for every $y \in O(x)$. In particular, $f(x) \in A \setminus O(x)$. Since
x is the first element of $A \setminus O(x)$ we conclude that $x \leq f(x)$. Thus $x \in B$.
□

Corollary 4.1.9 *If well-ordered sets A and B are isomorphic, then the
isomorphism $f\colon A \to B$ is unique.*

Proof Let $g\colon A \to B$ be another order isomorphism. We will show that
$f = g$. So let $x \in A$. The function $g^{-1} \circ f\colon A \to A$ is also an isomorphism.
Using Theorem 4.1.8 applied to $g^{-1} \circ f$ we conclude that $x \leq g^{-1}(f(x))$
and, applying function g to both sides, $g(x) \leq f(x)$. Similarly we argue for
$f(x) \leq g(x)$. Thus $f(x) = g(x)$. Since x was arbitrary, $f = g$.   □

Corollary 4.1.10 *No well-ordered set is isomorphic to any of its proper
initial segments.*

Proof If a well-ordered set A were isomorphic to an initial segment $O(x)$
for $x \in A$, and $f\colon A \to O(x)$ were an isomorphism establishing this fact,
then we would have $f(x) \in O(x)$, that is, $f(x) < x$, contradicting Theorem 4.1.8.   □

Corollary 4.1.11 *No two distinct initial segments of a well-ordered set
are isomorphic.*

Theorem 4.1.12 *If A and B are well-ordered sets then either*

(i)   *A and B are isomorphic,*

(ii) *A is isomorphic to a proper initial segment of B, or*

(iii) *B is isomorphic to a proper initial segment of A.*

Proof Let

$$Z = \{x \in A \colon \exists y_x \in B \ (O(x) \text{ and } O(y_x) \text{ are order isomorphic})\}.$$

Notice that if $x \in Z$ then, by Corollary 4.1.11, y_x from the definition of the set Z is unique. Define $f \colon Z \to B$ by putting $f(x) = y_x$, that is,

$$f = \{\langle x, y \rangle \in A \times B \colon O(x) \text{ and } O(y) \text{ are order isomorphic}\}.$$

First we will show that Z is an initial segment of A. To see it, let $x < z$ and $z \in Z$. We have to show that $x \in Z$. Since $z \in Z$, there is an isomorphism h between $O(z)$ and $O(f(z))$. The restriction of h to $O(x)$ gives an isomorphism, whose range must be a proper initial segment S of $O(f(z))$, and, by Theorem 4.1.5, $S = O(y)$ for some $y \in O(f(z))$. Thus $x \in Z$ and we conclude that Z is an initial segment of A.

This argument shows also that f is strictly increasing, since $x < z$ and $z \in Z$ imply that $O(x)$ is isomorphic to $O(y)$ for some $y \in O(f(z))$, that is, $f(x) = y < f(z)$.

Similarly as for Z we can also prove that $f[Z]$ is an initial segment of B. (Just note that $f[Z] = \{y \in B \colon \exists x \in A \ (O(y) \text{ and } O(x) \text{ are isomorphic})\}$.) So $f \colon Z \to f[Z]$ is an order isomorphism.

If $Z = A$ then f establishes an isomorphism between A and either B or a proper initial segment of B. Similarly, if $f[Z] = B$ then f^{-1} shows that B is isomorphic to either A or a proper initial segment of A. Thus, it is enough to prove that either $Z = A$ or $f[Z] = B$.

To obtain a contradiction assume that $Z \neq A$ and $f[Z] \neq B$. Then Z is a proper initial segment of A and $f[Z]$ is a proper initial segment of B. Hence, by Theorem 4.1.5, $Z = O(x)$ for some $x \in A$ and $f[Z] = O(y)$ for some $y \in B$. But then, f is an isomorphism between $Z = O(x)$ and $O(y)$, contradicting the fact that $x \notin Z$.   □

We will finish this section with the following construction of an ordering relation on a product space.

Theorem 4.1.13 *Let $\{F_t\}_{t \in T}$ be an indexed family of nonempty linearly ordered sets $\langle F_t, \leq_t \rangle$ and let T be well ordered by $\leq$. If $\preceq$ is a binary relation on the Cartesian product $P = \prod_{t \in T} F_t$ defined by the formula*

$$f \preceq g \Leftrightarrow f = g \ \vee \ \exists s \ (s = \min\{t \in T \colon f(t) \neq g(t)\} \ \& \ f(s) \leq_s g(s))$$

for every $f, g \in P$, then $\preceq$ is a linear order on P.

Moreover, if T is finite then P is well ordered by $\preceq$ if and only if all sets F_t are well ordered.

The proof is left as an exercise.

The relation $\preceq$ defined in Theorem 4.1.13 is called the *lexicographic order* of $\prod_{t \in T} F_t$ and is sometimes denoted by $\leq_{\text{lex}}$.

EXERCISES

1 Prove Proposition 4.1.3.

2 Prove Proposition 4.1.4.

3 Let $f \colon X \to Y$ be an isomorphism between partially ordered sets $\langle X, \leq \rangle$ and $\langle Y, \preceq \rangle$. Show that $S \subset X$ is a (proper) initial segment of X if and only if $f[S]$ is a (proper) initial segment of Y.

4 Complete the proof of Theorem 4.1.5 by showing that $S = O(\xi)$.

5 Prove Theorem 4.1.13.

6 Show that in the additional part of Theorem 4.1.13 the assumption that T is finite is essential by giving an example of well-ordered sets T and F for which the lexicographic order on F^T is not a well-ordering relation.

7 Let $\langle K, \leq \rangle$ and $\langle X, \leq \rangle$ be nonempty linearly ordered sets such that X has a fixed element $0 \in X$. For $f \colon K \to X$ let $\text{supp}(f) = \{k \in K \colon f(k) \neq 0\}$ and put

$$\mathcal{F}(K, X) = \{f \in X^K \colon \text{supp}(f) \text{ is finite}\}.$$

Define a binary relation $\preceq$ on $\mathcal{F}(K, X)$ by putting

$$f \preceq g \Leftrightarrow f = g \text{ or } f(m) \leq g(m), \text{ where } m = \max\{k \in K \colon f(k) \neq g(k)\},$$

for every $f, g \in \mathcal{F}(K, X)$. Prove that

(a) the relation $\preceq$ is a linear-order relation on $\mathcal{F}(K, X)$;

(b) if X has at least two elements and 0 is the minimal element of X then $\langle \mathcal{F}(K, X), \preceq \rangle$ is well ordered if and only if $\langle X, \leq \rangle$ and $\langle K, \leq \rangle$ are well ordered.

The relation $\preceq$ is usually called an *antilexicographic-order relation*.

4.2   Ordinal numbers

In the previous section we were informally talking about the "order type" of a given partially ordered set $\langle X, \leq \rangle$. The natural way of thinking about such an object is to consider it as a representative from the equivalence class $[X]_{\leq}$ of all ordered sets order isomorphic to $\langle X, \leq \rangle$. This equivalence class, however, is not a set! (The argument is similar to that for the class of all sets.) Moreover, if we want to talk about all order types, we find also that there are too many of them; that is, even if we discover a good way of choosing a representative from each equivalence class $[X]_{\leq}$, the class of all such representatives will not be a set.

One way to avoid this problem is to consider the sentences "$\langle Y, \leq \rangle$ has the order type τ of $\langle X, \leq \rangle$" and "$\langle Y, \leq \rangle$ and $\langle X, \leq \rangle$ have the same order type" just as abbreviations for the well-defined sentence "$\langle Y, \leq \rangle$ and $\langle X, \leq \rangle$ are order isomorphic." That is the way we will think about the order types of arbitrary partially ordered sets. For the well-ordered sets, however, we can indeed follow the path of our intuition described in the previous paragraph. This section is dedicated to fulfilling this goal by constructing the canonical order types of well-ordered sets, called *ordinal numbers*. The construction is due to von Neumann (1929).

The class of all ordinal numbers is still too big to be a set. Thus, instead of constructing it as a set, we will define a formula $\varphi(\alpha)$ that will represent an intuitive sentence "α is an ordinal number." The definition is as follows.

A set α is said to be an *ordinal number* (or just an *ordinal*) if it has the following properties:

(A_α) If $\beta \in \alpha$ then $\beta \subset \alpha$;

(B_α) If $\beta, \gamma \in \alpha$ then $\beta = \gamma$, $\beta \in \gamma$, or $\gamma \in \beta$;

(C_α) If $\emptyset \neq B \subset \alpha$ then there exists a $\gamma \in B$ such that $\gamma \cap B = \emptyset$.

It follows from Theorem 3.1.2 that if α stands for either any natural number or the set $\mathbb{N}$ of natural numbers then α satisfies conditions (A_α) and (B_α). Also, the well-ordering principle for the natural numbers implies condition (C_α) in this case. So every natural number is an ordinal number and so is $\mathbb{N}$. It is also easy to see that the set $\mathbb{N} \cup \{\mathbb{N}\}$ is an ordinal number. The sets $\mathbb{N}$ and $\mathbb{N} \cup \{\mathbb{N}\}$, when considered as ordinal numbers, are traditionally denoted as ω and $\omega + 1$, respectively.

We will start with the following basic property.

Theorem 4.2.1 *If α is an ordinal number and $\eta \in \alpha$ then η is also an ordinal number.*

Proof Assume that α satisfies (A_α)–(C_α) and let $\eta \in \alpha$. We have to show that η satisfies (A_η)–(C_η).

First notice that (A_α) implies $\eta \subset \alpha$. To see (C_η) let $\emptyset \neq B \subset \eta$. Then $\emptyset \neq B \subset \alpha$ and (C_η) follows from (C_α). To see (B_η) take $\beta, \gamma \in \eta$. Then $\beta, \gamma \in \alpha$ and (B_η) follows from (B_α).

Finally, to see (A_η) let $\beta \in \eta$. We have to show that $\beta \subset \eta$. So choose $\xi \in \beta$. The proof will be finished when we show that $\xi \in \eta$. But $\beta \in \eta \subset \alpha$. So, by (A_α), $\beta \subset \alpha$ and $\xi \in \alpha$, since $\xi \in \beta$. We have proved that $\xi, \eta \in \alpha$. Now, using (B_α), we conclude that $\xi \in \eta$, $\xi = \eta$, or $\eta \in \xi$. If $\xi \in \eta$ then we are done. We will prove that the other two cases are impossible.

To obtain a contradiction, assume first that $\eta \in \xi$. Then, in particular, $\eta \in \xi \in \beta \in \eta$. Put $B = \{\xi, \beta, \eta\}$. Then $\emptyset \neq B \subset \alpha$. But then (C_α) fails for this B, since $\eta \in \xi \cap B$, $\xi \in \beta \cap B$, and $\beta \in \eta \cap B$, that is, the sets $\xi \cap B$, $\beta \cap B$, and $\eta \cap B$ are nonempty.

So assume $\xi = \eta$. Then $\eta = \xi \in \beta \in \eta$. Choosing $B = \{\beta, \eta\}$ gives a contradiction similar to that in the previous case. $\qquad\square$

Now notice that for an ordinal number α and for arbitrary $\beta, \gamma \in \alpha$

$$\gamma \subset \beta \text{ if and only if } \gamma = \beta \vee \gamma \in \beta. \tag{4.3}$$

The implication $\Leftarrow$ follows immediately from (A_β), which holds by Theorem 4.2.1. To see the other implication let $\gamma \subset \beta$. Since $\beta, \gamma \in \alpha$, condition (B_α) implies that $\gamma = \beta$, $\gamma \in \beta$, or $\beta \in \gamma$. If either of the first two conditions holds, we are done. But if $\beta \in \gamma$ then, by (A_γ), $\beta \subset \gamma$, and combining this with $\gamma \subset \beta$ we obtain $\gamma = \beta$.

Theorem 4.2.2 *If α is an ordinal number then the relation $\subset$ is a well-ordering relation on α.*

Proof Inclusion $\subset$ is clearly a partial-order relation on any family of sets, so it is on α. It is a linear order on α by (B_α) and (4.3).

To see that α is well ordered by $\subset$, let $\emptyset \neq B \subset \alpha$ and let $\gamma \in B$ be as in (C_α), that is, such that $\gamma \cap B = \emptyset$. We will show that γ is minimal with respect to $\subset$. So choose arbitrary $\beta \in B$. Then, by (B_α) and (4.3), either $\gamma \subset \beta$ or $\beta \in \gamma$. But $\beta \in \gamma$ implies $\gamma \cap B \neq \emptyset$, which is impossible. Thus $\gamma \subset \beta$. Since β was an arbitrary element of B we conclude that γ is a minimal element in B. $\qquad\square$

It is customary that the relation $\subset$ on an ordinal number is denoted by $\leq$. It is also easy to see that, according to our general agreement, the relation $<$ on an ordinal number is identical to $\in$.

Notice also that by Theorem 4.1.5 every proper initial segment of an ordinal number α is of the form $O(\beta)$ for some $\beta \in \alpha$ and that $O(\beta) = \beta$. Thus every initial segment of an ordinal number is an ordinal number.

The basic properties of ordinal numbers are as follows.

Theorem 4.2.3 *For every ordinal numbers α and β*

(i)   *if α and β are order isomorphic then $\alpha = \beta$;*

(ii)   *$\alpha = \beta$, $\alpha < \beta$, or $\beta < \alpha$;*

(iii) *if $\mathcal{F}$ is a nonempty family of ordinal numbers then $\bigcup \mathcal{F}$ is also an ordinal number.*

Proof (i) Let $f \colon \alpha \to \beta$ be an order isomorphism and let

$$Z = \{\gamma \in \alpha \colon f(\gamma) = \gamma\}.$$

We will show, by transfinite induction, that $Z = \alpha$. This will finish the proof, since $Z = \alpha$ implies that f is the identity function on α, and thus $\alpha = f[\alpha] = \beta$.

To see that $Z = \alpha$ let $\gamma \in \alpha$ be such that $O(\gamma) \subset Z$. Then $f[O(\gamma)] = O(f(\gamma))$ and $f(\delta) = \delta$ for all $\delta < \gamma$. Hence

$$f(\gamma) = O(f(\gamma)) = f[O(\gamma)] = \{f(\delta) \colon \delta < \gamma\} = \{\delta \colon \delta < \gamma\} = \gamma.$$

Thus we have proved that for every $\gamma \in \alpha$ condition $O(\gamma) \subset Z$ implies that $\gamma \in Z$. Hence $Z = \alpha$. Condition (i) has been proved.

(ii) By Theorem 4.1.12 either α and β are isomorphic, or α is isomorphic to a proper initial segment of β, or β is isomorphic to a proper initial segment of α. By (i), this implies (ii).

(iii) We have to check that $\delta = \bigcup \mathcal{F}$ satisfies (A_δ)–(C_δ).

(A_δ): If $\beta \in \bigcup \mathcal{F}$ then there is an $\alpha \in \mathcal{F}$ such that $\beta \in \alpha$. Then $\beta \subset \alpha \subset \bigcup \mathcal{F}$.

(B_δ): If $\beta, \gamma \in \bigcup \mathcal{F}$ then there are $\alpha_1, \alpha_2 \in \mathcal{F}$ such that $\beta \in \alpha_1$ and $\gamma \in \alpha_2$. By (ii) either $\alpha_1 \leq \alpha_2$ or $\alpha_2 \leq \alpha_1$. Let α be the greater of these two. Then $\beta, \gamma \in \alpha$ and, by (B_α), $\beta = \gamma$, $\beta \in \gamma$, or $\gamma \in \beta$.

(C_δ): Let $\emptyset \neq B \subset \bigcup \mathcal{F}$ and let $\alpha \in B$. Then there is an $\eta \in \mathcal{F}$ such that $\alpha \in \eta$. Hence, by Theorem 4.2.1, α is an ordinal. If $\alpha \cap B = \emptyset$ put $\gamma = \alpha$. Otherwise, put $\gamma = \min(\alpha \cap B)$, which exists by (C_α). It is easy to see that $\gamma \cap B = \emptyset$. $\qquad\square$

It is not difficult to see that for an ordinal number α the set $\alpha \cup \{\alpha\}$ is also an ordinal number. We will denote it by $\alpha + 1$ and call it the *(ordinal immediate) successor* of α. Number α is also called the *(ordinal immediate) predecessor* of $\alpha+1$. It is also easy to check that $\alpha+1$ is the smallest ordinal number greater than α. Thus every ordinal number has its successor. Not every number, however, has an immediate predecessor; for example, 0 and ω do not have one. The ordinal numbers that have immediate predecessors (i.e., those in the form $\alpha + 1$) are called *ordinal successors*. Those that do not have immediate predecessors are called *limit ordinals*.

The next theorem justifies our intuition of considering ordinal numbers as representatives of the abstract classes of all order-isomorphic well-ordered sets.

Theorem 4.2.4 *For every well-ordered set $\langle W, \leq \rangle$ there exists precisely one ordinal number α that is order isomorphic to W.*

Proof The uniqueness of α follows immediately from Theorem 4.2.3(i).

For $w \in W$ let $O[w] = O(w) \cup \{w\}$ and define

$$Z = \{w \in W : O[w] \text{ is order isomorphic to some ordinal number } \alpha_w\}.$$

Notice that by Theorem 4.2.3(i) and Corollary 4.1.9 the ordinal number α_w and the isomorphism $f_w : O[w] \to \alpha_w$ are unique.

Now, if $v, w \in Z$ and $v \leq w$ then $f_w[O[v]]$ is an ordinal number α, as an initial segment of α_w. In particular, $f_w|_{O[v]} : O[v] \to \alpha$ is an order isomorphism and, by the preceding uniqueness remark, $f_w|_{O[v]} = f_v$. So we have shown that

$$f_v \subset f_w \quad \text{for every } v, w \in Z, \ v \leq w. \tag{4.4}$$

We will prove, by transfinite induction, that $Z = W$.

So let $O(w) \subset Z$. Then, by (4.4), $f = \bigcup_{v \in O(w)} f_v$ is a strictly increasing function from $\bigcup_{v \in O(w)} O[v]$ onto $\bigcup_{v \in O(w)} \alpha_v$. But $\bigcup_{v \in O(w)} O[v] = O(w)$ and, by Theorem 4.2.3(iii), $\alpha = \bigcup_{v \in O(w)} \alpha_v$ is an ordinal number. So $f : O(w) \to \alpha$ is an order isomorphism. Extend f to a function $F : O[w] \to \alpha + 1$ by putting $F(w) = \alpha$. Then the existence of F proves that $w \in Z$. We have shown that $Z = W$.

Now, by (4.4), the function $f = \bigcup_{w \in W} f_w$ is an isomorphism between $W = \bigcup_{w \in W} O[w]$ and an ordinal number $\bigcup_{w \in W} \alpha_w$. $\qquad\square$

In what follows it will be convenient to use the following theorems.

Theorem 4.2.5 *If α and β are ordinal numbers and $f : \alpha \to \beta$ is a strictly increasing function then $\alpha \leq \beta$. Moreover, $\xi \leq f(\xi)$ for every $\xi < \alpha$.*

Proof If $\beta < \alpha$ then $\beta \subset \alpha$ and f is a strictly increasing function from α into α. But then $f(\beta) \in \beta$, that is, $f(\beta) < \beta$, contradicting Theorem 4.1.8. So, by Theorem 4.2.3(ii), $\alpha \leq \beta$.

The additional part follows from Theorem 4.1.8. $\qquad\square$

For a well-ordered set W let $\mathrm{Otp}(W)$ stand for the *order type* of W, that is, the unique ordinal number that is order isomorphic to W.

Corollary 4.2.6 *If W is well ordered and $B \subset W$ then $\mathrm{Otp}(B) \leq \mathrm{Otp}(W)$.*

Proof Let $f\colon \mathrm{Otp}(B) \to B$ and $g\colon W \to \mathrm{Otp}(W)$ be order isomorphisms. Then $g \circ f\colon \mathrm{Otp}(B) \to \mathrm{Otp}(W)$ is strictly increasing, and by Theorem 4.2.5 $\mathrm{Otp}(B) \leq \mathrm{Otp}(W)$. □

We will also introduce the following arithmetic for ordinal numbers. For ordinal numbers α and β we define the *sum* $\alpha + \beta$ of α and β as the order type of the well-ordered set $(\{0\} \times \alpha) \cup (\{1\} \times \beta)$ ordered by

$$\langle i, \xi \rangle \leq \langle j, \zeta \rangle \Leftrightarrow i < j \ \lor \ (i = j \ \& \ \xi \leq \zeta).$$

Thus, we "append" β to α. It is easy to see that the relation so defined is indeed a well-ordering relation.

Similarly, we define the *product* $\alpha\beta$ of ordinal numbers α and β as the order type of the set $\beta \times \alpha$ ordered lexicographically. By Theorem 4.1.13 it is well ordered.

It is not difficult to see that, in general, $\alpha + \beta \neq \beta + \alpha$ and $\alpha\beta \neq \beta\alpha$ (see Exercise 1). We have the following monotonic laws for these operations.

Theorem 4.2.7 *For arbitrary ordinal numbers* α, β, γ

(a) $\alpha < \beta$ *implies* $\gamma + \alpha < \gamma + \beta$;

(b) $\alpha \leq \beta$ *implies* $\alpha + \gamma \leq \beta + \gamma$;

(c) $\alpha < \beta$ *and* $\gamma > 0$ *imply* $\gamma\alpha < \gamma\beta$;

(d) $\alpha \leq \beta$ *implies* $\alpha\gamma \leq \beta\gamma$.

Moreover, the inequalities $\leq$ *in (b) and (d) cannot be replaced by* $<$ *even if* $\alpha < \beta$.

Proof To see (a) let $f\colon (\{0\} \times \gamma) \cup (\{1\} \times \alpha) \to (\{0\} \times \gamma) \cup (\{1\} \times \beta)$ be the identity map. Notice that f is an isomorphism between $(\{0\} \times \gamma) \cup (\{1\} \times \alpha)$ and the initial segment $O(\langle 1, \alpha \rangle)$ of $(\{0\} \times \gamma) \cup (\{1\} \times \beta)$. Thus $\gamma + \alpha$ is isomorphic to a proper initial segment of $\gamma + \beta$, so $\gamma + \alpha < \gamma + \beta$.

To see (b) let $f\colon (\{0\} \times \alpha) \cup (\{1\} \times \gamma) \to (\{0\} \times \beta) \cup (\{1\} \times \gamma)$ be the identity map. Notice that f is strictly increasing. Thus, via an appropriate isomorphism, f can be transformed into a strictly increasing function from $\alpha + \gamma$ into $\beta + \gamma$. But then, by Theorem 4.2.5, $\alpha + \gamma \leq \beta + \gamma$.

To see that equality can hold even if $\alpha < \beta$, notice that $1 + \omega = \omega = 0 + \omega$ (see Exercise 1).

Parts (c) and (d) are left as exercises. □

For an ordinal number α any function a on α is called a *transfinite sequence* and is usually denoted by $\{a_\xi\}_{\xi < \alpha}$ or $\langle a_\xi \colon \xi < \alpha \rangle$, where $a_\xi = a(\xi)$. If α is a domain of a transfinite sequence a then we also often say that a is an *α-sequence*.

EXERCISES

1 Show that for ordinal numbers α and β we might have $\alpha + \beta \neq \beta + \alpha$ and $\alpha\beta \neq \beta\alpha$ by proving that

(a) $1 + \omega = \omega$, so $1 + \omega \neq \omega + 1$;

(b) $2\omega = \omega$, $\omega 2 = \omega + \omega$, and $\omega \neq \omega + \omega$.

2 Prove parts (c) and (d) of Theorem 4.2.7.

4.3   Definitions by transfinite induction

Most mathematicians take for granted that if we have a procedure associating a number a_{n+1} to a number a_n and we have defined a_0, then the sequence $\{a_n\}_{n\in\mathbb{N}}$ constructed in such a way exists. The proof of this fact for a finite portion $\{a_n\}_{n\leq m}$, $m \in \mathbb{N}$, of such a sequence does not require any special axioms. The proof of existence of the infinite sequence $\{a_n\}_{n\in\mathbb{N}}$, however, requires some form of argument, and the axiom of choice is needed to prove any procedure justifying such an argument. (For example, if we stay only within the framework of Peano arithmetic, we have no way to prove the existence of an infinite object $\{a_n\}_{n\in\mathbb{N}}$, since we operate in a world with finite objects only.) This argument is given in the next theorem.

Theorem 4.3.1 (Recursion theorem) *Let Z be a set, α be an ordinal number, and $\mathcal{F}$ be the family of all ξ-sequences for $\xi < \alpha$ with values in Z, that is, $\mathcal{F} = \bigcup_{\xi<\alpha} Z^\xi$. Then for each function $h\colon \mathcal{F} \to Z$ there exists precisely one function $f\colon \alpha \to Z$ such that*

$$f(\xi) = h(f|_\xi) \quad \text{for all } \xi < \alpha. \tag{4.5}$$

The function h from the theorem tells us how to choose $f(\xi)$ knowing the values of the sequence $\langle f(\zeta)\colon \zeta < \xi\rangle$ chosen so far. Because of this, we will sometimes call it an "oracle" function.

Proof of Theorem 4.3.1 First we prove the uniqueness of f. So let g be another function satisfying (4.5) and let $W = \{\xi < \alpha\colon f(\xi) = g(\xi)\}$. It is enough to show that $W = \alpha$. But if $\xi = O(\xi) \subset W$ then $f|_\xi = g|_\xi$ and $f(\xi) = h(f|_\xi) = h(g|_\xi) = g(\xi)$. Hence $\xi \in W$, and by a transfinite induction argument $W = \alpha$. The uniqueness of f has been proved.

To prove the main body of the theorem assume, to obtain a contradiction, that the theorem fails for some ordinal number α. We may assume that α is the smallest ordinal number for which the theorem fails, since if it is not, we can replace it by the minimum

$$\min\{\beta < \alpha\colon \text{ theorem fails for } \beta\}.$$

Let h be a function for which the theorem fails for ordinal α. For every $\xi < \alpha$ there is a function $f_\xi\colon \xi \to Z$ such that

$$f_\xi(\gamma) = h(f_\xi|_\gamma) \text{ for all } \gamma < \xi. \tag{4.6}$$

But for any $\xi \le \zeta < \alpha$ the function $f_\zeta|_\xi$ also satisfies (4.6). Thus, by the uniqueness part that we have already proved, $f_\zeta|_\xi = f_\xi$. In particular,

$$f_\xi(\gamma) = f_\zeta(\gamma) \text{ for every } \gamma < \xi \le \zeta < \alpha.$$

Now, if α is a successor ordinal, say $\alpha = \beta + 1$, then define $f\colon \alpha \to Z$ as an extension of f_β by putting $f(\beta) = h(f_\beta)$. Then f satisfies (4.5) since $f|_\xi = f_\beta|_\xi$ for every $\xi < \alpha$. This contradicts the choice of α.

Thus assume that α is a limit ordinal. Then $\xi + 1 < \alpha$ for every $\xi < \alpha$. Define $f\colon \alpha \to Z$ by

$$f(\xi) = f_{\xi+1}(\xi) \text{ for } \xi < \alpha$$

and notice that such a function satisfies (4.5), since

$$f(\xi) = f_{\xi+1}(\xi) = h(f_{\xi+1}|_\xi) = h(f|_\xi)$$

for every $\xi < \alpha$. This contradicts the choice of α. $\qquad\square$

Most commonly, the oracle function h is defined as a choice function C on some family of sets. In such a case a particular value $h(\xi)$ is chosen as a value of C on some set, whose definition depends on the previously chosen values of f, that is, it is of the form

$$h(p) = C(\{z \in Z\colon \varphi(z,p)\}),$$

where $C\colon \mathcal{P}(Z) \to Z$ is a choice function and $\varphi(z,p)$ is some formula. Of course, such a function h is not defined for the values of $p \in \mathcal{F}$ for which the set $\{z \in Z\colon \varphi(z,p)\}$ is empty. Formally, to use the recursion theorem we have to define it for these values of p as well; however, in most cases we are not interested in such values anyway. Thus, we usually assume that h is arbitrarily defined somehow for such values and we don't even bother to mention it. (The same is true for the other values for the oracle function in which we are not interested.) Also, it is customary in such situations to simply write that we choose in the inductive step

$$f(\xi) \in \{z \in Z\colon \varphi(z, f|_\xi)\}$$

or simply
$$f(\xi) = z \text{ such that } z \text{ satisfies } \varphi(z, f|_\xi)$$
without mentioning explicitly the oracle function at all.

In the applications using the recursion theorem we often separately specify the value of the function h for the empty sequence by $h(0) = A$ and for successor ordinals $\xi + 1$ by $h(\phi) = F(\phi(\xi))$ for some function $F \in Z^Z$, where $\phi \in Z^{\xi+1}$. Thus the value of h on successor ordinal $\xi + 1$ depends only on the value $\phi(\xi)$ of ϕ on its last element ξ. The α-sequence f obtained that way satisfies the conditions

$$f(0) = A, \quad f(\xi + 1) = F(f(\xi)) \text{ for any ordinal } \xi < \alpha,$$

$$f(\lambda) = h(f|_\lambda) \text{ for any limit ordinal } \lambda < \alpha.$$

Notice that to define the function f only on $\alpha = \omega$ we don't need the very last condition. Then the first two conditions can be rewritten as

$$f(0) = A, \quad f(n + 1) = F(f(n)) \text{ for any } n < \omega.$$

The existence of such a sequence is usually deduced from the standard (not transfinite) recursion theorem.

There is probably no need to explain to the reader the importance of this last theorem. Without this theorem it is sometimes hard to prove even very intuitive theorems. As an example we can state here the following characterization of well orderings.

Theorem 4.3.2 *A linearly ordered set $\langle X, \le \rangle$ is well ordered if and only if it does not contain a subset of order type $\omega^\star$ (i.e., a strictly decreasing ω-sequence).*

Proof $\Rightarrow$: If there exists a strictly decreasing sequence $\langle x_n : n < \omega \rangle$ then the set $\{x_n : n < \omega\}$ does not have a smallest element.

$\Leftarrow$: Assume that X is not well ordered and construct, by induction on $n < \omega$, a strictly decreasing sequence $\langle x_n : n < \omega \rangle$. The details are left as an exercise. $\qquad \square$

The next theorem is one of the most fundamental theorems in set theory.

Theorem 4.3.3 (Well-ordering or Zermelo's theorem) *Every nonempty set X can be well ordered.*

Proof Let C be a choice function on the family $\mathcal{P}(X) \setminus \{\emptyset\}$. Choose an arbitrary $p \notin X$, for example, $p = X$, and define $F : \mathcal{P}(X \cup \{p\}) \to X \cup \{p\}$ by

$$F(Z) = \begin{cases} C(Z) & \text{for } Z \in \mathcal{P}(X) \setminus \{\emptyset\}, \\ p & \text{otherwise.} \end{cases}$$

Let $\mathcal{G}$ be the family of all ordinal numbers ξ such that there is a one-to-one ξ-sequence in X. Notice that $\mathcal{G}$ is a set by the replacement axiom, since

$$\mathcal{G} = \{\mathrm{Otp}(\langle Y, \leq\rangle) \colon Y \in \mathcal{P}(X) \ \& \ \leq \in \mathcal{P}(X \times X) \ \& \ \langle Y, \leq\rangle \text{ is well ordered}\}.$$

Let α be an ordinal number that does not belong to $\mathcal{G}$. For example, take $\alpha = (\bigcup \mathcal{G}) + 1$. By the recursion theorem there exists a function $f \colon \alpha \to X \cup \{p\}$ such that

$$f(\xi) = F(X \setminus \{f(\zeta) \colon \zeta < \xi\}) \tag{4.7}$$

for all $\xi < \alpha$. (The function h is defined by $h(s) = F(X \setminus \mathrm{range}(s))$.) Notice that for every $\beta \leq \alpha$

$$\text{if } f[\beta] \subset X \text{ then } f|_\beta \text{ is one-to-one.} \tag{4.8}$$

This is the case since, by (4.7), $f(\xi) \in X \setminus \{f(\zeta) \colon \zeta < \xi\}$ for $\xi < \beta$. In particular, $f[\alpha] \not\subset X$ since there is no one-to-one α-sequence in X. So the set $S = \{\beta \leq \alpha \colon f(\beta) = p\}$ is nonempty. Let $\beta = \min S$. Then $f|_\beta$ is one-to-one and it establishes a well ordering of X of type β by

$$x \preceq y \Leftrightarrow f^{-1}(x) \leq f^{-1}(y). \qquad \qquad \square$$

Remark It is clear that we used the axiom of choice in the proof of Theorem 4.3.3. It is also easy to prove in ZF that Theorem 4.3.3 implies the axiom of choice.[1] To argue for this it is enough to show that Theorem 4.3.3 implies the existence of a choice function for every family $\mathcal{F}$ of nonempty sets. But if $\leq$ is a well ordering of $\bigcup \mathcal{F}$ then the function $f(F) = \min F$ for $F \in \mathcal{F}$ is a choice function defined only with help of the comprehension schema.

The combination of the recursion and well-ordering theorems gives a very strong proving technique. However, for its use usually some kind of cardinal argument is needed, a tool that we still have not developed. Thus, the direct use of this technique will be postponed until subsequent chapters. On the other hand, the next theorem will give us a way to prove the results that require some kind of transfinite induction argument without doing it explicitly. To state it we need the following definitions.

For a partially ordered set $\langle P, \leq\rangle$ we say that a set $S \subset P$ is a *chain* in P if S is linearly ordered by $\leq$. An element $b \in P$ is said to be an *upper bound* of a set $S \subset P$ if $s \leq b$ for every $s \in S$. Similarly, $b \in P$ is a *lower bound* of $S \subset P$ if $b \leq s$ for every $s \in S$.

---

[1] ZF stands for the ZFC axioms from which the axiom of choice AC has been removed. See Appendix A.

Notice that not all subsets of a partially ordered set must have upper or lower bounds. For example, $(0, \infty) \subset \mathbb{R}$ does not have an upper bound and every number $r \leq 0$ is a lower bound of $(0, \infty)$.

Let us also recall that $m \in P$ is a maximal element of $\langle P, \leq \rangle$ if there is no $p \in P$ such that $m < p$.

Theorem 4.3.4 (Zorn's lemma[2]) *If $\langle P, \leq \rangle$ is a partially ordered set such that every chain in P has an upper bound then P has a maximal element m.*

Proof The proof is similar to that of Theorem 4.3.3. Let C be a choice function on the family $\mathcal{P}(P) \setminus \{\emptyset\}$. Choose an arbitrary $x \notin P$, for example, $x = P$, and define $F \colon \mathcal{P}(P \cup \{x\}) \to P \cup \{x\}$ by

$$F(Z) = \begin{cases} C(Z) & \text{for } Z \in \mathcal{P}(P) \setminus \{\emptyset\}, \\ x & \text{otherwise.} \end{cases}$$

Let $\mathcal{G}$ be the family of all ordinal numbers ξ such that there is a strictly increasing sequence $\langle p_\zeta \colon \zeta < \xi \rangle$ in P. Let α be an ordinal number that does not belong to $\mathcal{G}$. For example, take $\alpha = (\bigcup \mathcal{G}) + 1$.

Define a function h on the family of all ξ-sequences $\langle p_\zeta \colon \zeta < \xi \rangle$ in $P \cup \{x\}$, $\xi < \alpha$, by putting

$$h(\langle p_\zeta \colon \zeta < \xi \rangle) = F(B \setminus \{p_\zeta \colon \zeta < \xi\}), \tag{4.9}$$

where $B = \{b \in P \colon b \text{ is an upper bound of } \{p_\zeta \colon \zeta < \xi\}\}$. (Notice that $B = \emptyset$ if $\{p_\zeta \colon \zeta < \xi\} \not\subset P$.) By the recursion theorem, there is a function f such that

$$f(\xi) = h(f|_\xi) \quad \text{for all } \xi < \alpha.$$

Notice that for every $\beta \leq \alpha$

$$\text{if } f[\beta] \subset P \text{ then } f|_\beta \text{ is strictly increasing.} \tag{4.10}$$

This is the case, since then, by (4.9), $f(\xi) \in B \setminus \{p_\zeta \colon \zeta < \xi\}$ for $\xi < \beta$. In particular, $f[\alpha] \not\subset P$ since there is no strictly increasing α-sequence in P. So the set $S = \{\beta \leq \alpha \colon f(\beta) = x\}$ is nonempty. Let $\beta = \min S$. Then $f|_\beta$ is strictly increasing, so $B = \{b \in P \colon b \text{ is an upper bound of } \{p_\xi \colon \xi < \beta\}\}$ is nonempty, $\{p_\xi \colon \xi < \beta\}$ being a chain. Let $m \in B$. It is enough to argue that m is a maximal element of P. But if there is a $y \in P$ with $y > m$, then $y > m \geq p_\xi$ for every $\xi < \beta$. In particular, $y \in B \setminus \{p_\xi \colon \xi < \beta\}$, so

---

[2] M. Zorn proved this lemma in 1935 and published it in *Bulletin AMS*. The same theorem was proved in 1922 by K. Kuratowski and published in *Fundamenta Mathematicae*. Thus priority for this theorem belongs without any doubt to Kuratowski. However, in essentially all published sources the name Zorn is associated to this theorem and it seems that the battle for historical justice has been lost.

$B \setminus \{p_\xi \colon \xi < \beta\} \neq \emptyset$. Thus, by (4.9), $f(\beta) \in P$, contradicting the choice of β.   □

Zorn's lemma is used most often in the situation when P is a family of subsets of a set A ordered by inclusion and such that for every chain $S \subset P$ its union $\bigcup S$ is also in P. (Notice that $\bigcup S$ is an upper bound for S with respect to $\subset$.) In this particular case Zorn's lemma has an especially nice form, stated as the following corollary.

Corollary 4.3.5 (Hausdorff maximal principle) *If $\mathcal{F}$ is a nonempty family of subsets of a set A such that for every chain $S \subset \mathcal{F}$ its union $\bigcup S$ belongs to $\mathcal{F}$ then $\mathcal{F}$ has an $\subset$-maximal element.*

Remark Zorn's lemma and the Hausdorff maximal principle are equivalent to the axiom of choice within the ZF theory. To see this first notice that Zorn's lemma follows from the axiom of choice (Theorem 4.3.4) and that the Hausdorff maximal principle follows from Zorn's lemma (Corollary 4.3.5). To see that the Hausdorff maximal principle implies the axiom of choice let $\mathcal{G}$ be a nonempty family of pairwise-disjoint nonempty sets. We have to find a selector S for $\mathcal{G}$. So let $\mathcal{F}$ be the family of all $T \subset \bigcup \mathcal{G}$ such that $T \cap G$ has at most one element for every $G \in \mathcal{G}$. Notice that $\mathcal{F}$ satisfies the assumptions of the Hausdorff maximal principle. Let $S \in \mathcal{F}$ be maximal. We will show that S is a selector for $\mathcal{G}$. But if S is not a selector for $\mathcal{G}$ then $G \cap S = \emptyset$ for some $G \in \mathcal{G}$. Then there is a $g \in G$ and $S \cup \{g\}$ is in $\mathcal{F}$, contradicting the maximality of S.

EXERCISE

1 Complete the details of the proof of Theorem 4.3.2.

4.4   Zorn's lemma in algebra, analysis, and topology

In this section we will see three standard applications of Zorn's lemma in three main branches of mathematics: algebra, analysis, and topology. Each of these areas will be represented, respectively, by the theorem that every linear space has a basis, the Hahn–Banach theorem, and the Tychonoff theorem. Going through these proofs should help the reader to appreciate the power of Zorn's lemma and, implicitly, the axiom of choice. The Hahn–Banach theorem and the Tychonoff theorem will not be used in the remaining part of this text.

Algebraic application To state the algebraic example let us recall the following definitions. A set G with a binary operation $+ \colon G \times G \to G$ is a *group* if $+$ is associative (i.e., $(u+v)+w = u+(v+w)$ for all $u, v, w \in G$), G has the identity element 0 (i.e., $0 + v = v + 0 = v$ for every $v \in G$), and every element of G has an inverse element (i.e., for every $v \in G$ there exists $-v \in G$ such that $(-v) + v = v + (-v) = 0$). A group $\langle G, + \rangle$ is *Abelian* if $a + b = b + a$ for every $a, b \in G$. A set K with two binary operations $+$ and $\cdot$ on K is a *field* if $\langle K, + \rangle$ is an Abelian group with 0 as an identity element, $\langle K \setminus \{0\}, \cdot \rangle$ is an Abelian group with 1 as an identity element, and $a(b+c) = ab + ac$ for every $a, b, c \in K$. In what follows we will mainly be concerned with the fields $\langle \mathbb{R}, +, \cdot \rangle$ of real numbers and $\langle \mathbb{Q}, +, \cdot \rangle$ of rational numbers.

An Abelian group $\langle V, + \rangle$ is said to be a *linear space* (or *vector space*) over a field K if there is an operation from $K \times V$ to V, $\langle k, v \rangle \to kv$, such that $(k + l)v = kv + lv$, $k(v + w) = kv + kw$, $k(lv) = (kl)v$, $0v = 0$, and $1v = v$ for every $k, l \in K$ and $v, w \in V$. A subset S of a vector space V over a field K is a *linear subspace* of V if it is a linear space when considered with the same operations. The main examples of vector spaces considered in this text will be $\mathbb{R}^n$ over either $\mathbb{R}$ or $\mathbb{Q}$.

A subset S of a vector space V over a field K is *linearly independent* if for every finite number of distinct elements $v_1, \ldots, v_n$ of S and every $k_1, \ldots, k_n \in K$ the condition $k_1 v_1 + \cdots + k_n v_n = 0$ implies $k_1 = \cdots = k_n = 0$. A subset S of a vector space V over a field K *spans* V if every $v \in V$ can be represented as $v = k_1 v_1 + \cdots + k_n v_n$ for some $v_1, \ldots, v_n \in S$ and $k_1, \ldots, k_n \in K$. A *basis* of V is a linearly independent subset of V that spans V.

It is easy to see that if B is a basis of V then every $v \in V$ has a unique representation $v = k_1 v_1 + \cdots + k_n v_n$, where $v_1, \ldots, v_n \in B$ are different and $k_1, \ldots, k_n \in K$ (and where we ignore the v_is for which $k_i = 0$). This is the case since if $v = l_1 v_1 + \cdots + l_n v_n$ is a different representation of v, then $(l_1 - k_1)v_1 + \cdots + (l_n - k_n)v_n = v - v = 0$ so $l_1 - k_1 = \cdots = l_n - k_n = 0$. Thus $l_i = k_i$ for $i \in \{1, \ldots, n\}$.

Theorem 4.4.1 *If S_0 is a linearly independent subset of a vector space V over K then there exists a basis $\mathcal{B}$ of V that contains S_0.*

In particular, every vector space has a basis.

Proof The additional part follows from the main part, since the empty set is linearly independent in any vector space.

To prove the main part of the theorem, let

$$\mathcal{F} = \{S \subset V \colon S_0 \subset S \text{ and } S \text{ is linearly independent in } V\}.$$

Notice first that $\mathcal{F}$ satisfies the assumptions of the Hausdorff maximal principle.

Clearly $\mathcal{F}$ is nonempty, since $S_0 \in \mathcal{F}$. To check the main assumption, let $\mathcal{G} \subset \mathcal{F}$ be a chain in $\mathcal{F}$ with respect to $\subset$. We will show that $\bigcup \mathcal{G}$ is linearly independent in V. So let $v_1, \ldots, v_n$ be different elements of $\bigcup \mathcal{G}$ and choose $k_1, \ldots, k_n \in K$ such that $k_1 v_1 + \cdots + k_n v_n = 0$. For every $i \in \{1, \ldots, n\}$ let $G_i \in \mathcal{G}$ be such that $v_i \in G_i$. Since $\{G_1, \ldots, G_n\}$ is a finite subset of a linearly ordered set $\mathcal{G}$, we can find the largest element, say G_j, in this set. Then $v_i \in G_i \subset G_j$ for all $i \in \{1, \ldots, n\}$. Hence all v_is are in a linearly independent set G_j. Thus $k_1 v_1 + \cdots + k_n v_n = 0$ implies $k_1 = \cdots = k_n = 0$. So we can use the Hausdorff maximal principle on $\mathcal{F}$.

Let $\mathcal{B}$ be a maximal element in $\mathcal{F}$. We will show that it is a basis of V. It is linearly independent, since it belongs to $\mathcal{F}$. So it is enough to prove that $\mathcal{B}$ spans V.

To obtain a contradiction assume that there is a $v \in V$ such that

$$v \neq k_1 v_1 + \cdots + k_n v_n \tag{4.11}$$

for every $v_1, \ldots, v_n \in V$ and $k_1, \ldots, k_n \in K$. We will show that this implies that $\mathcal{B} \cup \{v\}$ is linearly independent in V, which contradicts the maximality of $\mathcal{B}$ in $\mathcal{F}$. So choose different elements $v_1, \ldots, v_n$ from $\mathcal{B}$ and $k_0, \ldots, k_n \in K$ such that $k_0 v + k_1 v_1 + \cdots + k_n v_n = 0$. There are two cases.

Case 1: $k_0 = 0$. Then $k_1 v_1 + \cdots + k_n v_n = 0$ and $v_1, \ldots, v_n$ are from $\mathcal{B}$. Hence, also $k_1 = \cdots = k_n = 0$.

Case 2: $k_0 \neq 0$. Then $v = -(k_1/k_0)v_1 + \cdots - (k_n/k_0)v_n$, contradicting (4.11). □

Analytic application For the next theorem the term vector space will be used for the vector spaces over the field $\mathbb{R}$ of real numbers. Recall that a function f is said to be a *linear functional* on a vector space V if $f : V \to \mathbb{R}$ is such that $f(ax + by) = af(x) + bf(y)$ for every $x, y \in V$ and $a, b \in \mathbb{R}$.

Theorem 4.4.2 (Hahn–Banach theorem) *Let V be a vector space and $p : V \to \mathbb{R}$ be such that $p(x + y) \leq p(x) + p(y)$ and $p(ax) = ap(x)$ for all $x, y \in V$ and $a \geq 0$. If f is a linear functional on a linear subspace S of V such that $f(s) \leq p(s)$ for all $s \in S$, then there exists a linear functional F on V such that F extends f and $F(x) \leq p(x)$ for all $x \in V$.*

Proof Let $\mathcal{F}$ be the family of all functionals g on a linear subspace T of V such that g extends f and $g(x) \leq p(x)$ for all $x \in T$. First we will show that $\mathcal{F}$ satisfies the assumptions of the Hausdorff maximal principle.

So let $\mathcal{G} \subset \mathcal{F}$ be a chain in $\mathcal{F}$ with respect to $\subset$. We will show that $\bigcup \mathcal{G} \in \mathcal{F}$. Clearly $f \subset \bigcup \mathcal{G}$. To see that $g = \bigcup \mathcal{G}$ is a functional, notice first that g is a function since for $\langle x, s_0 \rangle, \langle x, s_1 \rangle \in g$ we can find $g_0, g_1 \in \mathcal{G}$ such that $\langle x, s_i \rangle \in g_i$ for $i < 2$. Then, for $g_0 \subset g_1$, $s_0 = g_0(x) = g_1(x) = s_1$. Similarly, if $g_1 \subset g_0$ then $s_1 = g_1(x) = g_0(x) = s_0$.

In a similar way we show that the domain of g is a linear subspace and that g is a linear functional on its domain. Since also $g(x) \leq p(x)$ for all $x \in \text{dom}(g)$, we conclude that $g \in \mathcal{F}$, that is, $\mathcal{F}$ satisfies the assumptions of the Hausdorff maximal principle.

Now let $F \colon T \to \mathbb{R}$ be a maximal element of $\mathcal{F}$. Then F is a linear functional extending f such that $F(x) \leq p(x)$ for every $x \in T$. It is enough to prove that $T = V$.

To obtain a contradiction assume that $V \neq T$ and let $w \in V \setminus T$. We will show that this contradicts the maximality of F by constructing a $G \in \mathcal{F}$ that extends F and is defined on a linear subspace $V_0 = \text{Span}(T \cup \{w\}) = \{aw + t \colon a \in \mathbb{R}, \ t \in T\}$ of V.

First notice that for every $s, t \in T$

$$F(s) + F(t) = F(s+t) \leq p(s+t) = p(s-w+t+w) \leq p(s-w) + p(t+w)$$

and so

$$-p(s-w) + F(s) \leq p(t+w) - F(t).$$

Hence, in particular,

$$\sup_{t \in T}[-p(t-w) + F(t)] \leq \inf_{t \in T}[p(t+w) - F(t)].$$

Choose α such that $\sup[-p(t-w) + F(t)] \leq \alpha \leq \inf[p(t+w) - F(t)]$. Then

$$\alpha \leq \inf[p(t+w) - F(t)] \ \& \ -\alpha \leq \inf[p(t-w) - F(t)]. \tag{4.12}$$

Now for $a \in \mathbb{R}$ and $t \in T$ define

$$G(aw + t) = a\alpha + F(t).$$

Clearly G is a functional on V_0 that extends F. We have to show only that $G(s) \leq p(s)$ for all $s \in V_0$. So let $s = aw + t$ with $t \in T$, $a \in \mathbb{R}$.

If $a = 0$ then $G(aw + t) = F(t) \leq p(t) = p(aw + t)$.

If $a > 0$ then, by the first part of (4.12),

$$
\begin{aligned}
G(aw + t) &= a\alpha + F(t) \\
&= a[\alpha + F(t/a)] \\
&\leq a[[p(t/a + w) - F(t/a)] + F(t/a)] \\
&= ap(t/a + w) \\
&= p(t + aw).
\end{aligned}
$$

If $a = -b < 0$ then, by the second part of (4.12),

$$
\begin{aligned}
G(aw + t) &= -b\alpha + F(t) \\
&= b[-\alpha + F(t/b)] \\
&\leq b[[p(t/b - w) - F(t/b)] + F(t/b)] \\
&= bp(t/b - w) \\
&= p(t + aw).
\end{aligned}
$$

This finishes the proof of Theorem 4.4.2.   □

Topological application Let us start by recalling some definitions. A family $\mathcal{F}$ of subsets of a set X is said to be a *cover* of a set $A \subset X$ if $A \subset \bigcup \mathcal{F}$. If $\mathcal{F}$ is a cover of A then a family $\mathcal{G} \subset \mathcal{F}$ is said to be a *subcover* of $\mathcal{F}$ if $A \subset \bigcup \mathcal{G}$. If $\langle X, \tau \rangle$ is a topological space then $\mathcal{F}$ is said to be an *open cover* of $A \subset X$ if $A \subset \bigcup \mathcal{F}$ and $\mathcal{F} \subset \tau$. A topological space $\langle X, \tau \rangle$ is *compact* if every open cover of X has a finite subcover.

For a topological space $\langle X, \tau \rangle$ a family $\mathcal{B} \subset \tau$ is said to be a *base* for X if for every $U \in \tau$ and every $x \in U$ there exists a $B \in \mathcal{B}$ such that $x \in B \subset U$. A family $\mathcal{S} \subset \tau$ is said to be a *subbase* of topological space $\langle X, \tau \rangle$ if the family

$$
\mathcal{B}(\mathcal{S}) = \left\{ \bigcap_{i < n} S_i \in \mathcal{P}(X) \colon n \in \mathbb{N} \ \& \ \forall i < n \ (S_i \in \mathcal{S}) \right\}
$$

forms a base for X. It is easy to see that for any family $\mathcal{S}$ of subsets of a set X the family

$$
\mathcal{T}(\mathcal{S}) = \left\{ \bigcup \mathcal{G} \colon \mathcal{G} \subset \mathcal{B}(\mathcal{S}) \right\}
$$

forms a topology on X. This topology is said to be *generated* by $\mathcal{S}$. The family $\mathcal{S}$ is also a subbase for the topology $\mathcal{T}(\mathcal{S})$.

Finally, let us recall that for a family $\{\langle X_\alpha, \tau_\alpha \rangle\}_{\alpha \in \mathcal{A}}$ of topological spaces their *Tychonoff product* is a topological space defined on the product set $X = \prod_{\alpha \in \mathcal{A}} X_\alpha$ with topology generated by the subbase

$$
\{p_\alpha^{-1}(U) \colon \alpha \in \mathcal{A} \ \& \ U \in \tau_\alpha\},
$$

where $p_\alpha \colon X \to X_\alpha$ is the projection of X onto its αth coordinate X_α.

Theorem 4.4.3 (Tychonoff theorem) *The product of an arbitrary family of compact spaces is compact.*

The proof will be based on the following lemma.

Lemma 4.4.4 (Alexander subbase theorem) *If there is a subbase $\mathcal{S}$ of a topological space X such that every open cover $\mathcal{V} \subset \mathcal{S}$ of X has a finite subcover then X is compact.*

Proof Let $\mathcal{S}$ be a subbase of X such that any cover $\mathcal{V} \subset \mathcal{S}$ of X has a finite subcover. Let $\mathcal{B} = \mathcal{B}(\mathcal{S})$ be a base for X generated by $\mathcal{S}$ and, to obtain a contradiction, assume that X is not compact. Then there is a subfamily $\mathcal{U}_0$ of $\mathcal{B}(\mathcal{S})$ that covers X but does not have a finite subcover. (We use here an elementary fact that for any base $\mathcal{B}$ of a topological space X, the space X is compact if and only if every cover $\mathcal{U} \subset \mathcal{B}$ of X has a finite subcover.)

Define $\mathcal{F}$ as

$$\left\{\mathcal{U} \subset \mathcal{B} \colon \mathcal{U}_0 \subset \mathcal{U}, \ X \subset \bigcup \mathcal{U}, \text{ and there is no finite subcover } \mathcal{V} \subset \mathcal{U} \text{ of } X\right\}.$$

To see that $\mathcal{F}$ satisfies the assumptions of the Hausdorff maximal principle let $\mathcal{G} \subset \mathcal{F}$ be a chain. If $\mathcal{T} \subset \bigcup \mathcal{G}$ is finite then there is a $\mathcal{U} \in \mathcal{G}$ such that $\mathcal{T} \subset \mathcal{U}$. So $\mathcal{T}$ cannot cover X, since $\mathcal{U} \in \mathcal{F}$. Thus $\bigcup \mathcal{G}$ does not have a finite subcover, that is, $\bigcup \mathcal{G} \in \mathcal{F}$.

So, by the Hausdorff maximal principle, there is a maximal element $\mathcal{U} \in \mathcal{F}$. Now let $U \in \mathcal{U}$. Then $U = \bigcap_{i<n} U_i$ for some $U_i \in \mathcal{S}$. We claim that

$$\text{there is a } j < n \text{ such that } U_j \in \mathcal{U}. \tag{4.13}$$

Otherwise, for every $i < n$ we can find a finite $\mathcal{U}_i \subset \mathcal{U}$ such that $\mathcal{U}_i \cup \{U_i\}$ covers X, that is, $\mathcal{U}_i$ covers $X \setminus U_i$. But then $\bigcup_{i<n} \mathcal{U}_i$ covers $\bigcup_{i<n} (X \setminus U_i) = X \setminus \bigcap_{i<n} U_i = X \setminus U$. So $\bigcup_{i<n} \mathcal{U}_i \cup \{U\} \subset \mathcal{U}$ is a finite subcover of X, contradicting the fact that $\mathcal{U} \in \mathcal{F}$.

But (4.13) implies that for every $U \in \mathcal{U}$ there is a $V = U_j \in \mathcal{U} \cap \mathcal{S}$ such that $U \subset V$. So $\mathcal{V} = \mathcal{U} \cap \mathcal{S}$ is also a cover of X. Hence, by our assumption, $\mathcal{V}$ has a finite subcover of X. This contradiction finishes the proof. □

Proof of the Tychonoff theorem Let

$$\mathcal{S} = \{p_\alpha^{-1}(U) \colon \alpha \in \mathcal{A} \ \& \ U \in \tau_\alpha\}$$

be a subbase of $X = \prod_{\alpha \in \mathcal{A}} X_\alpha$ and let $\mathcal{U} \subset \mathcal{S}$ be a cover of X. By the Alexander subbase theorem it is enough to find a finite subcover of $\mathcal{U}$. For $\alpha \in \mathcal{A}$ let

$$\mathcal{U}_\alpha = \{U \subset X_\alpha \colon U \text{ is open in } X_\alpha \text{ and } p_\alpha^{-1}(U) \in \mathcal{U}\}.$$

If for some $\alpha \in \mathcal{A}$ the family $\mathcal{U}_\alpha$ covers X_α then, by the compactness of X_α, there is a finite subcover $\mathcal{V} \subset \mathcal{U}_\alpha$ that covers X_α. But then

$\mathcal{U}' = \{p_\alpha^{-1}(U)\colon U \in \mathcal{V}\} \subset \mathcal{U}$ is a finite subcover of X. So, to obtain a contradiction, assume that no $\mathcal{U}_\alpha$ covers X_α and let $x(\alpha) \in X_\alpha \setminus \bigcup \mathcal{U}_\alpha$ for every $\alpha \in A$. Then $x \in X \setminus \bigcup_{\alpha \in A} \bigcup\{p_\alpha^{-1}(U)\colon U \in \mathcal{U}_\alpha\} = X \setminus \bigcup \mathcal{U}$. This finishes the proof. $\qquad\square$

EXERCISES

1 Prove that every partial-order relation $\preceq$ on a set X can be extended to a linear-order relation $\leq$ on X. (Here $\leq$ extends $\preceq$ if $\preceq \subset \leq$.)

2 We say that a subset $A \subset \mathbb{R}$ is algebraically independent if for every nonzero polynomial $p(x_1, \ldots, x_n)$ of n variables with rational coefficients and any sequence $a_1, \ldots, a_n$ of different elements from A, $p(a_1, \ldots, a_n) \neq 0$. Show that there exists an algebraically independent subset $\mathcal{A}$ of $\mathbb{R}$ such that if $\mathbb{Q}(\mathcal{A})$ is a field generated by $\mathbb{Q}$ and $\mathcal{A}$ then for every $b \in \mathbb{R} \setminus \mathbb{Q}(\mathcal{A})$ there exists a nonzero polynomial $p(x)$ with coefficients in $\mathbb{Q}(\mathcal{A})$ such that $p(b) = 0$. (A family $\mathcal{A}$ with this property is called a *transcendental basis* of $\mathbb{R}$ over $\mathbb{Q}$.)

3 A *filter* on a set X is a nonempty family $\mathcal{F}$ of subsets of X such that (1) $A \cap B \in \mathcal{F}$ provided $A, B \in \mathcal{F}$; and (2) if $A \subset B \subset X$ and $A \in \mathcal{F}$ then $B \in \mathcal{F}$. A filter $\mathcal{F}$ on X is *proper* if $\mathcal{F} \neq \mathcal{P}(X)$, and it is *prime* if for every $A \subset X$ either $A \in \mathcal{F}$ or $X \setminus A \in \mathcal{F}$. Show that every proper filter on a set X can be extended to a proper prime filter.

4 A *graph* on a (finite or infinite) set V is an ordered pair $\langle V, E \rangle$ such that $E \subset [V]^2$, where $[V]^2$ is the set of all two-element subsets of V. For $E_0 \subset E$ a graph $\langle V, E_0 \rangle$ is a *forest* in graph $\langle V, E \rangle$ if it does not contain any cycle, that is, if there is no sequence $v_0, v_1, \ldots, v_n = v_0$ with $n > 2$ such that $\{v_i, v_{i+1}\} \in E_0$ for $i \in n$. For $E_0 \subset E$ a graph $\langle V, E_0 \rangle$ *spans* $\langle V, E \rangle$ if for every $v \in V$ there is a $w \in V$ such that $\{v, w\} \in E_0$.

Let $\langle V, E \rangle$ be a graph that spans itself. Show that there exists a forest $\langle V, E_0 \rangle$ that spans $\langle V, E \rangle$.

Chapter 5

Cardinal numbers

5.1 Cardinal numbers and the continuum hypothesis

We say that the sets A and B have the *same cardinality* and write $A \approx B$ if there exists a bijection $f\colon A \to B$.

It is easy to see that for every A, B, and C

- $A \approx A$;

- if $A \approx B$ then $B \approx A$; and

- if $A \approx B$ and $B \approx C$ then $A \approx C$.

Thus the "relation" $\approx$ of having the same cardinality is an equivalence "relation." We put the word *relation* in quotation marks since our definition does not specify any field for $\approx$. This means that the "field" of this "relation" is the class of all sets. However, the restriction of $\approx$ to any set X is an equivalence relation on X. Thus we will use the term *relation* for $\approx$ in this sense.

By Zermelo's theorem (Theorem 4.3.3) and Theorem 4.2.4, for every set A there exists an ordinal number α such that $A \approx \alpha$. The smallest ordinal number with this property is called the *cardinality* of A and is denoted by $|A|$. Thus

$$|A| = \min\{\alpha\colon\ \alpha \text{ is an ordinal number and } A \approx \alpha\}.$$

In particular, $A \approx |A|$ for every A. It is not difficult to see that $A \approx B$ if

61

and only if $|A| = |B|$. Thus the terminology "sets A and B have the same cardinality" can be used for $A \approx B$ as well as for $|A| = |B|$.

Notice also that

$$|\alpha| \leq \alpha \ \text{ for every ordinal number } \ \alpha,$$

since $\alpha \approx \alpha$ and $|\alpha|$ is the smallest ordinal β such that $\alpha \approx \beta$. This implies also that

$$|\alpha| \leq |\beta| \text{ for all ordinal numbers } \alpha \leq \beta.$$

An ordinal number κ is said to be a *cardinal number* (or just a *cardinal*) provided $\kappa = |A|$ for some set A. We can distinguish the cardinal numbers from all other ordinal numbers by using the following properties.

Proposition 5.1.1 *For an ordinal number α the following conditions are equivalent:*

(i)   *α is a cardinal number;*

(ii)   *$|\alpha| = \alpha$;*

(iii) *$\beta < |\alpha|$ for every $\beta < \alpha$;*

(iv)   *$|\beta| < |\alpha|$ for every $\beta < \alpha$;*

(v)   *$|\beta| \neq |\alpha|$ for every $\beta < \alpha$.*

Proof (i)$\Rightarrow$(ii): If α is a cardinal number then there exists a set A such that $|A| = \alpha$. This means, in particular, that $A \approx \alpha$, that is, $|A| = |\alpha|$. So $|\alpha| = |A| = \alpha$.

(ii)$\Rightarrow$(iii): It is obvious.

(iii)$\Rightarrow$(iv): It follows from the fact that $|\beta| \leq \beta$ for every ordinal β.

(iv)$\Rightarrow$(v): It is obvious.

(v)$\Rightarrow$(i): Condition (v) implies that $\beta \not\approx \alpha$ for every $\beta < \alpha$. Thus α is the smallest ordinal γ such that $\alpha \approx \gamma$. But this means, by the definition of cardinality, that α is a cardinal number, as then $\alpha = |\alpha|$.   □

Property (v) of Proposition 5.1.1 explains why we often say that the cardinal numbers are the *initial ordinals*, that is, ordinals that are the smallest of a given cardinality.

Let us also note the following theorem.

Theorem 5.1.2 *Let A and B be arbitrary sets. The following conditions are equivalent:*

(i) $|A| \leq |B|$;

(ii)   *there exists a one-to-one function $g \colon A \to B$.*

Moreover, if $A \neq \emptyset$ then these conditions are equivalent to the condition that

(iii)   *there exists a function $f \colon B \to A$ from B onto A.*

Proof For $A = \emptyset$ the equivalence of (i) and (ii) is obvious. So assume that $A \neq \emptyset$.

(i)$\Rightarrow$(iii): Let $h \colon B \to |B|$ and $g \colon |A| \to A$ be bijections. Pick $a \in A$. Notice that $|A| \subset |B|$ and $h^{-1}(|A|) \subset B$.

Define $f \colon B \to A$ by

$$f(b) = \begin{cases} g(h(b)) & \text{for } b \in h^{-1}(|A|), \\ a & \text{otherwise.} \end{cases}$$

Then

$$A \supset f[B] \supset f[h^{-1}(|A|)] = (g \circ h)[h^{-1}(|A|)] = g[|A|] = A.$$

So f is onto A.

(iii)$\Rightarrow$(ii): Let $f \colon B \to A$ be onto. Let G be a choice function for the family $\{f^{-1}(a) \colon a \in A\}$. Then $g(a) = G(f^{-1}(a))$ is a one-to-one function from A into B.

(ii)$\Rightarrow$(i): Let $g \colon A \to B$ be a one-to-one function and $h \colon B \to |B|$ be a bijection. Then $(h \circ g) \colon A \to |B|$ is a one-to-one function from A onto $(h \circ g)[A] \subset |B|$. Hence $|A| = |(h \circ g)[A]|$ and, by Corollary 4.2.6, $\mathrm{Otp}((h \circ g)[A]) \leq |B|$. So

$$|A| = |(h \circ g)[A]| = |\mathrm{Otp}((h \circ g)[A])| \leq \mathrm{Otp}((h \circ g)[A]) \leq |B|. \qquad \Box$$

Corollary 5.1.3 *If $A \subset B$ then $|A| \leq |B|$.*

Proof Since the identity function from A into B is one-to-one, the corollary follows immediately from Theorem 5.1.2. $\qquad \Box$

Corollary 5.1.4 *If $A \subset B \subset C$ and $|A| = |C|$ then $|B| = |A|$.*

Proof By Corollary 5.1.3 we have $|A| \leq |B| \leq |C| = |A|$. So $|B| = |A|$. $\Box$

Remark It is often the case that the relation $A \preceq B$ for a set A having cardinality less than or equal to the cardinality of B, $|A| \leq |B|$, is defined

as in Theorem 5.1.2(ii), that is, by saying that there exists a one-to-one
function $f: A \to B$. Then it is necessary to prove that $A \preceq B$ and $B \preceq A$
imply that there exists a bijection from A onto B. This fact is known as the
Schröder–Bernstein theorem and follows immediately from our definition
and Theorem 5.1.2.

Theorem 5.1.5 ω *is a cardinal number and every* $n \in \omega$ *is a cardinal
number.*

Proof First we show that every $n \in \omega$ is a cardinal number. So let $n \in \omega$.
By Proposition 5.1.1(iv) it is enough to prove that $|k| < |n|$ for every $k < n$.
Thus it is enough to show that the inequality $|n| \leq |k|$ is false, that is, by
Theorem 5.1.2, that

$$\text{there is no one-to-one function } f: n \to k \text{ for every } k < n. \quad (5.1)$$

We will prove (5.1) by induction on $n < \omega$.

Notice that for $n = 0$ condition (5.1) is true, since there is no $k < n$.
So assume that it is true for some $n < \omega$. We will show that this implies
(5.1) for $n + 1$.

To obtain a contradiction assume that for some $k < n+1$ there is a one-
to-one function $f: (n+1) \to k$. If $k < n$ then $f|_n: n \to k$ contradicts (5.1).
So $k = n$, $f: (n+1) \to n$, and $f[n+1] \not\subset (n-1)$, that is, $n-1 \in f[n+1]$.
(Notice that $n - 1$ exists, since the existence of $f: (n+1) \to n$ implies that
$n \neq 0$.)

Let $i < n+1$ be such that $f(i) = n-1$. Define $g: n \to n-1$ by putting
$g(j) = f(j)$ for $j < n$, $j \neq i$, and $g(i) = f(n)$. (If $i = n$ this last condition is
redundant and can simply be ignored.) Notice that the values of g indeed
belong to $n - 1$ and that g is one-to-one. Thus g contradicts (5.1) for n.
This contradiction finishes the proof of (5.1) and the fact that every $n \in \omega$
is a cardinal number.

To see that ω is a cardinal number notice that $n \subset (n+1) \subset \omega$ for every
$n < \omega$. Hence, by Corollary 5.1.3, $|n| \leq |n+1| \leq |\omega|$. But by what we have
already proved, $|n| = n < n+1 = |n+1|$. So $|n| < |n+1| \leq |\omega|$, that is, we
conclude that $|n| < |\omega|$ for every $n < \omega$. Therefore, by Proposition 5.1.1,
we deduce that ω is a cardinal number. $\qquad\square$

The sets with cardinality less than ω, that is, equal to some $n < \omega$, are
called *finite sets*. A set A is *countable* if $|A| = \omega$. A set is *infinite* if it is
not finite. A set is *uncountable* if it is infinite and not countable.

We have already established a convention that general infinite ordinal
numbers are denoted by the greek letters $\alpha, \beta, \gamma, \zeta, \eta, \xi$. The infinite cardinal numbers will usually be denoted by the letters κ, μ, λ.

The natural numbers and ω are cardinal numbers. Are there any other
(i.e., uncountable) cardinal numbers? The next theorem gives us a tool to
construct a lot of uncountable cardinal numbers.

Theorem 5.1.6 (Cantor's theorem) $|X| < |\mathcal{P}(X)|$ *for every set* X.

Proof Fix a set X. First notice that $|X| \leq |\mathcal{P}(X)|$, since the function $f \colon X \to \mathcal{P}(X)$ defined by $f(x) = \{x\}$ for $x \in X$ is one-to-one.

To finish the proof it is enough to show that $|X| \neq |\mathcal{P}(X)|$. So let $f \colon X \to \mathcal{P}(X)$. We will prove that f is not a bijection by showing that f is not onto $\mathcal{P}(X)$. To see it, put

$$Y = \{x \in X : x \notin f(x)\}.$$

Then $Y \in \mathcal{P}(X)$ is not in the range of f, since if there were a $z \in X$ such that $f(z) = Y$ then we would have

$$z \in Y \Leftrightarrow z \in \{x \in X : x \notin f(x)\} \Leftrightarrow z \notin f(z) \Leftrightarrow z \notin Y,$$

a contradiction. $\square$

Cantor's theorem tell us, in particular, that for any cardinal number κ there is a cardinal number λ larger than κ, namely, $\lambda = |\mathcal{P}(\kappa)|$. We will denote this cardinal number by 2^κ. That way, the symbol 2^κ will stand for two objects: a cardinal number and the family of all functions from κ into 2. Although these two object are definitely different, this notation is consistent in the sense that

$$|\mathcal{P}(X)| = |2^X| \text{ for every set } X. \tag{5.2}$$

This last equation is established by a bijection $\chi \colon \mathcal{P}(X) \to 2^X$ defined by $\chi(A) = \chi_A$, where χ_A is the *characteristic function* of A, that is, $\chi_A(x) = 1$ for $x \in A$ and $\chi_A(x) = 0$ for $x \in X \setminus A$.

The notation 2^κ is also consistent with the cardinal exponentiation operation that will be introduced in the next section.

Notice also that for every cardinal number κ there is the smallest cardinal number λ greater than κ, namely, a cardinal equal to the minimum of the set

$$K = \{\mu \in 2^{2^\kappa} : \mu \text{ is a cardinal and } \kappa < \mu\}.$$

The minimum exists, since K is a nonempty ($2^\kappa \in K$) subset of a well-ordered set 2^{2^κ}. This unique cardinal number is denoted by κ^+ and is called the *cardinal successor of* κ.

It is also worthwhile to note the following property of cardinal numbers.

Proposition 5.1.7 *If $\mathcal{F}$ is a family of cardinal numbers then $\alpha = \bigcup \mathcal{F}$ is also a cardinal number.*

Proof By Theorem 4.2.3(iii) α is an ordinal number. Moreover, if $\beta < \alpha$ then $\beta \in \alpha$ and there exists a $\kappa \in \mathcal{F}$ such that $\beta \in \kappa$. So $\beta < \kappa = |\kappa| \leq |\alpha|$. Thus, by Proposition 5.1.1(iii), α is a cardinal number. $\square$

With the use of Proposition 5.1.7 and the operation of cardinal successor we can construct for every ordinal number α a cardinal number ω_α by induction on $\xi \leq \alpha$ in the following way:

$$\omega_0 = \omega;$$

$$\omega_{\xi+1} = (\omega_\xi)^+ \text{ for every } \xi < \alpha;$$

$$\omega_\lambda = \bigcup_{\xi<\lambda} \omega_\xi \text{ for every limit } \lambda \leq \alpha.$$

It is easy to see that ω_α is the αth infinite cardinal number. It is also worthwhile to mention that very often ω_α is denoted by $\aleph_\alpha$. ($\aleph$ is a Hebrew letter pronounced *aleph*.)

Similarly, for every ordinal number α we can construct a cardinal number $\beth_\alpha$ by induction on $\xi \leq \alpha$ as follows:

$$\beth_0 = \omega;$$

$$\beth_{\xi+1} = 2^{\beth_\xi} \text{ for every } \xi < \alpha;$$

$$\beth_\lambda = \bigcup_{\xi<\lambda} \beth_\xi \text{ for every limit } \lambda \leq \alpha.$$

($\beth$ is a Hebrew letter, which we read *bet*.) In particular, the cardinality of $\mathcal{P}(\omega)$ is equal to $\beth_1 = 2^\omega$. This number is usually denoted by $\mathfrak{c}$, a Hebrew letter c, which we read *continuum*.

Obviously, $\omega_0 = \omega = \beth_0$ and, by Cantor's theorem, $\omega_1 \leq \beth_1 = \mathfrak{c}$. It is also not difficult to prove that $\omega_\alpha \leq \beth_\alpha$ for every ordinal number α. Two natural questions that arise in this context are the following: Is $\omega_1 = \mathfrak{c}$? Is $\omega_\alpha = \beth_\alpha$ for every ordinal number α? Surprisingly, both these questions cannot be decided within ZFC set theory. The statement that $\omega_1 = \mathfrak{c}$ is called the *continuum hypothesis* (usually abbreviated by CH) and is independent of ZFC set theory (see Section 1.1). Similarly, the statement "$\omega_\alpha = \beth_\alpha$ for every ordinal number α" is called the *generalized continuum hypothesis* (usually abbreviated by GCH) and is also independent of ZFC set theory.

The independence of CH from the ZFC axioms will be proved in Chapter 9. Meanwhile, we will often use CH as an additional assumption to deduce other properties of interest. That way, we will know that the deduced properties are consistent with ZFC set theory, that is, that we have no way to prove their negations using only ZFC axioms.

The last theorem of this section gives some rules on comparing the cardinalities of sets.

Theorem 5.1.8 *If* $|A| = |B|$ *and* $|C| = |D|$ *then*

(a) $|A \times C| = |B \times D|$;

(b) $|A \cup C| = |B \cup D|$, *provided* $A \cap C = \emptyset = B \cap D$;

(c) $\left|A^C\right| = \left|B^D\right|$.

Proof Let $f \colon A \to B$ and $g \colon C \to D$ be bijections. Define the bijections for (a)–(c) as follows:

   (a) $F \colon A \times C \to B \times D$, $F(a, c) = \langle f(a), g(c) \rangle$.

   (b) $F \colon A \cup C \to B \cup D$, $F = f \cup g$.

   (c) $F \colon A^C \to B^D$, and $F(h) \in B^D$ for $h \in A^C$ is defined by the formula $F(h)(d) = f(h(g^{-1}(d)))$ for every $d \in D$.

   It is left as an exercise to show that the functions F so defined are indeed bijections.   □

We will finish this section with the following remark. The only cardinal numbers that were defined in a "natural" way (i.e., without an essential use of the axiom of choice) are the natural numbers and ω. The infinite ordinal numbers that we can easily construct, that is, construct with the use of ordinal number operations, are of the form $\omega + 1$, $\omega + 2$, $\omega + \omega$, $\omega\omega$, and so forth, and are not cardinal numbers. (For example, notice that $|\omega\omega| \leq \omega$ by defining the one-to-one function $f \colon \omega\omega \to \omega$ with the formula $f(n, m) = 2^n 3^m$, where we identify $\omega\omega$ with the set $\omega \times \omega$ from the definition of $\omega\omega$.) In fact, there is no way to prove without the axiom of choice the existence of uncountable ordinal numbers. This might seem strange, since the proof of Cantor's theorem evidently did not use the axiom of choice. However, we in fact proved there only that $|X| \neq |\mathcal{P}(X)|$. The proof that the set $\mathcal{P}(X)$ can indeed be well ordered requires the use of Zermelo's theorem, which is equivalent to the axiom of choice.

EXERCISES

1 Let $n \in \omega$. Show that if α is an ordinal number such that $|\alpha| = n$ then $\alpha = n$.

2 Prove that $\omega_\alpha \leq \beth_\alpha$ for every ordinal number α.

3 Complete the proof of Theorem 5.1.8 by showing that all functions F defined in this proof are bijections.

5.2 Cardinal arithmetic

For cardinal numbers κ and λ define their *cardinal sum* by

$$\kappa \oplus \lambda = |(\kappa \times \{0\}) \cup (\lambda \times \{1\})|,$$

their *cardinal product* by

$$\kappa \otimes \lambda = |\kappa \times \lambda|,$$

and a *cardinal exponentiation* operation by

$$\kappa^\lambda = \left|\kappa^\lambda\right|.$$

Notice that the exponentiation operation for $\kappa = 2$ gives us another definition of 2^λ. However, it is consistent with the previous definition, since $\left|2^\lambda\right| = 2^\lambda$. Notice also that by the foregoing definitions and Theorem 5.1.8, for all sets A and B,

$$|A \cup B| = |A| \oplus |B| \quad \text{for } A \text{ and } B \text{ disjoint} \tag{5.3}$$

and

$$|A \times B| = |A| \otimes |B|, \qquad \left|A^B\right| = |A|^{|B|}. \tag{5.4}$$

Proposition 5.2.1 *For all cardinal numbers κ and λ*

(i) $\kappa \oplus \lambda = \lambda \oplus \kappa,$

(ii) $\kappa \otimes \lambda = \lambda \otimes \kappa.$

Proof From Theorem 5.1.8 we have

$$\kappa \oplus \lambda = |(\kappa \times \{0\}) \cup (\lambda \times \{1\})| = |(\lambda \times \{0\}) \cup (\kappa \times \{1\})| = \lambda \oplus \kappa,$$

where the second equation follows from the fact that $|\kappa \times \{0\}| = |\kappa \times \{1\}|$ and $|\lambda \times \{1\}| = |\lambda \times \{0\}|$. Also,

$$\kappa \otimes \lambda = |\kappa \times \lambda| = |\lambda \times \kappa| = \lambda \otimes \kappa,$$

since $f(\zeta, \xi) = \langle \xi, \zeta \rangle$ is a bijection between $\kappa \times \lambda$ and $\lambda \times \kappa$. □

Thus, unlike the ordinal operations $+$ and $\cdot$, the cardinal operations $\oplus$ and $\otimes$ are commutative.

Proposition 5.2.2 *For every* $m, n \in \omega$

(i) $m \oplus n = m + n < \omega$,

(ii) $m \otimes n = mn < \omega$.

The proof is left as an exercise.

Lemma 5.2.3 *Every infinite cardinal number is a limit ordinal number.*

Proof First notice that
$$|\omega \cup \{x\}| = |\omega|.$$

If $x \in \omega$ then this is obvious. Otherwise this is justified by a function $f \colon \omega \cup \{x\} \to \omega$ defined by $f(x) = 0$ and $f(n) = n + 1$ for $n < \omega$. Hence, by Theorem 5.1.8(ii), for every infinite ordinal number α

$$|\alpha + 1| = |\alpha \cup \{\alpha\}| = |(\alpha \setminus \omega) \cup (\omega \cup \{\alpha\})| = |(\alpha \setminus \omega) \cup \omega| = |\alpha|.$$

Therefore, by Proposition 5.1.1, an ordinal number $\alpha + 1$ is not a cardinal number for any infinite α.   □

Theorem 5.2.4 *If* κ *is an infinite cardinal then* $\kappa \otimes \kappa = \kappa$.

Proof Let κ be an infinite cardinal number. We will prove by transfinite induction on $\omega \leq \alpha \leq \kappa$ that

$$|\alpha| \otimes |\alpha| = |\alpha|. \tag{5.5}$$

So let $\omega \leq \lambda \leq \kappa$ be such that (5.5) holds for every $\omega \leq \alpha < \lambda$. We will show that (5.5) holds for λ as well. This will finish the proof.
  If there is an $\alpha < \lambda$ such that $|\alpha| = |\lambda|$ then

$$|\lambda| \otimes |\lambda| = |\alpha| \otimes |\alpha| = |\alpha| = |\lambda|.$$

Hence we may assume that λ is a cardinal number.
  Now notice that for every $\alpha < \lambda$

$$|\alpha \times \alpha| < \lambda. \tag{5.6}$$

It follows from Proposition 5.2.2(ii) for a finite α and from (5.5) for an infinite α. Define a well ordering $\preceq$ on $\lambda \times \lambda$ by putting $\langle \alpha, \beta \rangle \preceq \langle \gamma, \delta \rangle$ if and only if
$$\max\{\alpha, \beta\} < \max\{\gamma, \delta\}$$

or

$$\max\{\alpha, \beta\} = \max\{\gamma, \delta\} \quad \text{and} \quad \langle \alpha, \beta \rangle \leq_{\text{lex}} \langle \gamma, \delta \rangle,$$

where $\leq_{\text{lex}}$ stands for the lexicographic order on $\kappa \times \kappa$ (see Theorem 4.1.13). The proof that $\preceq$ is indeed a well ordering is left as an exercise.

Now let $\langle \gamma, \delta \rangle \in \lambda \times \lambda$ and let $\varepsilon = \max\{\gamma + 1, \delta + 1\}$. Then $\varepsilon < \lambda$ by Lemma 5.2.3. Moreover, the initial segment $O_{\preceq}(\langle \gamma, \delta \rangle)$ (generated by $\langle \gamma, \delta \rangle$ with respect to $\preceq$) is a subset of $\varepsilon \times \varepsilon$, since $\langle \alpha, \beta \rangle \preceq \langle \gamma, \delta \rangle$ implies that $\max\{\alpha, \beta\} \leq \max\{\gamma, \delta\} < \varepsilon$. Hence, by (5.6),

$$|\operatorname{Otp}(O_{\preceq}(\langle \gamma, \delta \rangle))| = |O_{\preceq}(\langle \gamma, \delta \rangle)| \leq |\varepsilon \times \varepsilon| = |\varepsilon| < \lambda$$

and so also

$$\operatorname{Otp}(O_{\preceq}(\langle \gamma, \delta \rangle)) < \lambda.$$

(Otherwise, we would have $\lambda = |\lambda| \leq |\operatorname{Otp}(O_{\preceq}(\langle \gamma, \delta \rangle))|$.) Since every ordinal number $\beta \in \operatorname{Otp}(\lambda \times \lambda, \preceq)$ is of the form $\operatorname{Otp}(O_{\preceq}(\langle \gamma, \delta \rangle))$ we can conclude that $\operatorname{Otp}(\lambda \times \lambda, \preceq) \leq \lambda$. So

$$|\lambda \times \lambda| = |\operatorname{Otp}(\lambda \times \lambda, \preceq)| \leq \operatorname{Otp}(\lambda \times \lambda, \preceq) \leq \lambda = |\lambda|.$$

Thus $|\lambda \times \lambda| = |\lambda|$, since the inequality $|\lambda| \leq |\lambda \times \lambda|$ is obvious. $\square$

Corollary 5.2.5 *If* λ *and* κ *are infinite cardinals then*

$$\kappa \oplus \lambda = \kappa \otimes \lambda = \max\{\kappa, \lambda\}.$$

Proof We may assume that $\lambda \leq \kappa$. Then

$$\kappa \oplus \lambda = |(\kappa \times \{0\}) \cup (\lambda \times \{1\})| \leq |\kappa \times 2| \leq |\kappa \times \lambda| = \kappa \otimes \lambda$$

and

$$\kappa \otimes \lambda = |\kappa \times \lambda| \leq |\kappa \times \kappa| = \kappa.$$

Since evidently $\kappa \leq \kappa \oplus \lambda$ we obtain

$$\kappa \leq \kappa \oplus \lambda \leq \kappa \otimes \lambda \leq \kappa. \qquad \square$$

Corollary 5.2.6 *A countable union of countable sets is countable.*

Proof Let $\mathcal{F}$ be a countable family of countable sets, $\mathcal{F} = \{F_n : n < \omega\}$. Then for every $n < \omega$ there is a bijection $f_n \colon \omega \to F_n$. Define function $f \colon \omega \times \omega \to \bigcup \mathcal{F}$ by putting $f(n, m) = f_n(m)$. Notice that f is onto. So, by Theorem 5.1.2, $|\bigcup \mathcal{F}| \leq |\omega \times \omega| = \omega$. Hence $|\bigcup \mathcal{F}| = \omega$. $\square$

Similarly, we can prove the following.

Corollary 5.2.7 *If* κ *is an infinite cardinal and* $|X_\alpha| \leq \kappa$ *for all* $\alpha < \kappa$ *then* $\left|\bigcup_{\alpha < \kappa} X_\alpha\right| \leq \kappa$.

The proof is left as an exercise.

Corollary 5.2.8 *A finite union of finite sets is finite.*

The proof is similar to that of Corollary 5.2.6 (use Proposition 5.2.2).

For a set A let $A^{<\omega} = \bigcup_{n<\omega} A^n$. Thus $A^{<\omega}$ is the set of all finite sequences with values in A.

Corollary 5.2.9 *If κ is an infinite cardinal, then $|\kappa^{<\omega}| = \kappa$.*

Proof It is easy to prove, by induction on $n < \omega$, that $|\kappa^n| = \kappa$ for every $0 < n < \omega$. So, by Corollary 5.2.7,

$$\kappa \leq |\kappa^{<\omega}| = \left| \bigcup_{n<\omega} \kappa^n \right| \leq \kappa. \qquad \square$$

Corollary 5.2.10 $|\mathbb{Q}| = |\mathbb{Z}| = \omega$.

Proof Clearly $\omega = |\mathbb{N}| \leq |\mathbb{Z}| \leq |\mathbb{Q}|$. To see that $|\mathbb{Q}| \leq \omega$ it is enough to show that $|\mathbb{Q}| \leq |2 \times \omega \times \omega|$, since $|2 \times \omega \times \omega| = 2 \otimes \omega \otimes \omega = \omega$. So let $f \colon 2 \times \omega \times \omega \to \mathbb{Q}$ be defined by $f(i, m, n) = (-1)^i \frac{m}{n+1}$. It is easy to see that f is onto $\mathbb{Q}$. $\qquad \square$

Theorem 5.2.11 $|\mathbb{R}| = |2^\omega| = |\mathcal{P}(\omega)| = \mathfrak{c}$.

Proof $|2^\omega| = 2^\omega = \mathfrak{c}$ by the definition of $\mathfrak{c}$. $|2^\omega| = |\mathcal{P}(\omega)|$ was proved in (5.2). The equation $|\mathbb{R}| = |2^\omega|$ is proved by two inequalities.

The first inequality $|2^\omega| \leq |\mathbb{R}|$ is established by a one-to-one function $f \colon 2^\omega \to \mathbb{R}$ defined by

$$f(\langle a_0, a_1, \ldots \rangle) = \sum_{n<\omega} \frac{2a_n}{3^{n+1}}.$$

(The range of f is Cantor's classical "ternary" set; see Section 6.2.)

To see that $|\mathbb{R}| \leq |2^\omega|$ define a one-to-one function $g \colon \mathbb{R} \to \mathcal{P}(\mathbb{Q})$ by $g(r) = \{q \in \mathbb{Q} \colon q < r\}$ (i.e., $g(r)$ is the Dedekind cut associated with r). So

$$|\mathbb{R}| \leq |\mathcal{P}(\mathbb{Q})| = |2^{\mathbb{Q}}| = 2^{|\mathbb{Q}|} = 2^\omega. \qquad \square$$

Theorem 5.2.12 *If λ and κ are cardinal numbers such that $\lambda \geq \omega$ and $2 \leq \kappa \leq \lambda$ then $\kappa^\lambda = 2^\lambda$.*

In particular, $\lambda^\lambda = 2^\lambda$ for every infinite cardinal number λ.

Proof Notice that
$$2^\lambda \subset \kappa^\lambda \subset \lambda^\lambda \subset \mathcal{P}(\lambda \times \lambda).$$
Hence
$$2^\lambda \leq \kappa^\lambda \leq \lambda^\lambda \leq |\mathcal{P}(\lambda \times \lambda)| = \left|2^{\lambda \times \lambda}\right| = 2^{|\lambda \times \lambda|} = 2^\lambda. \qquad \square$$

Theorem 5.2.13 *If κ, λ, and μ are cardinals, then*
$$\kappa^{\lambda \oplus \mu} = \kappa^\lambda \otimes \kappa^\mu \qquad and \qquad (\kappa^\lambda)^\mu = \kappa^{\lambda \otimes \mu}.$$

Proof It follows immediately from the following properties:
$$A^{(B \cup C)} \approx (A^B) \times (A^C) \quad \text{for } B \cap C = \emptyset, \tag{5.7}$$
and
$$(A^B)^C \approx A^{B \times C}. \tag{5.8}$$
Their proof is left as an exercise. $\square$

For a set X and a cardinal number κ define
$$[X]^{\leq \kappa} = \{A \in \mathcal{P}(X) \colon |A| \leq \kappa\}, \quad [X]^{<\kappa} = \{A \in \mathcal{P}(X) \colon |A| < \kappa\},$$

$$[X]^\kappa = \{A \in \mathcal{P}(X) \colon |A| = \kappa\}.$$

Proposition 5.2.14 *For every infinite set X and nonzero cardinal $\kappa \leq |X|$*
$$\left|[X]^\kappa\right| = \left|[X]^{\leq \kappa}\right| = |X|^\kappa.$$

Proof Define $\psi \colon X^\kappa \to [X]^{\leq \kappa}$ by $\psi(f) = f[\kappa]$. Since ψ is clearly onto, and $[X]^\kappa \subset [X]^{\leq \kappa}$, we have
$$\left|[X]^\kappa\right| \leq \left|[X]^{\leq \kappa}\right| \leq |X|^\kappa.$$
To finish the proof it is enough to show that $|X|^\kappa \leq \left|[X]^\kappa\right|$.

To prove it define $\psi \colon X^\kappa \to [\kappa \times X]^\kappa$ by $\psi(f) = f$. Clearly ψ is one-to-one. Hence

$$
\begin{aligned}
|X|^\kappa &= |X^\kappa| \\
&\leq |[\kappa \times X]^\kappa| \\
&= |[|\kappa \times X|]^\kappa| \tag{5.9} \\
&= |[|X|]^\kappa| \\
&= |[X]^\kappa|, \tag{5.10}
\end{aligned}
$$

where equations (5.9) and (5.10) follow from the fact that $|Y| = |Z|$ implies $|[Y]^\kappa| = |[Z]^\kappa|$, which is left as an exercise. $\square$

We will finish this section with the following important fact.

Theorem 5.2.15 *The family* $\mathcal{C}(\mathbb{R}) = \{f \in \mathbb{R}^{\mathbb{R}} : f$ *is continuous$\}$ has cardinality continuum.*

Proof Define a function $R \colon \mathcal{C}(\mathbb{R}) \to \mathbb{R}^{\mathbb{Q}}$ by $R(f) = f|_{\mathbb{Q}}$. By the density of $\mathbb{Q}$ in $\mathbb{R}$ and the continuity of our functions it follows that R is one-to-one (see Section 3.3). Hence

$$|\mathcal{C}(\mathbb{R})| \leq |\mathbb{R}^{\mathbb{Q}}| = |\mathbb{R}|^{|\mathbb{Q}|} = |2^{\omega}|^{\omega} = (2^{\omega})^{\omega} = 2^{\omega \otimes \omega} = 2^{\omega}.$$

The other inequality is proved by a one-to-one function $c \colon \mathbb{R} \to \mathcal{C}(\mathbb{R})$, where $c(a) \colon \mathbb{R} \to \mathbb{R}$ is a constant function with value a. $\square$

EXERCISES

1 Use the definitions of cardinal arithmetic operations and Theorem 5.1.8 to prove (5.3) and (5.4).

2 Prove Proposition 5.2.2. *Hint:* Show by induction on n that $m + n < \omega$ and $mn < \omega$. Then use Exercise 1 from Section 5.1.

3 Prove that the relation $\preceq$ defined in the proof of Theorem 5.2.4 is a well-ordering relation on $\lambda \times \lambda$.

4 Prove Corollary 5.2.7.

5 Complete the proof of Theorem 5.2.13 by proving (5.7) and (5.8).

6 Complete the proof of Proposition 5.2.14 by showing that $|Y| = |Z|$ implies $|[Y]^{\kappa}| = |[Z]^{\kappa}|$ for every sets Y and Z and any cardinal number κ.

7 A real number r is an *algebraic number* if there is a polynomial $p(x)$ with rational coefficients such that $p(r) = 0$. Show that the set $A \subset \mathbb{R}$ of all algebraic numbers is countable. Conclude that the set $\mathbb{R} \setminus A$ of all nonalgebraic real numbers has cardinality continuum.

8 A function $f \colon \mathbb{R} \to \mathbb{R}$ is a Baire class-one function if there exists a sequence $f_n \colon \mathbb{R} \to \mathbb{R}$ of continuous functions such that f_n converges to f pointwise, that is, such that $\lim_{n \to \infty} f_n(x) = f(x)$ for every $x \in \mathbb{R}$. Find the cardinality of the family $\mathcal{B}_1$ of all Baire class-one functions.

5.3   Cofinality

We say that a subset A of an ordinal number α is *unbounded in* α if A has no *strict* upper bound in α, that is, when there is no $\gamma \in \alpha$ with $\xi < \gamma$ for all $\xi \in A$. It is also easy to see that $A \subset \alpha$ is unbounded in α when $\alpha = \bigcup\{\xi + 1 : \xi \in A\}$ or when

$$\forall \gamma < \alpha \ \exists \xi \in A \ (\gamma \leq \xi).$$

For an ordinal number α let $\mathrm{cf}(\alpha)$ be the smallest ordinal number β such that there exists a function $f \colon \beta \to \alpha$ with the property that $f[\beta]$ is unbounded in α. Such a function f is called a *cofinal map (in α)*. The number $\mathrm{cf}(\alpha)$ is called the *cofinality* of α. Thus $f \colon \beta \to \alpha$ is cofinal in α if

$$\forall \xi \in \alpha \ \exists \eta \in \beta \ (\xi \leq f(\eta))$$

and

$$\mathrm{cf}(\alpha) = \min\{\beta \colon \text{ there is a cofinal map } f \colon \beta \to \alpha\}.$$

Notice that

$$\mathrm{cf}(\alpha) \leq \alpha \ \text{ for every ordinal number } \alpha,$$

since the identity map is a cofinal map. Also,

$$\mathrm{cf}(\alpha + 1) = 1 \ \text{ for every ordinal number } \alpha,$$

since $f \colon 1 \to \alpha + 1$, $f(0) = \alpha$, is cofinal in $\alpha + 1$. It is also easy to see that if $\mathrm{cf}(\alpha) = 1$ then α has a greatest element. Thus $\mathrm{cf}(\alpha) = 1$ if and only if $\mathrm{cf}(\alpha) < \omega$ if and only if α is a successor ordinal.

We will be interested mainly in cofinalities of limit ordinal numbers.

Lemma 5.3.1 *For every ordinal number α there is a strictly increasing cofinal map $f \colon \mathrm{cf}(\alpha) \to \alpha$.*

Proof   For α a successor ordinal the map just shown works.

If α is a limit ordinal and $g \colon \mathrm{cf}(\alpha) \to \alpha$ is a cofinal map, define by transfinite induction on $\eta < \mathrm{cf}(\alpha)$

$$f(\eta) = \max\left\{g(\eta), \bigcup_{\xi < \eta}(f(\xi) + 1)\right\}.$$

Clearly $f(\xi) < f(\xi) + 1 \leq f(\eta)$ for every $\xi < \eta < \mathrm{cf}(\alpha)$. Thus f is strictly increasing.

The fact that

$$f(\eta) \in \alpha \quad \text{for every} \quad \eta < \text{cf}(\alpha) \tag{5.11}$$

is proved by induction on $\eta < \text{cf}(\alpha)$. If for some $\eta < \text{cf}(\alpha)$ condition (5.11) holds for all $\xi < \eta$ then $f(\eta) = \max\{g(\eta), \bigcup_{\xi<\eta}(f(\xi)+1)\} \leq \alpha$. But $f(\eta)$ can't be equal to α since otherwise the restriction of f to η would be a cofinal map in α and this would contradict the minimality of $\text{cf}(\alpha)$. Thus $\eta < \text{cf}(\alpha)$. Condition (5.11) has been proved.

Finally, f is cofinal in α since for every $\xi \in \alpha$ there is an $\eta < \text{cf}(\alpha)$ with $\xi \leq g(\eta) \leq f(\eta)$. $\square$

Corollary 5.3.2 $\text{cf}(\text{cf}(\alpha)) = \text{cf}(\alpha)$ *for every ordinal number α.*

Proof Let $f\colon \text{cf}(\text{cf}(\alpha)) \to \text{cf}(\alpha)$ and $g\colon \text{cf}(\alpha) \to \alpha$ be strictly increasing cofinal maps. Then $g \circ f\colon \text{cf}(\text{cf}(\alpha)) \to \alpha$ is also cofinal, since for every $\xi < \alpha$ there exist $\eta \in \text{cf}(\alpha)$ with $\xi \leq g(\eta)$ and $\zeta \in \text{cf}(\text{cf}(\alpha))$ with $\eta \leq f(\zeta)$, so

$$\xi \leq g(\eta) \leq g(f(\zeta)) = (g \circ f)(\zeta).$$

Thus, by the minimality of $\text{cf}(\alpha)$, we have $\text{cf}(\text{cf}(\alpha)) \geq \text{cf}(\alpha)$. So $\text{cf}(\text{cf}(\alpha)) = \text{cf}(\alpha)$. $\square$

An ordinal number α is *regular* if it is a limit ordinal and $\text{cf}(\alpha) = \alpha$.

Theorem 5.3.3 *If α is a regular ordinal number then α is a cardinal number.*

Proof Put $\beta = |\text{cf}(\alpha)|$, and choose a bijection $f\colon \beta \to \text{cf}(\alpha)$ and a cofinal function $g\colon \text{cf}(\alpha) \to \alpha$. Then $(g \circ f)\colon \beta \to \alpha$ is cofinal in α, since $(g \circ f)[\beta] = g[f[\beta]] = g[\text{cf}(\alpha)]$. Hence, by minimality of $\text{cf}(\alpha)$, $\text{cf}(\alpha) \leq \beta = |\text{cf}(\alpha)|$. $\square$

Proposition 5.3.4 ω *is regular.*

Proof It follows easily from Corollary 5.2.8. $\square$

Theorem 5.3.5 κ^+ *is regular for every infinite cardinal number κ.*

Proof Let $\alpha < \kappa^+$ and $f\colon \alpha \to \kappa^+$. Then $|\alpha| \leq \kappa$ and $|f(\xi)| \leq \kappa$ for every $\xi < \alpha$. Hence $|\bigcup_{\xi<\alpha}[f(\xi)+1]| \leq \kappa$ since a union of $\leq \kappa$ sets of cardinality $\leq \kappa$ has cardinality $\leq \kappa$ (see Corollary 5.2.7). Thus $\bigcup_{\xi<\alpha}[f(\xi)+1] \neq \kappa^+$, that is, $\text{cf}(\kappa^+) \neq \alpha$ for every $\alpha < \kappa^+$. So $\text{cf}(\kappa^+) = \kappa^+$. $\square$

One of the most useful properties of regular cardinals is given in the following proposition.

Proposition 5.3.6 *If λ is a regular cardinal, $A \subset \lambda$, and $|A| < \lambda$ then there is an $\alpha < \lambda$ such that $A \subset \alpha$.*

Proof Let $\beta = \mathrm{Otp}(A)$ and let $f \colon \beta \to A \subset \lambda$ be an order isomorphism. Notice that $\beta < \mathrm{cf}(\lambda)$, since $|\beta| = |A| < \lambda = \mathrm{cf}(\lambda)$. So the set $A = f[\beta]$ cannot be unbounded in λ. This implies the existence of an $\alpha < \lambda$ such that $\xi < \alpha$ for all $\xi \in A$. Thus $A \subset \alpha$. $\square$

Another useful fact is the following.

Proposition 5.3.7 *If $\alpha < \omega_1$ is a limit ordinal, then $\mathrm{cf}(\alpha) = \omega$.*

The proof is left as an exercise.
We will finish this section with the following cardinal inequality.

Theorem 5.3.8 *For every infinite cardinal κ*

$$\kappa^{\mathrm{cf}(\kappa)} > \kappa.$$

Proof Let $f \colon \mathrm{cf}(\kappa) \to \kappa$ be a cofinal map and let $G \colon \kappa \to \kappa^{\mathrm{cf}(\kappa)}$. We will show that G is not onto $\kappa^{\mathrm{cf}(\kappa)}$.
Define $h \colon \mathrm{cf}(\kappa) \to \kappa$ by

$$h(\alpha) = \min(\kappa \setminus \{G(\xi)(\alpha) \colon \xi < f(\alpha)\})$$

for $\alpha < \mathrm{cf}(\kappa)$. Such a definition makes sense, since $|\{G(\xi)(\alpha) \colon \xi < f(\alpha)\}| \le |f(\alpha)| < \kappa$. But for every $\xi < \kappa$ there exists an $\alpha < \mathrm{cf}(\kappa)$ such that $\xi < f(\alpha)$, that is, such that $h(\alpha) \ne G(\xi)(\alpha)$. Hence $h \ne G(\xi)$ for every $\xi < \kappa$ and $h \notin G[\kappa]$. $\square$

Corollary 5.3.9 $\mathrm{cf}(\mathfrak{c}) > \omega$.

Proof Assume, to obtain a contradiction, that $\mathrm{cf}(\mathfrak{c}) = \omega$. Then $\mathfrak{c}^{\mathrm{cf}(\mathfrak{c})} = (2^\omega)^\omega = 2^\omega = \mathfrak{c}$, contradicting Theorem 5.3.8. $\square$

EXERCISE

1 Prove Proposition 5.3.7.

Part III

The power of recursive definitions

Chapter 6

Subsets of $\mathbb{R}^n$

6.1 Strange subsets of $\mathbb{R}^n$ and the diagonalization argument

In this section we will illustrate some typical constructions by transfinite induction. We will do so by constructing recursively some subsets of $\mathbb{R}^n$ with strange geometric properties. The choice of geometric descriptions of these sets is not completely arbitrary here – we still have not developed the basic facts concerning "nice" subsets of $\mathbb{R}^n$ that are necessary for most of our applications. This will be done in the remaining sections of this chapter.

To state the next theorem, we will need the following notation. For a subset A of the plane $\mathbb{R}^2$ the horizontal section of A generated by $y \in \mathbb{R}$ (or, more precisely, its projection onto the first coordinate) will be denoted by A^y and defined as $A^y = \{x \in \mathbb{R} \colon \langle x, y \rangle \in A\}$. The vertical section of A generated by $x \in \mathbb{R}$ is defined by $A_x = \{y \in \mathbb{R} \colon \langle x, y \rangle \in A\}$.

We will start with the following example.

Theorem 6.1.1 *There exists a subset A of the plane with every horizontal section A^y being dense in $\mathbb{R}$ and with every vertical section A_x having precisely one element.*

Proof We will define the desired set by induction. To do so, we will first reduce our problem to the form that is most appropriate for a recursive construction.

The requirement that every horizontal section of A is dense in $\mathbb{R}$ tells us that A is "reasonably big." More precisely, the condition says that $A^y \cap (a, b) \neq \emptyset$ for every $a < b$ and $y \in \mathbb{R}$. Therefore if we put $\mathcal{F} =$

$\{(a,b) \times \{y\} \colon a,b,y \in \mathbb{R}\ \&\ a < b\}$ then we can restate it as

$$A \cap J \neq \emptyset \quad \text{for every } J \in \mathcal{F}. \tag{6.1}$$

The requirement that every vertical section of A has precisely one point tells us that A is "reasonably small." It can be restated as

$$|A \cap [\{x\} \times \mathbb{R}]| = 1 \quad \text{for every } x \in \mathbb{R}. \tag{6.2}$$

The idea of constructing A is to "grow it bit by bit" to obtain at the end a set satisfying (6.1), while keeping approximations "reasonably small." More precisely, by transfinite induction on $\xi < \mathfrak{c}$, we will define a sequence $\langle\langle x_\xi, y_\xi\rangle \colon \xi < \mathfrak{c}\rangle$ such that $A_0 = \{\langle x_\xi, y_\xi\rangle \colon \xi < \mathfrak{c}\}$ satisfies (6.1) and has the property that $|A_0 \cap [\{x\} \times \mathbb{R}]| \leq 1$ for every $x \in \mathbb{R}$. It is enough, since then the set $A = A_0 \cup \{\langle x, 0\rangle \colon A_0 \cap [\{x\} \times \mathbb{R}] = \emptyset\}$ satisfies the theorem.

To describe our construction of the sequence, notice first that $|\mathcal{F}| = \mathfrak{c}$, since $\mathfrak{c} = |\mathbb{R}| \leq |\mathcal{F}| \leq |\mathbb{R}^3| = \mathfrak{c}$. Let $\{J_\xi \colon \xi < \mathfrak{c}\}$ be an enumeration of the family $\mathcal{F}$. If for some $\xi < \mathfrak{c}$ the sequence $\langle\langle x_\zeta, y_\zeta\rangle \colon \zeta < \xi\rangle$ is already defined, choose

$$\langle x_\xi, y_\xi\rangle \in J_\xi \setminus \left[\bigcup_{\zeta < \xi} \{x_\zeta\} \times \mathbb{R}\right]. \tag{6.3}$$

The choice can be made, since for every $\xi < \mathfrak{c}$ we have $|J_\xi| = |\mathbb{R}| = \mathfrak{c} > |\xi| \geq \left|J_\xi \cap \left[\bigcup_{\zeta < \xi} \{x_\zeta\} \times \mathbb{R}\right]\right|$. So, by the recursion theorem, the sequence $\langle\langle x_\xi, y_\xi\rangle \colon \xi < \mathfrak{c}\rangle$ exists. (More precisely, we define $\langle x_\xi, y_\xi\rangle = f(\xi) = h(f|_\xi)$ where the oracle function h is defined implicitly by formula (6.3). Explicitly, $h(\langle\langle x_\zeta, y_\zeta\rangle \colon \zeta < \xi\rangle)$ is defined as $C(J_\xi \setminus \{\langle x_\zeta, y_\zeta\rangle \colon \zeta < \xi\})$, where C is a choice function from the family $\mathcal{P}(\mathbb{R}^2) \setminus \{\emptyset\}$.)

It is clear that $A_0 = \{\langle x_\xi, y_\xi\rangle \colon \xi < \mathfrak{c}\}$ satisfies (6.1), since for every $J \in \mathcal{F}$ there exists a $\xi < \mathfrak{c}$ such that $J = J_\xi$ and $\langle x_\xi, y_\xi\rangle \in A_0 \cap J_\xi = A_0 \cap J$. It is also easy to see that the choice as in (6.3) implies that every vertical section of A_0 has at most one point. $\qquad\square$

The proof of Theorem 6.1.1 is a typical example of a diagonalization technique. In order to prove the theorem we had to find for every $J \in \mathcal{F}$ a point $\langle x_J, y_J\rangle \in J$ that would belong to a future set A. The choice was supposed to be done while *preserving* at every step the property that no two points chosen so far belong to the same vertical line. In order to prove the theorem we listed *all* elements of $\mathcal{F}$ and made our construction, taking care of each of its elements one at a time, using the fact that we still had "enough room" to make our choice. This technique of making inductive constructions by listing all important objects for the desired property and then taking care of each of them one at a time is called the *diagonalization technique*. It is typical for such constructions that the inductive step is possible because of some cardinal argument similar to the one given earlier. The *preservation* part, like that mentioned before, is usually the hard part of the argument.

The next theorem, due to Mazurkiewicz (1914), is very similar in character to Theorem 6.1.1.

Theorem 6.1.2 *There exists a subset A of the plane $\mathbb{R}^2$ that intersects every straight line in exactly two points.*

Proof Let $\{L_\xi \colon \xi < \mathfrak{c}\}$ be an enumeration of all straight lines in the plane $\mathbb{R}^2$. By transfinite induction on $\xi < \mathfrak{c}$ we will construct a sequence $\{A_\xi \colon \xi < \mathfrak{c}\}$ of subsets of $\mathbb{R}^2$ such that for every $\xi < \mathfrak{c}$

(I) A_ξ has at most two points;

(P) $\bigcup_{\zeta \le \xi} A_\zeta$ does not have three collinear points;

(D) $\bigcup_{\zeta \le \xi} A_\zeta$ contains precisely two points of L_ξ.

Then the set $A = \bigcup_{\xi < \mathfrak{c}} A_\xi$ will have the desired property. This is the case since the *preservation* condition (P) implies that every line contains at most two points of A, while the *diagonal* condition (D) makes sure that every line L contains at least two points from A. Thus, it is enough to show that we can choose a set A_ξ satisfying (I), (P), and (D) for every $\xi < \mathfrak{c}$. This will be proved by induction on $\xi < \mathfrak{c}$.

So assume that for some $\xi < \mathfrak{c}$ the sequence $\{A_\zeta \colon \zeta < \xi\}$ is already constructed. By condition (I) the set $B = \bigcup_{\zeta < \xi} A_\zeta$ has cardinality $< \mathfrak{c}$, being a union of $|\xi| < \mathfrak{c}$ many finite sets. Similarly, the family $\mathcal{G}$ of all lines containing two points from B has cardinality $\le |B^2| < \mathfrak{c}$. Notice that by the inductive assumption (P) the set $B \cap L_\xi$ has at most two points. If it has precisely two points put $A_\xi = \emptyset$ and notice that (P) and (D) are satisfied. If $B \cap L_\xi$ contains less than two points then L_ξ intersects every line from $\mathcal{G}$ in at most one point. Thus $L_\xi \cap \bigcup \mathcal{G} = \bigcup_{L \in \mathcal{G}} L_\xi \cap L$ has at most $|\mathcal{G}| < \mathfrak{c}$ many points.

Choose $A_\xi \subset L_\xi \setminus \bigcup \mathcal{G}$ to have two elements if $B \cap L_\xi = \emptyset$ and to have one element if $B \cap L_\xi$ has one point. It is easy to see that conditions (P) and (D) are satisfied with either choice of A_ξ. The construction has now been completed. $\qquad\Box$

The next theorem is yet another example of a geometric problem solved with help of the recursion theorem. In this theorem a *circle* in $\mathbb{R}^3$ is understood to be a set of points forming any nontrivial circle in any plane in $\mathbb{R}^3$.

Theorem 6.1.3 $\mathbb{R}^3$ *is a union of disjoint circles.*

Proof We have to construct a family $\mathcal{C}$ of subsets of $\mathbb{R}^3$ such that (1) each $C \in \mathcal{C}$ is a circle, (2) the family $\mathcal{C}$ covers $\mathbb{R}^3$, and (3) different elements of $\mathcal{C}$ are disjoint. We will construct the family $\mathcal{C}$ as $\{C_\xi \colon \xi < \mathfrak{c}\}$ by induction on $\xi < \mathfrak{c}$. For this, the natural approach is the following. Choose an enumeration $\{p_\xi \colon \xi < \mathfrak{c}\}$ of $\mathbb{R}^3$, and for each p_ξ choose a circle C_ξ such that $p_\xi \in C_\xi$. This certainly would take care of (2); however, then we could not keep circles disjoint: For $p_\xi \in C_0 \setminus \{p_0\}$ the circles C_0 and C_ξ would not be disjoint. Thus we have to settle for a slightly weaker condition: We will choose C_ξ such that

(D) $\quad p_\xi \in \bigcup_{\zeta \leq \xi} C_\zeta$.

This is the "diagonal" condition that will imply that $\mathcal{C}$ covers $\mathbb{R}^3$. However, we have to make the choice in such a way that circles in $\mathcal{C}$ are pairwise disjoint. We will do this by assuming that for each $\xi < \mathfrak{c}$ the circles constructed so far are pairwise disjoint and we will choose C_ξ preserving this property. More precisely, we will choose C_ξ such that the following "preservation" condition is satisfied:

(P) $\quad C_\xi \cap \bigcup_{\zeta < \xi} C_\zeta = \emptyset$.

Evidently, if we could construct $\mathcal{C} = \{C_\xi \colon \xi < \mathfrak{c}\}$ such that conditions (D) and (P) are satisfied for every $\xi < \mathfrak{c}$, then $\mathcal{C}$ would be the desired family of circles.

On the other hand, the recursion theorem tells us that we can find such a sequence $\langle C_\xi \colon \xi < \mathfrak{c} \rangle$ as long as the family of all circles C_ξ satisfying (D) and (P) is nonempty for every $\xi < \mathfrak{c}$. (Since then the choice function will work as an oracle function.)

So assume that for some $\xi < \mathfrak{c}$ the sequence $\{C_\zeta \colon \zeta < \xi\}$ is already constructed satisfying (D) and (P). We will prove that there exists a circle C_ξ satisfying (D) and (P).

If $p_\xi \notin \bigcup_{\zeta < \xi} C_\zeta$ define $p = p_\xi$. Otherwise, choose an arbitrary point $p \in \mathbb{R}^3 \setminus \bigcup_{\zeta < \xi} C_\zeta$. This can be done since for every straight line L in $\mathbb{R}^3$ the set $L \cap \bigcup_{\zeta < \xi} C_\zeta = \bigcup_{\zeta < \xi}(L \cap C_\zeta)$ has cardinality less than continuum, being a union of less than continuum many finite sets $L \cap C_\zeta$. We will choose C_ξ containing p and satisfying (P). To do this, take a plane P containing p that does not contain any of the circles C_ζ for $\zeta < \xi$. This can be done since there are continuum many planes passing through p and there are only $\leq |\xi| < \mathfrak{c}$ many planes containing circles from $\{C_\zeta \colon \zeta < \xi\}$. Now notice that the plane P intersects each circle C_ζ in at most two points. Thus the set $S = P \cap \bigcup_{\zeta < \xi} C_\zeta$ has cardinality less than continuum. Fix a line L in P containing p and let $\mathcal{C}_0$ be the family of all circles in P containing p and tangent to L. Notice that different circles from $\mathcal{C}_0$ intersect only at the point p. Thus there is a circle $C_\xi \in \mathcal{C}_0$ disjoint from S. This finishes the inductive construction and the proof. $\square$

If in the previous three theorems the reader got the impression that we can construct a subset or a partition of $\mathbb{R}^n$ with almost arbitrary paradoxical geometric properties, the next theorem will prove that this impression is wrong.

Theorem 6.1.4 *The plane $\mathbb{R}^2$ is not a union of disjoint circles.*

Proof Assume, to obtain a contradiction, that there is a family $\mathcal{F}$ of disjoint circles in $\mathbb{R}^2$ such that $\mathbb{R}^2 = \bigcup \mathcal{F}$. Construct, by induction on $n < \omega$, a sequence $\{C_n : n < \omega\}$ of circles from $\mathcal{F}$ in the following way: Start with an arbitrary circle $C_0 \in \mathcal{F}$ and in step $n + 1$ choose a circle $C_{n+1} \in \mathcal{F}$ that contains the center c_n of circle C_n.

Notice that if r_n is the radius of C_n then $|c_{n+1} - c_n| = r_{n+1} < r_n/2$. Thus $\langle c_n : n < \omega \rangle$ is a Cauchy sequence. Let $p = \lim_{n \to \infty} c_n$. Then p belongs to the closed disk D_n bounded by C_n for every $n < \omega$. So p cannot belong to any circle C_n since it belongs to D_{n+1}, which is disjoint from C_n.

Now let $C \in \mathcal{F}$ be such that $p \in C$. Then $C \neq C_n$ for every $n < \omega$. But if $n < \omega$ is such that r_n is smaller than the radius of C then $C \cap C_n \neq \emptyset$. This contradicts the choice of $\mathcal{F}$. $\square$

The next theorem shows that a simple diagonalization may lead to messy multiple-case considerations. In the theorem $\mathbb{R}^2$ will be considered as a vector space. In particular, $A + r$ will stand for the algebraic sum of a set $A \subset \mathbb{R}^2$ and $r \in \mathbb{R}^2$, that is, $A + r = \{a + r : a \in A\}$.

Theorem 6.1.5 *There is a subset A of $\mathbb{R}^2$ that intersects every one of its translations in a singleton, that is, such that the set $(A+r) \cap A$ has precisely one element for every $r \in \mathbb{R}^2$, $r \neq 0$.*

Proof Enumerate $\mathbb{R}^2 \setminus \{0\}$ as $\{r_\xi : \xi < \mathfrak{c}\}$. We will define, by induction on $\xi < \mathfrak{c}$, the sequences $\langle a_\xi \in \mathbb{R}^2 : \xi < \mathfrak{c} \rangle$, $\langle b_\xi \in \mathbb{R}^2 : \xi < \mathfrak{c} \rangle$, and $\langle A_\xi : \xi < \mathfrak{c} \rangle$ such that for every $\xi < \mathfrak{c}$

(I) $A_\xi = \{a_\zeta : \zeta \leq \xi\} \cup \{b_\zeta : \zeta \leq \xi\}$;

(D) $b_\xi - a_\xi = r_\xi$;

(P)   if $a, b, a', b' \in A_\xi$ are such that $b - a = b' - a' \neq 0$ then $a = a'$ and $b = b'$.

Before we describe the construction notice that $A = \bigcup_{\xi < \mathfrak{c}} A_\xi$ has the desired properties: $(A + r) \cap A$ has at least one element by the "diagonal" condition (D) for $r = r_\xi$, since $b_\xi \in (A + r_\xi) \cap A$; it has at most one element by the "preservation" condition (P), where we take $r = b - a = b' - a'$.

To make an inductive step assume that for some $\xi < \mathfrak{c}$ the sequences are already constructed. The difficulty will be to find points a_ξ and b_ξ satisfying (D) while preserving (P).

Let $B = \bigcup_{\zeta < \xi} A_\zeta$. Notice that B satisfies (P).

If $B \cap (B + r_\xi) \neq \emptyset$ and $b \in B \cap (B + r_\xi)$ then $a = b - r_\xi \in B$ and we can define $a_\xi = a$ and $b_\xi = b$. Condition (D) is satisfied by the choice of a and b, and condition (P) holds, since $A_\xi = B$.

So assume that $B \cap (B + r_\xi) = \emptyset$ and notice that if we find $a_\xi = x$ then $b_\xi = a_\xi + r_\xi = x + r_\xi$ would be uniquely defined. Thus, it is enough to find x guaranteeing satisfaction of (P). In order to find it we have to avoid the situation

$$b - a = b' - a' \neq 0 \quad \text{and} \quad b \neq b', \tag{6.4}$$

where $a, b, a', b' \in A_\xi = B \cup \{x, x + r_\xi\}$. So take x, a, b, a', b' that satisfy (6.4).

There are several cases to consider. First notice that by our inductive hypothesis $\{a, b, a', b'\} \cap \{x, x + r_\xi\} \neq \emptyset$.

In the first group of cases we assume that all elements a, b, a', b' are different. Then there are the following possibilities:

- $\{a, b, a', b'\} \cap \{x, x + r_\xi\} = \{x\}$. Then $x = c + d - e$ for some $c, d, e \in B$. To avoid this situation it is enough to take x outside the set $S_1 = \{c + d - e \colon c, d, e \in B\}$.

- $\{a, b, a', b'\} \cap \{x, x + r_\xi\} = \{x + r_\xi\}$. Then $x = c + d - e - r_\xi$ for some $c, d, e \in B$. To avoid this situation it is enough to take x outside the set $S_2 = \{c + d - e - r_\xi \colon c, d, e \in B\}$.

- $\{a, b, a', b'\} \cap \{x, x + r_\xi\} = \{x, x + r_\xi\}.$

  If $\{a, b\} = \{x, x + r_\xi\}$ then $\pm r_\xi = b' - a'$. However, this would contradict the fact that $B \cap (B + r_\xi) = \emptyset$. Similarly $\{x, x + r_\xi\}$ cannot be equal to $\{a', b'\}$, $\{a, a'\}$, or $\{b, b'\}$.

  If $\{x, x + r_\xi\} = \{a, b'\}$ or $\{x, x + r_\xi\} = \{a', b\}$ then $2x = c + d - r_\xi$ for some $c, d \in B$. To avoid this situation it is enough to take x outside the set $S_3 = \{(c + d - r_\xi)/2 \colon c, d \in B\}$.

The second group of cases consists of the situations in which some of the elements a, b, a', b' are equal to each other. It is easy to see that there are only two possibilities for such equations: $b = a'$ and $a = b'$. Moreover, these equations cannot happen at the same time. Thus assume that $a' = b$

and that the other elements are different (the case when $a = b'$ is identical). Then $2b = a + b'$.

All the cases when $b \notin \{x, x + r_\xi\}$ are identical to those previously considered. So assume that $b \in \{x, x + r_\xi\}$. Then there are the following possibilities:

- $b = x$. Then $2x = a + b'$.

  If $x + r_\xi \notin \{a, b'\}$ then $2x = c + d$ for some $c, d \in B$. To avoid this case it is enough to take x outside the set $S_4 = \{(c + d)/2 \colon c, d \in B\}$.

  If $x + r_\xi \in \{a, b'\}$ then $2x = c + x + r_\xi$ for some $c \in B$. To avoid this situation it is enough to take x outside the set $S_5 = \{c + r_\xi \colon c \in B\}$.

- $b = x + r_\xi$. Then $2x = a + b' - 2r_\xi$.

  If $x \notin \{a, b'\}$ then $2x = c + d - 2r_\xi$ for some points $c, d \in B$. To avoid this particular situation it is enough to choose x outside the set $S_6 = \{(c + d - 2r_\xi)/2 \colon c, d \in B\}$.

  If $x \in \{a, b'\}$ then $2x = c + x - 2r_\xi$ for some $c \in B$. To avoid this case it is enough to take x outside the set $S_7 = \{c - 2r_\xi \colon c \in B\}$.

It is easy to see that each of the sets S_i for $i = 1, \ldots, 7$ has cardinality $\leq |\xi|^3 < \mathfrak{c}$. Thus we can take x from the set $\mathbb{R}^2 \setminus \bigcup_{i=1}^{7} S_i$. $\quad\square$

In all of the previous examples of this section we used preservation conditions of finite character such as noncollinearity, finite intersections of different circles, and so forth. This made obvious the fact that these conditions were preserved when we took a union of previously constructed sets. This approach, however, does not always work. For example, the next theorem will state that $\mathbb{R}^2$ is a union of countably many sets S_i such that no set S_i contains two different points of rational distance apart. If we try to prove this theorem starting with an arbitrary well ordering $\{p_\xi \colon \xi < \mathfrak{c}\}$ of $\mathbb{R}^2$ and at step ξ try to add the point p_ξ to some set S_i, then already at step ω we might find ourselves in trouble – it might happen that for every $i < \omega$ there is already an element $p_n \in S_i$, $n < \omega$, of rational distance from p_ω. If so, we could not assign p_ω to any S_i. To solve these difficulties we will proceed in a different manner, to be described subsequently.

We will start with the following easy but very useful lemma that will also be used very often in the rest of this text. To state it, we need the following definitions. Let $F \colon X^k \to [X]^{\leq \omega}$ for some $k < \omega$. We say that a subset Y of X is *closed under the action of* F if $F(y_1, \ldots, y_k) \subset Y$ for every $y_1, \ldots, y_k \in Y$. If $f \colon X^k \to X$ then $Y \subset X$ is *closed under the action of* f if it is closed under the action of F, where $F(x) = \{f(x)\}$. Finally, if $\mathcal{F}$ is a family of functions from finite powers of X into either $[X]^{\leq \omega}$ or X then $Y \subset X$ is *closed under the action of* $\mathcal{F}$ if it is closed under the action of F for every $F \in \mathcal{F}$.

Lemma 6.1.6 *Let $\mathcal{F}$ be an at most countable family of functions from finite powers of X into $[X]^{\leq \omega}$ or X. Then*

(a) *for every $Z \subset X$ there exists a smallest subset Y of X closed under the action of $\mathcal{F}$ and containing Z; this set, denoted by $\mathrm{cl}_{\mathcal{F}}(Z)$, has cardinality less than or equal to $|Z| + \omega$; in particular, $|\mathrm{cl}_{\mathcal{F}}(Z)| = |Z|$ for every infinite Z;*

(b) *if $|X| = \kappa > \omega$ then there exists an increasing sequence $\langle X_\alpha : \alpha < \kappa \rangle$ of subsets of X closed under the action of $\mathcal{F}$ such that $X = \bigcup_{\alpha < \kappa} X_\alpha$, $|X_\alpha| < \kappa$ for all $\alpha < \kappa$, and $X_\lambda = \bigcup_{\alpha < \lambda} X_\alpha$ for every limit ordinal $\lambda < \kappa$.*

Proof In the proof we will identify functions $f \colon X \to X$ from $\mathcal{F}$ with $F \colon X \to [X]^1$, $F(x) = \{f(x)\}$.

(a) Let $\mu = |Z| + \omega$. Construct, by induction on $n < \omega$, an increasing sequence $\langle Z_n \in [X]^\mu : n < \omega \rangle$ by putting $Z_0 = Z$ and defining Z_{n+1} as

$$Z_n \cup \bigcup \{ F(z_1, \ldots, z_m) \colon F \in \mathcal{F} \ \& \ \mathrm{dom}\, F = X^m \ \& \ \langle z_1, \ldots, z_m \rangle \in Z_n^m \}.$$

Notice that $|Z_{n+1}| \leq \mu$. This is the case since $|Z_n| \leq \mu$ and the second set is a union of a family of sets of cardinality $\leq \omega$ indexed by a set of cardinality $\leq |\mathcal{F}| \otimes |Z_n|^{<\omega} \leq \mu$.

Define $\mathrm{cl}_{\mathcal{F}}(Z) = \bigcup_{n < \omega} Z_n$. Obviously $Z \subset \mathrm{cl}_{\mathcal{F}}(Z) \subset X$ and $|\mathrm{cl}_{\mathcal{F}}(Z)| \leq \mu$, as $\mathrm{cl}_{\mathcal{F}}(Z)$ is a union of countably many sets of cardinality at most μ.

To see that $\mathrm{cl}_{\mathcal{F}}(Z)$ is closed under the action of $\mathcal{F}$ take $F \in \mathcal{F}$. If X^m is the domain of F and $z_1, \ldots, z_m \in \mathrm{cl}_{\mathcal{F}}(Z)$ then there is an $n < \omega$ such that $z_1, \ldots, z_m \in Z_n$, and so $F(z_1, \ldots, z_m) \subset Z_{n+1} \subset \mathrm{cl}_{\mathcal{F}}(Z)$.

In order to prove that $\mathrm{cl}_{\mathcal{F}}(Z)$ is the smallest subset of X closed under the action of $\mathcal{F}$ and containing Z let Y be another such set. Then $Z_0 = Z \subset Y$ and by an easy induction we can prove that $Z_n \subset Y$ for every $n < \omega$. So $\mathrm{cl}_{\mathcal{F}}(Z) = \bigcup_{n < \omega} Z_n \subset Y$.

(b) Enumerate X as $\{x_\xi : \xi < \kappa\}$ and define $X_\alpha = \mathrm{cl}_{\mathcal{F}}(\{x_\xi : \xi < \alpha\})$. Then the equation $X = \bigcup_{\alpha < \kappa} X_\alpha$ is obvious and $|X_\alpha| \leq |\alpha| + \omega < \kappa$ for $\alpha < \kappa$ follows immediately from (a). The two other conditions are simple consequences of the fact that $\mathrm{cl}_{\mathcal{F}}(Z)$ is the smallest subset of X containing Z and being closed under the action of $\mathcal{F}$.   □

The next theorem is due to Erdős and Hajnal (Erdős 1969).

Theorem 6.1.7 *There is a countable partition $\{S_i : i < \omega\}$ of $\mathbb{R}^2$ such that the distance between any two different points of the same set S_i is irrational.*

Proof By induction on $\kappa = |X|$ we are going to prove that every set $X \subset \mathbb{R}^2$ can be decomposed as described in the theorem.

So let $X \subset \mathbb{R}^2$ be such that $|X| = \kappa$ and suppose that the foregoing statement is true for every $Y \subset \mathbb{R}^2$ of cardinality $< \kappa$. We have to prove the statement for X.

If $\kappa \leq \omega$ then the statement is obvious, since we can put every element of X in a different set S_i. So assume $\kappa > \omega$.

For every $p, q \in \mathbb{Q}$ define $F_{pq} \colon (\mathbb{R}^2)^2 \to [\mathbb{R}^2]^{<\omega}$ by

$$F_{pq}(x, y) = \{z \in \mathbb{R}^2 \colon |x - z| = p \,\&\, |y - z| = q\}.$$

Notice that indeed $F_{pq}(x, y) \in [\mathbb{R}^2]^{<\omega}$ since it has at most two points. Let $\mathcal{F} = \{F_{pq} \colon p, q \in \mathbb{Q}\}$ and let $X = \bigcup_{\alpha < \kappa} X_\alpha$, where $\langle X_\alpha \colon \alpha < \kappa \rangle$ is a sequence of subsets closed under the action of $\mathcal{F}$ as in Lemma 6.1.6(b).

We will define a decomposition of X into sets S_i by defining $g \colon X \to \omega$ and $S_i = g^{-1}(i)$. (Thus the function g tells us that an element $x \in X$ is put into $S_{g(x)}$.) The function g must have the property that

$$\text{if } g(x) = g(y) \text{ for different } x, y \in X \text{ then } |x - y| \notin \mathbb{Q}. \qquad (6.5)$$

We will define the function g inductively on the sets X_α for $\alpha < \kappa$.

Assume that for some $\beta < \kappa$ the function g is already defined on each X_α for all $\alpha < \beta$. If β is a limit ordinal, then g is already defined on $X_\beta = \bigcup_{\alpha < \beta} X_\alpha$ and it is easy to see that it satisfies property (6.5). Assume that β is a successor ordinal, say $\beta = \alpha + 1$. Then g is defined on X_α. We have to extend our definition to the set $Z = X_{\alpha+1} \setminus X_\alpha$.

Since $|Z| \leq |X_{\alpha+1}| < \kappa$ we can find a function $h \colon Z \to \omega$ satisfying (6.5). We might try to define g on Z as h. However, this might not work, since for $z \in Z$ there might be an $x \in X_\alpha$ such that $|z - x| \in \mathbb{Q}$. But X_α is closed under the action of $\mathcal{F}$. Thus for every $z \in Z$ there is at most one $x_z \in X_\alpha$ such that $|z - x_z| \in \mathbb{Q}$. So we will define g on Z such that for every $z \in Z$ and $h(z) = n$ we have $g(z) \in \{2n, 2n + 1\}$. Then $g|_Z$ will satisfy (6.5), as h did. Now it is enough to choose $g(z)$ different from $g(x_z)$, if x_z exists, and arbitrarily otherwise. □

It can be proved also that $\mathbb{R}^n$ can be decomposed as in Theorem 6.1.7 for all $n \in \mathbb{N}$. However, the proof for $n > 2$ is considerably more difficult.

The last theorem of this section is due to Sierpiński (1919). Although its proof is essentially easier that those of previous theorems it has its own flavor – the set constructed in the theorem exists if and only if the continuum hypothesis is assumed.

Theorem 6.1.8 *The continuum hypothesis is equivalent to the existence of a subset A of $\mathbb{R}^2$ with the property that*

$$|A^y| \leq \omega \quad and \quad \left|\left(\mathbb{R}^2 \setminus A\right)_x\right| \leq \omega$$

for every $x, y \in \mathbb{R}$.

Proof $\Rightarrow$: First assume the continuum hypothesis $\mathfrak{c} = \omega_1$ and let $\preceq$ be a well ordering of $\mathbb{R}$ of type ω_1. (To find it, take a bijection $f\colon \mathbb{R} \to \omega_1$ and define $x \preceq y$ if and only if $f(x) \leq f(y)$.) Define $A = \{\langle x, y\rangle\colon x \preceq y\}$.

Now $A^y = \{x \in \mathbb{R}\colon \langle x, y\rangle \in A\} = \{x \in \mathbb{R}\colon x \preceq y\}$ is an initial segment of a set with order type ω_1, so it has cardinality $< \omega_1$, that is, $\leq \omega$. Similarly,

$$\begin{aligned}
(\mathbb{R}^2 \setminus A)_x &= (\{\langle x, y\rangle \in \mathbb{R}^2\colon \langle x, y\rangle \notin A\})_x \\
&= (\{\langle x, y\rangle \in \mathbb{R}^2\colon y \prec x\})_x \\
&= \{y \in \mathbb{R}\colon y \prec x\}
\end{aligned}$$

is an initial segment of a set with order type ω_1, that is, has cardinality $\leq \omega$.

$\Leftarrow$: Assume that $|\mathbb{R}| = \mathfrak{c} > \omega_1$ and let $A \subset \mathbb{R}^2$ be such that $|A^y| \leq \omega$ for every $y \in \mathbb{R}$. We will show that the complement of A has an uncountable vertical section.

Let $Y \subset \mathbb{R}$ be such that $|Y| = \omega_1$ (if $f\colon \mathfrak{c} \to \mathbb{R}$ is a bijection, take $Y = f[\omega_1]$). Let $X = \bigcup_{y \in Y} A^y$. Then $|X| \leq \omega_1$ as it is a union of ω_1 sets of cardinality $< \omega_1$. Take $x \in \mathbb{R} \setminus X$. Then $\langle x, y\rangle \notin A$ for every $y \in Y$, since $x \notin A^y$. Hence $\{x\} \times Y \subset \mathbb{R}^2 \setminus A$ and so $Y \subset (\mathbb{R}^2 \setminus A)_x$. Therefore $\left|(\mathbb{R}^2 \setminus A)_x\right| \geq \omega_1 > \omega$. $\qquad\square$

EXERCISES

1 Complete the proof of Lemma 6.1.6(b) by showing that for every limit ordinal $\lambda < \omega_1$ the set $X_\lambda = \bigcup_{\alpha < \lambda} X_\alpha$ is closed under the action of $\mathcal{F}$ as long as every X_α is closed under the action of $\mathcal{F}$ for $\alpha < \lambda$.

2 Prove that $\mathbb{R}^3 \setminus \mathbb{Q}^3$ is a union of disjoint lines.

3 Generalize Theorem 6.1.3 by proving that $\mathbb{R}^3$ is a union of disjoint circles of radius 1.

4 Modify Mazurkiewicz's theorem (Theorem 6.1.2) by proving that there exists a subset A of the plane $\mathbb{R}^2$ that intersects every circle in exactly three points.

5 Modify Sierpiński's theorem (Theorem 6.1.8) by proving that the continuum hypothesis is equivalent to the following statement: There exist sets $A_0, A_1, A_2 \subset \mathbb{R}^3$ such that $\mathbb{R}^3 = A_0 \cup A_1 \cup A_2$ and

$$|\{\langle x_0, x_1, x_2 \rangle \in A_i \colon x_i = x\}| \leq \omega$$

for every $i < 3$ and $x \in \mathbb{R}$.

    Try to generalize this and Theorem 6.1.8 to $\mathbb{R}^n$ for all $1 < n < \omega$. Prove your claim.

6 (Challenging) Prove the following theorem of Ceder: There is a decomposition $\{S_i\}_{i<\omega}$ of $\mathbb{R}^2$ with no S_i spanning an equilateral triangle. *Remark:* It is also possible to find a decomposition $\{S_i\}_{i<\omega}$ of $\mathbb{R}^2$ with no S_i spanning an isosceles triangle. This has been proved recently by Schmerl.

6.2 Closed sets and Borel sets

In what follows we will concentrate on the topological structure of $\mathbb{R}^n$. Recall that $\mathbb{Q}^n$ is a countable dense subset of $\mathbb{R}^n$ and that the family $\mathcal{B} = \{B(p, \varepsilon) \colon p \in \mathbb{Q}^n \ \& \ \varepsilon \in \mathbb{Q}\}$ of open balls forms a countable base for $\mathbb{R}^n$, that is, for every open set U in $\mathbb{R}^n$ and every $p \in U$ there is a $B \in \mathcal{B}$ such that $p \in B \subset U$.

    A point p of a subset P of $\mathbb{R}^n$ is an *isolated point* of P if there is an open set U (in $\mathbb{R}^n$ or in P) such that $U \cap P = \{p\}$. A subset S of $\mathbb{R}^n$ is *discrete* if every point of S is isolated in S. A nonempty closed subset F of $\mathbb{R}^n$ is said to be *perfect* if it has no isolated points.

    We will start this section with a study of the structure of closed subsets of $\mathbb{R}^n$. For this we will need a few theorems that are of interest in their own right.

Theorem 6.2.1

   (i)   *Every family $\mathcal{U}$ of pairwise-disjoint open subsets of $\mathbb{R}^n$ is at most countable.*

   (ii)   *Every discrete subset S of $\mathbb{R}^n$ is at most countable.*

   (iii)   *If α is an ordinal number and $\{S_\xi \colon \xi < \alpha\}$ is a strictly increasing (decreasing) sequence of open (closed) sets in $\mathbb{R}^n$ then α is at most countable.*

Proof (i) Define $f \colon \mathcal{U} \setminus \{\emptyset\} \to \mathbb{Q}^n$ by choosing $f(U) \in U \cap \mathbb{Q}^n$ for every $U \in \mathcal{U}$. Then f is one-to-one, and so $|\mathcal{U}| \leq |\mathbb{Q}^n| + 1 = \omega$.

(ii) For every $s \in S$ let $r_s > 0$ be such that $S \cap B(s, 2r_s) = \{s\}$ and let $\mathcal{U} = \{B(s, r_s) : s \in S\}$. Notice that $B(s, r_s) \cap B(t, r_t) = \emptyset$ for distinct $s, t \in S$. Thus, by (i), $\mathcal{U}$ is at most countable. Hence $|S| \leq |\mathcal{U}| \leq \omega$.

(iii) We assume that $\mathcal{U} = \{S_\xi : \xi < \alpha\}$ is a strictly increasing sequence of open sets. The argument for the other case is essentially the same.

To obtain a contradiction assume that α is uncountable. Thus $\alpha \geq \omega_1$. For $\xi < \omega_1$ choose $x_\xi \in S_{\xi+1} \setminus S_\xi$ and let $B_\xi \in \mathcal{B}$ be such that $x_\xi \in B_\xi \subset S_{\xi+1}$. Then the set $X = \{x_\xi : \xi < \omega_1\}$ is uncountable, since $x_\xi \neq x_\zeta$ for $\zeta < \xi < \omega_1$. But the function $f : X \to \mathcal{B}$, $f(x_\xi) = B_\xi$, is one-to-one, since for $\zeta < \xi < \omega_1$, $x_\xi \in B_\xi \setminus B_\zeta$, that is, $B_\xi \neq B_\zeta$. However, this is impossible, since it would imply $\omega_1 = |X| \leq |\mathcal{B}| = \omega$.

This contradiction finishes the proof. $\qquad\qquad\qquad\qquad\qquad\qquad\square$

Theorem 6.2.2 *For an ordinal number α the following conditions are equivalent:*

(i) $\alpha < \omega_1$;

(ii)   *there exists an $S \subset \mathbb{R}$ well ordered by the standard relation $\leq$ of order type α.*

Proof (ii)$\Rightarrow$(i): Assume, to obtain a contradiction, that for some $\alpha \geq \omega_1$ there exists an $S \subset \mathbb{R}$ with order type α. Let $h \colon \alpha \to S$ be an order isomorphism and define $U_\xi = (h(\xi), h(\xi+1))$ for $\xi < \omega_1$. Then $\{U_\xi : \xi < \omega_1\}$ is an uncountable family of nonempty, open, and pairwise-disjoint subsets of $\mathbb{R}$. This contradicts Theorem 6.2.1(i).

(i)$\Rightarrow$(ii): For the other implication we will prove the following fact by induction on $\alpha < \omega_1$:

($\star$)   for every $a, b \in \mathbb{R}$ with $a < b$ there exists an $S \subset [a, b)$ with $\mathrm{Otp}(S) = \alpha$.

For $\alpha = 0$ the condition is clearly true. So let $0 < \alpha < \omega_1$ be such that ($\star$) holds for all $\beta < \alpha$. We will show that ($\star$) holds for α.

Take $a, b \in \mathbb{R}$ with $a < b$. If $\alpha = \beta + 1$ is an ordinal successor, pick c with $a < c < b$ and $S \subset [a, c)$ with $\mathrm{Otp}(S) = \beta$. This can be done by the inductive assumption. Then $S \cup \{c\} \subset [a, b)$ and $\mathrm{Otp}(S \cup \{c\}) = \beta + 1 = \alpha$.

If α is a limit ordinal, then $\mathrm{cf}(\alpha) \leq \alpha < \omega_1$ and $\mathrm{cf}(\alpha)$ is infinite. So $\mathrm{cf}(\alpha) = \omega$. Let $\{\alpha_n < \alpha : n < \omega\}$ be a strictly increasing sequence that is cofinal in α with $\alpha_0 = 0$. Find a strictly increasing sequence $a = a_0 < a_1 < \cdots < a_n < \cdots < b$ (for example, put $a_n = b - [(b-a)/(n+1)]$). For every $n < \omega$ choose $S_n \subset [a_n, a_{n+1})$ with $\mathrm{Otp}(S_n) = \mathrm{Otp}(\alpha_{n+1} \setminus \alpha_n)$. This can be done by the inductive hypothesis, since $\mathrm{Otp}(\alpha_{n+1} \setminus \alpha_n) \leq \alpha_{n+1} < \alpha$. Put $S = \bigcup_{n < \omega} S_n$. Then $S \subset [a, b)$. Moreover, if $f_n \colon (\alpha_{n+1} \setminus \alpha_n) \to S_n$ is an order isomorphism then $f = \bigcup_{n < \omega} f_n \colon \alpha \to S$ is also an order isomorphism, that is, $\mathrm{Otp}(S) = \alpha$. Condition ($\star$) has been proved for every $\alpha < \omega_1$.

Now ($\star$) clearly implies (ii). $\qquad\qquad\qquad\qquad\qquad\qquad\qquad\qquad\square$

Theorem 6.2.3 *Every perfect subset P of $\mathbb{R}^n$ has cardinality continuum.*

Proof Clearly $|P| \leq |\mathbb{R}^n| = \mathfrak{c}$. Thus it is enough to show that $|P| \geq \mathfrak{c}$. Before we prove it we notice the following easy fact.

(A) If $R \subset \mathbb{R}^n$ is perfect, $U \subset \mathbb{R}^n$ is open, and $R \cap U = R \cap \mathrm{cl}(U) \neq \emptyset$ then $R \cap \mathrm{cl}(U)$ is perfect.

Clearly $R \cap \mathrm{cl}(U)$ is closed. Also, $p \in R \cap \mathrm{cl}(U)$ cannot be isolated since otherwise there would exist an open set $W \subset \mathbb{R}^n$ with the property that $\{p\} = (R \cap \mathrm{cl}(U)) \cap W = (R \cap U) \cap W = R \cap (U \cap W)$ and the set $U \cap W$ would show that p is an isolated point of R.

Now let $p_i \colon \mathbb{R}^n \to \mathbb{R}$ be the projection onto the ith coordinate. If there exists an $i < n$ such that $|p_i[P]| = \mathfrak{c}$, then $|P| \geq \mathfrak{c}$. So, to obtain a contradiction, assume that

(B) $|p_i[P]| < \mathfrak{c}$ for every $i < n$.

Notice also that

(C) if $R \subset P$ is perfect, then there are disjoint perfect sets $R_0, R_1 \subset R$.

To see it, let $i < n$ be such that $p_i[R]$ contains two numbers $a < b$. By (B) we can find $r \in (a,b) \setminus p_i[R]$. Then, by (A), the sets $R_0 = R \cap p_i^{-1}((-\infty, r])$ and $R_1 = R \cap p_i^{-1}([r, \infty))$ are disjoint and perfect.

Now we construct a family $\{P_s \subset P \colon s \in 2^{<\omega}\}$ of perfect bounded sets by induction on the length $|s|$ of a sequence s.

We put $P_\emptyset = P \cap [a, b]^n$, where $a < b$ are chosen in such a way that $P_\emptyset \neq \emptyset$ and $a, b \in \mathbb{R} \setminus \bigcup_{i<n} p_i[P]$. This can be done by (B). So $P \cap (a, b)^n = P \cap [a, b]^n$. In particular, $P_\emptyset$ is compact and, by (A), perfect. We will continue the induction, maintaining the following inductive condition to be satisfied for every $s \in 2^{<\omega}$.

(I) P_{s0} and P_{s1} are disjoint perfect subsets of P_s,

where $s0$ and $s1$ stand for the sequences of length $|s| + 1$ extending s by 0 and 1, respectively. We can find such sets by (C).

Now notice that for $f \in 2^\omega$ the sets $P_{f|k}$ for $k < \omega$ form a decreasing sequence of nonempty compact sets. Thus, the sets $P_f = \bigcap_{k<\omega} P_{f|k}$ are nonempty. Notice also that the sets P_f and P_g are disjoint for different $f, g \in 2^\omega$. To see this, let $k = \min\{i \in \omega \colon f(i) \neq g(i)\}$. Then $\{f(k), g(k)\} = \{0, 1\}$ and

$$P_f \cap P_g \subset P_{f|k+1} \cap P_{g|k+1} = P_{s0} \cap P_{s1} = \emptyset,$$

where $s = f|k = g|k$.

Let $\psi\colon 2^\omega \to P$ be such that $\psi(f) \in P_f$ (this is a choice function). Then ψ is clearly one-to-one and so $|P| \geq \mathfrak{c}$. □

The induction constructions of the preceding proof are often referred to as *tree constructions*, since $T = 2^{<\omega}$ is called a tree. (We imagine every $s \in T$ as a "tree-branching point," where the tree is splitting into $s0$ and $s1$.)

For $F \subset \mathbb{R}^n$ we denote by F' the set of all *limit points* of F, that is, all points of $\mathrm{cl}(F)$ that are not isolated. Clearly $F \setminus F'$ is equal to the set of all isolated points of F.

Now we are ready to prove the following theorem characterizing closed subsets of $\mathbb{R}^n$.

Theorem 6.2.4 (Cantor–Bendixson theorem) *Every uncountable, closed subset F of $\mathbb{R}^n$ can be represented as a disjoint union of a perfect set P and an at most countable set C.*

Proof By induction on $\alpha < \omega_1$ we construct a sequence $\{F_\alpha : \alpha < \omega_1\}$ of closed subsets of $\mathbb{R}^n$ by defining:

$$F_0 = F, \qquad F_\lambda = \bigcap_{\xi < \lambda} F_\xi \ \text{ for } \lambda < \omega_1, \ \lambda \text{ a limit ordinal,}$$

and

$$F_{\alpha+1} = (F_\alpha)'.$$

The sequence $\{F_\alpha : \alpha < \omega_1\}$ is decreasing and formed of closed sets. It can't be strictly decreasing by Theorem 6.2.1(iii). Thus there exists an $\alpha < \omega_1$ such that $F_\alpha = F_{\alpha+1}$. This means that F_α does not have any isolated points, that is, F_α is either perfect or empty. But $F \setminus F_\alpha = \bigcup_{\beta < \alpha}(F_\beta \setminus F_{\beta+1})$ and every set $F_\beta \setminus F_{\beta+1} = F_\beta \setminus (F_\beta)'$ is discrete, so, by Theorem 6.2.1(ii), it is at most countable. Hence $F \setminus F_\alpha$ is a union of $|\alpha| \leq \omega$ sets, each of which is at most countable. Therefore $C = F \setminus F_\alpha$ is at most countable, and $P = F_\alpha = F \setminus C$ cannot be empty. So P is perfect. □

Corollary 6.2.5 *The cardinality of a closed subset of $\mathbb{R}^n$ is either equal to continuum or is at most countable.*

Theorems 6.2.3 and 6.2.4 tell us the properties that closed subsets of $\mathbb{R}^n$ must have. Can we tell anything more? Exercise 1 shows that we cannot expect much improvement in the case of Theorem 6.2.4. But what about Theorem 6.2.3? How can perfect sets look? They can certainly be formed as a union of some number of closed intervals. But for such sets Theorem 6.2.3 is obvious. Are there any other perfect sets? The affirmative answer is given by a set $C \subset \mathbb{R}$ to be constructed and known as the *Cantor set* (or *Cantor ternary set*). It is constructed as follows.

First, we define the family $\{I_s \subset P \colon s \in 2^{<\omega}\}$ of closed intervals by induction on the length $|s|$ of s. We put $I_\emptyset = [0, 1]$. If the interval $I_s = [a_s, b_s]$ is already constructed, we remove from it the middle third interval, and define I_{s0} and I_{s1} as the left interval and the right interval, respectively, of the remaining two intervals, that is, $I_{s0} = [a_s, a_s + (b_s - a_s)/3]$ and $I_{s1} = [b_s - (b_s - a_s)/3, b_s]$. Put $C_n = \bigcup\{I_s \colon |s| = n\}$. Thus C_n is the union of all 2^n internals of length $1/3^n$ constructed at step n. Notice that $[0, 1] = C_0 \supset C_1 \supset \cdots$ and that each C_n is closed, being a finite union of closed intervals. We define $C = \bigcap_{n<\omega} C_n$.

The set C is clearly closed and nonempty. To see that it is perfect, let $p \in C$ and let J be an open interval containing p. Then there is an $s \in 2^{<\omega}$ such that the interval I_s (of length $1/3^{|s|}$) contains p and is contained in J. Let $j < 2$ be such that $p \notin I_{sj}$. Since $C \cap I_{sj} = \bigcap_{n<\omega}(C_n \cap I_{sj})$ is nonempty, being the intersection of a decreasing sequence of nonempty compact sets, we conclude that $C \cap J \neq \{p\}$. So C is perfect.

Notice also that C does not contain any nonempty interval. This is so since for any interval J of length $\varepsilon > 0$ we can find an $n < \omega$ such that $1/3^n < \varepsilon$. But C_n is a union of intervals of length $1/3^n < \varepsilon$. So $J \not\subset C_n$, and $J \not\subset C$ since $C \subset C_n$.

Finally, it is worthwhile to mention that C is the set of all numbers from $[0, 1]$ whose ternary representations do not contain the digit 1, that is,

$$C = \left\{ \sum_{n=0}^{\infty} \frac{d(n)}{3^{n+1}} \colon d \in \{0, 2\}^\omega \right\}.$$

This will be left without a proof.

To define Borel subsets of topological spaces we need the following definitions. A family $\mathcal{A} \subset \mathcal{P}(X)$ is a *σ-algebra* on X if $\mathcal{A}$ is nonempty and closed under complements and under countable unions, that is, such that

(i) $\emptyset, X \in \mathcal{A}$;

(ii) if $A \in \mathcal{A}$ then $X \setminus A \in \mathcal{A}$;

(iii) if $A_k \in \mathcal{A}$ for every $k < \omega$ then $\bigcup_{k<\omega} A_k \in \mathcal{A}$.

Notice also that every σ-algebra is also closed under countable intersections, since

$$\bigcap_{k<\omega} A_k = X \setminus \bigcup_{k<\omega} (X \setminus A_k).$$

Examples For every nonempty set X the following families form σ-algebras:

(1) $\mathcal{A} = \{\emptyset, X\}$,

(2) $\mathcal{A} = \mathcal{P}(X)$,

(3) $\mathcal{A} = \{A \in \mathcal{P}(X) : |A| \leq \kappa \text{ or } |X \setminus A| \leq \kappa\}$ for any infinite cardinal κ.

Notice that for every $\mathcal{F} \subset \mathcal{P}(X)$ there is a smallest σ-algebra $\sigma[\mathcal{F}]$ on X containing $\mathcal{F}$, namely,

$$\sigma[\mathcal{F}] = \bigcap \{\mathcal{A} \subset \mathcal{P}(X) : \mathcal{F} \subset \mathcal{A} \ \& \ \mathcal{A} \text{ is a } \sigma\text{-algebra on } X\}. \quad (6.6)$$

In particular, for every topological space $\langle X, \tau \rangle$ we define the σ-algebra *Bor* of *Borel sets* by *Bor* $= \sigma[\tau]$.

It is obvious from the definition that the following sets are Borel:

- G_δ sets, that is, the countable intersections of open sets;

- F_σ sets, that is, the countable unions of closed sets;

- $G_{\delta\sigma}$ sets, that is, the countable unions of G_δ sets;

- $F_{\sigma\delta}$ sets, that is, the countable intersections of F_σ sets;

and so on.

Is an arbitrary subset of a topological space $\langle X, \tau \rangle$ Borel? The answer to this question depends on the topological space. For example, in a discrete space $\langle X, \mathcal{P}(X) \rangle$ every set is open and hence Borel. This is, however, quite an exceptional example and more often $\mathcal{P}(X) \neq$ *Bor*. We will not approach this question in general. However, we will address it in the case of $\mathbb{R}^n$.

To shed more light on the structure of Borel sets, define inductively the following hierarchy of sets for a topological space $\langle X, \tau \rangle$:

$$\Sigma_1^0 = \tau,$$

$$\Sigma_\alpha^0 = \left\{ \bigcup_{k<\omega} (X \setminus A_k) : A_k \in \bigcup_{0<\beta<\alpha} \Sigma_\beta^0 \text{ for all } k < \omega \right\} \quad \text{for } 1 < \alpha < \omega_1.$$

Define also

$$\Pi_\alpha^0 = \{X \setminus A : A \in \Sigma_\alpha^0\} \quad \text{for } 0 < \alpha < \omega_1.$$

Thus

$$\Sigma_\alpha^0 = \left\{ \bigcup_{k<\omega} A_k : A_k \in \bigcup_{0<\beta<\alpha} \Pi_\beta^0 \text{ for all } k < \omega \right\} \quad \text{for } 1 < \alpha < \omega_1.$$

It is very easy to prove, by induction on $0 < \alpha < \omega_1$, that

$$\Sigma^0_\alpha, \Pi^0_\alpha \subset \mathcal{B}or \quad \text{for every } 0 < \alpha < \omega_1. \tag{6.7}$$

Notice that Π^0_1 is equal to the family of all closed subsets of X, $\Sigma^0_2 = F_\sigma$, $\Pi^0_2 = G_\delta$, $\Sigma^0_3 = G_{\delta\sigma}$, and $\Pi^0_3 = F_{\sigma\delta}$.

From now on we will assume that $X = \mathbb{R}^n$. In particular, any closed subset F of $\mathbb{R}^n$ is G_δ, as

$$F = \bigcap_{k<\omega} \{x \in \mathbb{R}^n : \exists y \in F\ (d(y,x) < 1/(k+1))\} = \bigcap_{k<\omega} \bigcup_{y\in F} B(y, 1/(k+1)).$$

Similarly, every open set in $\mathbb{R}^n$ is F_σ. Hence

$$\Sigma^0_1 \subset \Sigma^0_2 \quad \text{and} \quad \Pi^0_1 \subset \Pi^0_2. \tag{6.8}$$

Using these inclusions it is not difficult to prove

Proposition 6.2.6 *For every* $0 < \beta < \alpha < \omega_1$

$$\Sigma^0_\beta \subset \Sigma^0_\alpha, \quad \Pi^0_\beta \subset \Pi^0_\alpha, \quad \Sigma^0_\beta \subset \Pi^0_\alpha, \quad \Pi^0_\beta \subset \Sigma^0_\alpha.$$

Proof Inclusion $\Pi^0_\beta \subset \Sigma^0_\alpha$ follows immediately from the definition of Σ^0_α. Now, to see $\Sigma^0_\beta \subset \Pi^0_\alpha$ notice that

$$A \in \Sigma^0_\beta \Rightarrow \mathbb{R}^n \setminus A \in \Pi^0_\beta \subset \Sigma^0_\alpha \Rightarrow A \in \Pi^0_\alpha.$$

We prove

$$\Sigma^0_\beta \subset \Sigma^0_\alpha, \quad \Pi^0_\beta \subset \Pi^0_\alpha \quad \text{for every } 0 < \beta < \alpha$$

by joint induction on α. So assume that for some $1 < \gamma < \omega_1$ this condition holds for every $\alpha < \gamma$ and let $0 < \delta < \gamma$. We have to prove that $\Sigma^0_\delta \subset \Sigma^0_\gamma$ and $\Pi^0_\delta \subset \Pi^0_\gamma$.

But then $\bigcup_{\beta<\delta} \Pi^0_\beta \subset \bigcup_{\beta<\gamma} \Pi^0_\beta$, and so

$$\Sigma^0_\delta = \left\{ \bigcup_{k<\omega} A_k : A_k \in \bigcup_{\beta<\delta} \Pi^0_\beta \right\} \subset \left\{ \bigcup_{k<\omega} A_k : A_k \in \bigcup_{\beta<\gamma} \Pi^0_\beta \right\} = \Sigma^0_\gamma.$$

Now $\Pi^0_\delta \subset \Pi^0_\gamma$ follows from

$$A \in \Pi^0_\delta \Rightarrow \mathbb{R}^n \setminus A \in \Sigma^0_\delta \subset \Sigma^0_\gamma \Rightarrow A \in \Pi^0_\gamma.$$

Proposition 6.2.6 has been proved. $\qquad\square$

Theorem 6.2.7 *The family $\mathcal{B}or$ of Borel subsets of $\mathbb{R}^n$ is equal to*

$$\mathcal{B}or = \bigcup_{0<\alpha<\omega_1} \Sigma^0_\alpha = \bigcup_{0<\alpha<\omega_1} \Pi^0_\alpha.$$

Proof First notice that Proposition 6.2.6 implies

$$\bigcup_{0<\alpha<\omega_1} \Sigma^0_\alpha = \bigcup_{0<\alpha<\omega_1} \Pi^0_\alpha.$$

Let $\mathcal{F}$ denote this set. Notice also that (6.7) implies that $\mathcal{F} \subset \mathcal{B}or$. Since $\mathcal{F}$ contains the topology τ of $\mathbb{R}^n$ it is enough to show that $\mathcal{F}$ is a σ-algebra, since $\mathcal{B}or = \sigma[\tau]$ is the smallest σ-algebra containing τ. Thus we have to prove that $\mathcal{F}$ is closed under complements and countable unions.

If $A \in \mathcal{F}$ then $A \in \Sigma^0_\alpha$ for some $0 < \alpha < \omega_1$. Hence $\mathbb{R}^n \setminus A \in \Pi^0_\alpha \subset \mathcal{F}$.

If $A_k \in \mathcal{F}$ for $k < \omega$ then for every $k < \omega$ there exists $0 < \alpha_k < \omega_1$ such that $A_k \in \Pi^0_{\alpha_k}$. But $\omega_1 = \omega^+$ is regular, so there exists an $\alpha < \omega_1$ such that $\alpha_k < \alpha$ for every $k < \omega$. Hence

$$A_k \in \Pi^0_{\alpha_k} \subset \bigcup_{0<\beta<\alpha} \Pi^0_\beta$$

for every $k < \omega$, and so $\bigcup_{k<\omega} A_k \in \Sigma^0_\alpha \subset \mathcal{F}$. $\qquad\qquad\square$

Theorem 6.2.8 *The family $\mathcal{B}or$ of all Borel subsets of $\mathbb{R}^n$ has cardinality continuum.*

Proof First we will prove by induction that

$$|\Pi^0_\alpha| = |\Sigma^0_\alpha| = \mathfrak{c} \quad \text{for every} \ \ 0 < \alpha < \omega_1. \tag{6.9}$$

Equation $|\Pi^0_\alpha| = |\Sigma^0_\alpha|$ is established by a bijection $f \colon \Sigma^0_\alpha \to \Pi^0_\alpha$ defined by $f(A) = X \setminus A$. Next we will prove that $|\Sigma^0_1| = \mathfrak{c}$. The inequality $|\Sigma^0_1| = |\tau| \le \mathfrak{c}$ is justified by a surjective function $f \colon \mathcal{P}(\mathcal{B}) \to \tau$, $f(\mathcal{U}) = \bigcup\mathcal{U}$, where $\mathcal{B}$ is a countable base for $\mathbb{R}^n$. The inequality $|\tau| \ge \mathfrak{c}$ follows from the facts that the base $\mathcal{B}$ contains an infinite subfamily $\mathcal{G}$ of pairwise-disjoint sets and that f restricted to $\mathcal{P}(\mathcal{G})$ is one-to-one, while $|\mathcal{P}(\mathcal{G})| = \mathfrak{c}$.

Now assume that for some $1 < \alpha < \omega_1$ we have $|\Pi^0_\beta| = |\Sigma^0_\beta| = \mathfrak{c}$ for all $0 < \beta < \alpha$. We have to prove $|\Sigma^0_\alpha| = \mathfrak{c}$.

But, by the inductive hypothesis, $|\bigcup_{\beta<\alpha} \Pi^0_\beta| = \mathfrak{c}$ since it is a union of $|\alpha| < \omega_1 \le \mathfrak{c}$ sets of cardinality $\mathfrak{c}$. Moreover, $F \colon \left(\bigcup_{\beta<\alpha} \Pi^0_\beta\right)^\omega \to \Sigma^0_\alpha$ defined by $F(\langle A_0, A_1, \ldots \rangle) = \bigcup_{n<\omega} A_n$ is onto Σ^0_α, so

$$|\Sigma^0_\alpha| \le \left| \left(\bigcup_{\beta<\alpha} \Pi^0_\beta \right)^\omega \right| = \left| \bigcup_{\beta<\alpha} \Pi^0_\beta \right|^\omega = (2^\omega)^\omega = \mathfrak{c}.$$

Since $\mathfrak{c} = |\Sigma_1^0| \leq |\Sigma_\alpha^0|$, condition (6.9) has been proved.

Now, by Theorem 6.2.7, $\mathcal{B}or = \bigcup_{0<\alpha<\omega_1} \Sigma_\alpha^0$ is a union of $\omega_1 \leq \mathfrak{c}$ sets of cardinality $\mathfrak{c}$, that is, it has cardinality continuum. $\square$

Corollary 6.2.9 *There is a non-Borel subset of $\mathbb{R}^n$.*

Proof By Theorem 6.2.8 we have

$$|\mathcal{P}(\mathbb{R}^n)| = 2^{|\mathbb{R}^n|} = 2^{\mathfrak{c}} > \mathfrak{c} = |\mathcal{B}or|,$$

that is, $\mathcal{P}(\mathbb{R}^n) \setminus \mathcal{B}or \neq \emptyset$. $\square$

We will construct some explicit examples of non-Borel sets in the next section.

EXERCISES

1 For a closed set $F \subset \mathbb{R}$ let $\{F_\alpha : \alpha < \omega_1\}$ be a sequence from the proof of Theorem 6.2.4. Show that for every $\alpha < \omega_1$ there exists a closed subset $F \subset \mathbb{R}$ for which $F_{\alpha+1} \neq F_\alpha$.

2 Prove that the family $\sigma[\mathcal{F}]$ defined by (6.6) is indeed a σ-algebra on X.

3 Show that Theorem 6.2.7 remains true for an arbitrary topological space.

4 Prove the following generalization of Lemma 6.1.6.

Let X be a set and let $\mathcal{F}$ be a family of functions of the form $f: X^\alpha \to X$, where $\alpha \leq \omega$. If $|\mathcal{F}| \leq \mathfrak{c}$ then for every $Z \subset X$ there exists a $Y \subset X$ such that

(i) $Z \subset Y$;

(ii) Y is closed under the action of $\mathcal{F}$, that is, for every $f: X^\alpha \to X$ from $\mathcal{F}$ and $S \in Y^\alpha$ we have $f(S) \in Y$;

(iii) $|Y| \leq |Z|^\omega + \mathfrak{c}$.

Notice that this fact (used with $X = \mathcal{P}(\mathbb{R}^n)$, $Z = \tau$, and $\mathcal{F}$ composed of countable unions and complements) implies Theorem 6.2.8.

5 Prove that every uncountable Borel subset of $\mathbb{R}$ contains a perfect subset.

6.3 Lebesgue-measurable sets and sets with the Baire property

To construct the most useful σ-algebras we need the following definitions.

Recall, from Section 2.1, that a nonempty family $\mathcal{I}$ of subsets of a set X is an *ideal* on X if it is closed under the subset operation and under finite unions, that is, such that

(i)   if $A, B \in \mathcal{I}$ then $A \cup B \in \mathcal{I}$;

(ii)   if $A \in \mathcal{I}$ and $B \subset A$ then $B \in \mathcal{I}$.

An ideal $\mathcal{I}$ is said to be a σ-*ideal* if it is closed under countable unions, that is, if

(i′) $\bigcup_{k \in \omega} A_k \in \mathcal{I}$ provided $A_k \in \mathcal{I}$ for all $k < \omega$.

The elements of an ideal on X are usually considered as "small" in some sense.

Examples 1. $\mathcal{I} = \{\emptyset\}$ is a σ-ideal on every set X.
2. For every $A \subset X$ the family $\mathcal{I} = \mathcal{P}(A)$ forms a σ-ideal on X. In the case when $A = X$ we obtain $\mathcal{I} = \mathcal{P}(X)$. This ideal does not agree with our intuitive notion of a family of small sets. Thus, usually we will work with the *proper* ideals on X, that is, the ideals that are not equal to $\mathcal{P}(X)$.
3. For every infinite cardinal number κ and every set X the family $\mathcal{I} = [X]^{<\kappa}$ is an ideal on X. For $\kappa = \omega$ this is the ideal of finite subsets of X. For $\mathrm{cf}(\kappa) > \omega$ it is also a σ-ideal. Notice also that for $\kappa = \omega_1$ this is the ideal $[X]^{\leq\omega}$ of at most countable subsets of X.
4. For every topological space $\langle X, \tau \rangle$ the family $\mathcal{ND}$ of all nowhere-dense subsets of X (i.e., the subsets S of X such that $\mathrm{int}(\mathrm{cl}(S)) = \emptyset$) is an ideal on X.
5. For a topological space $\langle X, \tau \rangle$ the family

$$\mathcal{M} = \left\{ \bigcup_{k<\omega} A_k \colon A_k \text{ is nowhere dense in } X \right\}$$

is a σ-ideal on X. If $X = \mathbb{R}^n$ (or, more generally, X is a complete metric space or a compact space) then, by the Baire category theorem, $\mathcal{M}$ is proper. If moreover X does not have any isolated points, then $[X]^{\leq\omega} \subset \mathcal{M}$. The ideal $\mathcal{M}$ is usually called the *ideal of Meager* (or *first-category*) *subsets* of X.
6. For n-dimensional Euclidean space $\mathbb{R}^n$ we say that $X \subset \mathbb{R}^n$ is a *(Lebesgue) measure-zero set* or a *null set* in $\mathbb{R}^n$ if for every $\varepsilon > 0$ there

is a family of open balls $\{B(x_k, r_k) \colon k < \omega\}$ such that $X \subset \bigcup_{k<\omega} B(x_k, r_k)$ and $\sum_{k<\omega} r_k^n < \varepsilon$. The family

$$\mathcal{N} = \{X \subset \mathbb{R}^n \colon X \text{ is a null set in } \mathbb{R}^n\}$$

forms a proper σ-ideal in $\mathbb{R}^n$ called the σ-*ideal of (Lebesgue) measure-zero sets* or *null sets*.

The ideals $\mathcal{M}$ and $\mathcal{N}$ on $\mathbb{R}^n$ are of main interest in real analysis. It is easy to see that every countable set belongs to both $\mathcal{M}$ and $\mathcal{N}$. Also, for $n = 1$, the Cantor set C (see Section 6.2) is in $\mathcal{M} \cap \mathcal{N}$. It belongs to $\mathcal{M}$ since it is closed and does not contain any nonempty open interval. It belongs to $\mathcal{N}$ since the total length of intervals forming $C_n \supset C$ is $2^n \frac{1}{3^n} = \left(\frac{2}{3}\right)^n$.

In what follows we will study σ-ideals $\mathcal{M}$ and $\mathcal{N}$ from the set-theoretic point of view. It is worthwhile to notice, however, that both these ideals measure "smallness" in a very different sense. Ideal $\mathcal{M}$ describes the smallness in a topological sense whereas $\mathcal{N}$ does so in a measure sense. The difference between these two "smallness" notions is best captured by the following example.

Proposition 6.3.1 *There exists a dense G_δ set $G \subset \mathbb{R}^n$ such that $G \in \mathcal{N}$ and $\mathbb{R}^n \setminus G \in \mathcal{M}$.*

Proof Recall that $\mathbb{Q}^n$ is a countable dense subset of $\mathbb{R}^n$. Let $\{q_k \colon k < \omega\}$ be an enumeration of $\mathbb{Q}^n$. For $m < \omega$ let

$$G_m = \bigcup_{k<\omega} B(q_k, 2^{-(m+k)/n})$$

and let $G = \bigcap_{m<\omega} G_m$.

$G \in \mathcal{N}$ since for every $\varepsilon > 0$ there exists an $m < \omega$ with $2^{-(m-1)} < \varepsilon$ and

$$G \subset G_m = \bigcup_{k<\omega} B(q_k, 2^{-(m+k)/n}),$$

while $\sum_{k<\omega} \left[2^{-(m+k)/n}\right]^n = 2^{-(m-1)} < \varepsilon$.

To see that $\mathbb{R}^n \setminus G \in \mathcal{M}$ notice that

$$\mathbb{R}^n \setminus G = \mathbb{R}^n \setminus \bigcap_{m<\omega} G_m = \bigcup_{m<\omega} (\mathbb{R}^n \setminus G_m)$$

and each of the sets $\mathbb{R}^n \setminus G_m$ is closed and nowhere dense, since G_m is open and dense. $\qquad\square$

To continue our journey through the special subsets of $\mathbb{R}^n$ we need the following definitions and constructions. For a σ-ideal $\mathcal{I}$ on a set X define a binary relation $\sim_\mathcal{I}$ on $\mathcal{P}(X)$ by

$$A \sim_\mathcal{I} B \Leftrightarrow A \triangle B \in \mathcal{I}.$$

Notice the following easy facts.

Proposition 6.3.2 *Let* $\mathcal{I}$ *be a* σ-*ideal on* X.

(i) *For every* $A, B \subset X$,

$$A \sim_\mathcal{I} B \quad \text{if and only if} \quad A \setminus D = B \setminus D \quad \text{for some} \quad D \in \mathcal{I}.$$

(ii) $\sim_\mathcal{I}$ *is an equivalence relation on* $\mathcal{P}(X)$.

(iii) *If* $A \sim_\mathcal{I} B$ *then* $X \setminus A \sim_\mathcal{I} X \setminus B$.

(iv) *If* $A_n \sim_\mathcal{I} B_n$ *for every* $n < \omega$ *then* $\bigcup_{n<\omega} A_n \sim_\mathcal{I} \bigcup_{n<\omega} B_n$.

Proof (i) If $A \sim_\mathcal{I} B$ then $D = A \triangle B \in \mathcal{I}$ and $A \setminus D = B \setminus D$. Conversely, if $A \setminus D = B \setminus D$ for some $D \in \mathcal{I}$ then $A \triangle B \subset D \in \mathcal{I}$.

(ii) It is easy to see that $A \sim_\mathcal{I} A$ and that $A \sim_\mathcal{I} B$ implies $B \sim_\mathcal{I} A$ for every $A, B \subset X$. Now, if $A \sim_\mathcal{I} B$ and $B \sim_\mathcal{I} C$ then $A \setminus D = B \setminus D$ and $B \setminus E = C \setminus E$ for some $D, E \in \mathcal{I}$. So $A \setminus (D \cup E) = B \setminus (D \cup E) = C \setminus (D \cup E)$ and $D \cup E \in \mathcal{I}$. So $A \sim_\mathcal{I} C$.

(iii) If $A \sim_\mathcal{I} B$ then $A \setminus D = B \setminus D$ for some $D \in \mathcal{I}$ and so $(X \setminus A) \setminus D = (X \setminus B) \setminus D$. Hence $X \setminus A \sim_\mathcal{I} X \setminus B$.

(iv) Let $A_n, B_n \subset X$ be such that $A_n \sim_\mathcal{I} B_n$ for every $n < \omega$ and let $D_n \in \mathcal{I}$ be such that $A_n \setminus D_n = B_n \setminus D_n$ for $n < \omega$. Then $D = \bigcup_{n<\omega} D_n \in \mathcal{I}$ and so $\bigcup_{n<\omega} A_n \setminus D = \bigcup_{n<\omega} B_n \setminus D$. Therefore $\bigcup_{n<\omega} A_n \sim_\mathcal{I} \bigcup_{n<\omega} B_n$. $\square$

Theorem 6.3.3 *Let* $\mathcal{I}$ *be a* σ-*ideal on* X *and let* $\mathcal{A} \subset \mathcal{P}(X)$ *be nonempty. The family*

$$\mathcal{A}[\mathcal{I}] = \{A \triangle D : A \in \mathcal{A} \ \& \ D \in \mathcal{I}\}$$

forms a σ-*algebra on* X *if and only if*

(1) $X \setminus A \in \mathcal{A}[\mathcal{I}]$ *for every* $A \in \mathcal{A}$, *and*

(2) $\bigcup_{n<\omega} A_n \in \mathcal{A}[\mathcal{I}]$ *provided* $A_n \in \mathcal{A}$ *for every* $n < \omega$.

In particular, if $\mathcal{A}$ *is a* σ-*algebra on* X *then* $\mathcal{A}[\mathcal{I}]$ *is a* σ-*algebra generated by* $\mathcal{A} \cup \mathcal{I}$.

Proof The implication $\Rightarrow$ as well as the additional part of the theorem are obvious. To prove the other implication we have to prove that $\mathcal{A}[\mathcal{I}]$ is closed under complements and under countable unions. First notice that

$$C \sim_{\mathcal{I}} B \; \& \; B \in \mathcal{A}[\mathcal{I}] \Rightarrow C \in \mathcal{A}[\mathcal{I}], \tag{6.10}$$

since $B \in \mathcal{A}[\mathcal{I}]$ implies that $B \sim_{\mathcal{I}} A$ for some $A \in \mathcal{A}$, so $C \sim_{\mathcal{I}} A$ and $C = A \triangle D \in \mathcal{A}[\mathcal{I}]$, where $D = A \triangle C \in \mathcal{I}$.

Now, if $B = A \triangle D$ for some $A \in \mathcal{A}$ and $D \in \mathcal{I}$ then $B \triangle A \in \mathcal{I}$ and, by Proposition 6.3.1(iii), $X \backslash B \sim_{\mathcal{I}} X \backslash A$. But, by our assumption, $X \backslash A \in \mathcal{A}[\mathcal{I}]$ and so, by (6.10), $X \setminus B \in \mathcal{A}[\mathcal{I}]$.

Similarly, if $B_n = A_n \triangle D_n$ for some $A_n \in \mathcal{A}$ and $D_n \in \mathcal{I}$ then $B_n \triangle A_n \in \mathcal{I}$ and, by Proposition 6.3.1(iv), $\bigcup_{n<\omega} B_n \sim_{\mathcal{I}} \bigcup_{n<\omega} A_n$. Since $\bigcup_{n<\omega} A_n \in \mathcal{A}[\mathcal{I}]$, condition (6.10) implies that $\bigcup_{n<\omega} B_n \in \mathcal{A}[\mathcal{I}]$. $\square$

The most important σ-algebras on $\mathbb{R}^n$ generated as in Theorem 6.3.3 are the σ-algebras $\mathcal{L} = \mathcal{B}or[\mathcal{N}]$ of *Lebesgue-measurable sets* and the σ-algebra $\mathcal{B}aire = \mathcal{B}or[\mathcal{M}]$ of *sets with the Baire property*, where $\mathcal{B}or$ stands for the σ-algebra of Borel subsets of $\mathbb{R}^n$. Both these families are very rich. In fact, without use of the axiom of choice we cannot prove the existence of a subset of $\mathbb{R}^n$ that does not have the Baire property. A similar theorem for the family $\mathcal{L}$ of all measurable sets can also be proved.

In what follows we prove that there is a subset of $\mathbb{R}^n$ that is nonmeasurable and does not have the Baire property. For this, however, we need some structural theorems about $\mathcal{B}aire$ and $\mathcal{L}$.

Theorem 6.3.4 $\mathcal{B}aire = \tau[\mathcal{M}] = \{U \triangle M : U \text{ is open in } \mathbb{R}^n \text{ and } M \in \mathcal{M}\}$.

Proof This follows immediately from Theorem 6.3.3 since a union of open sets is open and since for every open set A its complement $F = \mathbb{R}^n \setminus A$ is closed, and $F = \text{int}(F) \cup (F \setminus \text{int}(F))$, where $F \setminus \text{int}(F)$ is closed and nowhere dense. $\square$

The similar characterization for $\mathcal{L}$ is more complicated and is given by the next theorem.

Theorem 6.3.5 *For every $A \in \mathcal{L}$ there exists an F_σ set F and a G_δ set G such that $F \subset A \subset G$ and $G \setminus F \in \mathcal{N}$.*

In particular,

$$
\begin{aligned}
\mathcal{L} &= \{G \triangle N : G \text{ is } G_\delta \text{ in } \mathbb{R}^n \text{ and } N \in \mathcal{N}\} \\
&= \{F \triangle N : F \text{ is } F_\sigma \text{ in } \mathbb{R}^n \text{ and } N \in \mathcal{N}\}.
\end{aligned}
$$

The property described in this theorem is called the *regularity* of the family $\mathcal{L}$ and is a basic fact about Lebesgue measure. We will leave it here without proof (see, e.g., Royden 1988).

Theorem 6.3.6 *If $A \in \mathcal{L} \setminus \mathcal{N}$ or $A \in \mathcal{B}aire \setminus \mathcal{M}$ then there exists a perfect set P such that $P \subset A$.*

Proof First assume that $A \in \mathcal{L} \setminus \mathcal{N}$. Then, by Theorem 6.3.5, there exists an F_σ set F such that $F \subset A$ and $A \setminus F \in \mathcal{N}$. Thus $F \in \mathcal{L} \setminus \mathcal{N}$. Let $F = \bigcup_{n<\omega} F_n$, where the sets F_n are closed in $\mathbb{R}^n$. Notice that at least one of the sets F_n must be uncountable, since otherwise F would be at most countable and then it would belong to $\mathcal{N}$. So let F_n be uncountable. Then, by the Cantor–Bendixson theorem (Theorem 6.2.4), there is a perfect set $P \subset F_n \subset F \subset A$.

Now assume that $A \in \mathcal{B}aire \setminus \mathcal{M}$ and let $A = U \triangle S$ for some nonempty open set U and $S \in \mathcal{M}$. Let $S = \bigcup_{n<\omega} S_n$ for some $S_n \in \mathcal{ND}$.

As in Theorem 6.2.3 we will use a "tree-construction" argument to define a family $\{U_s \subset U : s \in 2^{<\omega}\}$ of nonempty open sets; that is, the construction will be done by induction on the length $|s|$ of a sequence s.

We choose a bounded $U_\emptyset$ such that $\operatorname{cl}(U_\emptyset) \subset U \setminus S_0$ and continue the induction, maintaining the following condition to be satisfied for every $s \in 2^{<\omega}$:

(I) U_{s0} and U_{s1} are open balls such that

$$\operatorname{cl}(U_{s0}) \cap \operatorname{cl}(U_{s1}) = \emptyset, \quad \operatorname{cl}(U_{s0}) \cup \operatorname{cl}(U_{s1}) \subset U_s \setminus S_{|s|}.$$

To make an inductive step assume that U_s is already constructed. Since $\operatorname{cl}(S_{|s|}) \in \mathcal{ND}$ we have $U_s \setminus \operatorname{cl}(S_{|s|}) \neq \emptyset$ and we can find two open disjoint balls in $U_s \setminus \operatorname{cl}(S_{|s|})$. Decreasing their radii, if necessary, we can satisfy condition (I). This completes the construction.

Now let $F_n = \bigcup \{\operatorname{cl}(U_s) : s \in 2^{<\omega}, \ |s| = n\}$ and $F = \bigcap_{n<\omega} F_n$. By our construction, $F \subset U$ and $F_n \cap S_n = \emptyset$ for every $n < \omega$. Hence, $F \subset U \setminus S \subset A$. Also, each of the sets F_n is compact, being a finite union of closed bounded sets. Thus F is compact. To finish the proof it is enough to show that F is uncountable, since then, by the Cantor–Bendixson theorem, there is a perfect set $P \subset F \subset A$.

To see this, for every $f \in 2^\omega$ consider a set $F_f = \bigcap_{n<\omega} \operatorname{cl}(U_{f|_n}) \subset F$. Notice that, by Theorem 3.3.1, each set F_f is nonempty. Note also that the sets $\{F_f : f \in 2^\omega\}$ are pairwise disjoint. Indeed, if $f, g \in 2^\omega$ are different, $n = \min\{i < \omega : f(i) \neq g(i)\}$, and $s = f|_n = g|_n$ then $F_f \cap F_g \subset \operatorname{cl}(U_{s0}) \cap \operatorname{cl}(U_{s1}) = \emptyset$. Now let $c : 2^\omega \to F$ be a choice function for the family $\{F_f : f \in 2^\omega\}$, that is, $c(f) \in F_f$ for every $f \in 2^\omega$. Then c is one-to-one and so $|2^\omega| \leq |F|$, proving that F is indeed uncountable.

This finishes the proof.   □

We conclude this section with the following two theorems.

Theorem 6.3.7 *There exists a set $B \subset \mathbb{R}^n$ such that neither B nor $\mathbb{R}^n \setminus B$ contains any perfect subset.*

Proof First notice that the family $\mathcal{F}$ of all perfect subsets of $\mathbb{R}^n$ has cardinality $\leq \mathfrak{c}$, since $\mathcal{F} \subset \mathcal{B}or$ and $|\mathcal{B}or| = \mathfrak{c}$.

Let $\mathcal{F} = \{P_\xi : \xi < \mathfrak{c}\}$. By transfinite induction on $\xi < \mathfrak{c}$ define the sequences $\langle a_\xi : \xi < \mathfrak{c} \rangle$ and $\langle b_\xi : \xi < \mathfrak{c} \rangle$ by choosing in step $\xi < \mathfrak{c}$ points $a_\xi \neq b_\xi$ from $\mathbb{R}^n$ such that

$$a_\xi, b_\xi \in P_\xi \setminus (\{a_\zeta : \zeta < \xi\} \cup \{b_\zeta : \zeta < \xi\}).$$

This can be done since

$$|\{a_\zeta : \zeta < \xi\} \cup \{b_\zeta : \zeta < \xi\}| = 2 \otimes |\xi| < \mathfrak{c}$$

and, by Theorem 6.2.3, $|P_\xi| = \mathfrak{c}$. This finishes the construction.

Now define $B = \{a_\xi : \xi < \mathfrak{c}\}$. Then for every perfect set P there is a $\xi < \mathfrak{c}$ such that $P = P_\xi$. Hence $a_\xi \in P \cap B$ and $b_\xi \in P \cap (\mathbb{R}^n \setminus B)$. Thus $P \not\subset B$ and $P \not\subset \mathbb{R}^n \setminus B$. $\qquad\square$

The set B from Theorem 6.3.7 is called a *Bernstein set*.

Theorem 6.3.8 *A Bernstein subset B of $\mathbb{R}^n$ neither is measurable nor has the Baire property.*

Proof To obtain a contradiction assume that $B \in \mathcal{L}$. Then, by Theorem 6.3.6, $B \in \mathcal{N}$. Similarly, $\mathbb{R}^n \setminus B \in \mathcal{L}$ and $\mathbb{R}^n \setminus B \in \mathcal{N}$. But then $\mathbb{R}^n = B \cup (\mathbb{R}^n \setminus B) \in \mathcal{N}$, which is false. So $B \notin \mathcal{L}$.

The proof that $B \notin \mathcal{B}aire$ is exactly the same. $\qquad\square$

EXERCISES

1 Show that there exists a Bernstein set B such that $B + B = \mathbb{R}$, where $B + B = \{b_0 + b_1 : b_0, b_1 \in B\}$.

2 Show that there exists a Bernstein set B such that $B + B$ is also a Bernstein set.

Chapter 7

Strange real functions

This chapter is designed to help the reader to master the technique of recursive definitions. Thus, most of the examples presented will involve constructions by transfinite induction.

7.1 Measurable and nonmeasurable functions

Let $\mathcal{B}$ be a σ-algebra on $\mathbb{R}^n$. A function $f\colon \mathbb{R}^n \to \mathbb{R}$ is said to be a $\mathcal{B}$-*measurable function* if $f^{-1}(U) \in \mathcal{B}$ for every open set $U \subset \mathbb{R}$. Notice that if f is $\mathcal{B}$-measurable then $f^{-1}(B) \in \mathcal{B}$ for every Borel set $B \subset \mathbb{R}$. This is the case since the family $\{B \subset \mathbb{R}\colon f^{-1}(B) \in \mathcal{B}\}$ is a σ-algebra containing all open sets.

We will use this notion mainly for the σ-algebras of Borel, Lebesgue-measurable, and Baire subsets of $\mathbb{R}^n$, respectively. In each of these cases $\mathcal{B}$-measurable functions will be termed, respectively, as *Borel functions* (or *Borel-measurable functions*), *measurable functions* (or *Lebesgue-measurable functions*), and *Baire functions* (or *Baire-measurable functions*). Clearly, every continuous function is Borel-measurable and every Borel-measurable function is measurable and Baire.

A function $f\colon \mathbb{R}^n \to \mathbb{R}$ is *non-Borel* (or *non–Borel-measurable*) if it is not Borel. Similarly, we define *non-Baire(-measurable)* functions and *non-(Lebesgue-)measurable* functions.

Also recall that the *characteristic function* χ_A of a subset A of a set X is defined by putting $\chi_A(x) = 1$ if $x \in A$ and $\chi_A(x) = 0$ for $x \in X \setminus A$.

The first theorem is a corollary to Theorems 6.3.7 and 6.3.8.

Theorem 7.1.1 *The characteristic function χ_B of a Bernstein subset B of $\mathbb{R}^n$ is neither a measurable nor a Baire function. In particular, there exists a nonmeasurable, non-Baire function from $\mathbb{R}^n$ to $\mathbb{R}$.*

Proof The set $(\chi_B)^{-1}(\{1\}) = B$ neither is measurable nor has the Baire property. ☐

Let us recall that the Fubini–Tonelli theorem says that for a measurable function $f\colon \mathbb{R}^2 \to \mathbb{R}$ that is either integrable or nonnegative we have

$$\iint f \, dm_2 = \int \left(\int f_x \, dm_1 \right) dm_1 = \int \left(\int f^y \, dm_1 \right) dm_1,$$

where m_2 and m_1 stand for the Lebesgue measures on $\mathbb{R}^2$ and $\mathbb{R}$, respectively, and $f_x, f^y\colon \mathbb{R} \to \mathbb{R}$ are defined by $f_x(y) = f(x,y)$ and $f^y(x) = f(x,y)$. The integrals $\int \left(\int f_x \, dm_1 \right) dm_1$ and $\int \left(\int f^y \, dm_1 \right) dm_1$ are called *iterated integrals*. Thus, the Fubini–Tonelli theorem tells us for measurable functions what for the continuous functions is taught in every multivariate calculus course: A two-dimensional integral is equal to both iterated integrals. But what if the function f is nonmeasurable? Then we can't talk about the integral $\iint f \, dm_2$. However, it is still possible that both iterated integrals exist. Must they be equal? The next theorem, due to Sierpiński (1920), gives a negative answer to this question.

Theorem 7.1.2 *If the continuum hypothesis holds then there exists a function* $f\colon [0,1]^2 \to [0,1]$ *such that* $\int f_x \, dm_1 = 1$ *and* $\int f^y \, dm_1 = 0$ *for all* $x, y \in [0,1]$. *In particular,*

$$\int \left(\int f_x \, dm_1 \right) dm_1 = 1 \quad and \quad \int \left(\int f^y \, dm_1 \right) dm_1 = 0.$$

Proof Let A be as in Theorem 6.1.8, put $B = A \cap [0,1]^2$, and define $f = \chi_B$. Notice that $|B^y| \leq \omega$ and $|([0,1]^2 \setminus B)_x| \leq \omega$ for all $x, y \in [0,1]$. Thus $f^y(x) = 0$ for all but countably many $x \in \mathbb{R}$, and so $\int f^y \, dm_1 = \int 0 \, dm_1 = 0$. Similarly, for every $x \in [0,1]$ we have $f_x(y) = 1$ for all but countably many $y \in [0,1]$. Thus $\int f_x \, dm_1 = \int_{[0,1]} 1 \, dm_1 = 1$. ☐

A set $B \subset [0,1]^2$ for which the function $f = \chi_B$ satisfies Theorem 7.1.2 is usually called a *0-1 set*. Its existence is not equivalent to the continuum hypothesis; that is, it might exist when the continuum hypothesis fails. On the other hand, the existence of the function f from Theorem 7.1.2 cannot be proved in ZFC, which has been shown by H. Friedman (1980) and, independently, by C. Freiling (1986).

EXERCISE

1 Prove that there exists a function $f\colon \mathbb{R} \to \mathbb{R}$ such that $f^{-1}(r)$ is a Bernstein set for every $r \in \mathbb{R}$.

7.2   Darboux functions

To motivate what follows let us recall one of the most fundamental theorems of calculus – the intermediate value theorem. It tells us that every continuous function $f\colon \mathbb{R} \to \mathbb{R}$ has the following property:

(DP)   For every $a < b$ and every number y between $f(a)$ and $f(b)$ there is an $x \in (a, b)$ such that $f(x) = y$.

The functions satisfying property (DP) form an important class of functions known as *Darboux functions*. In particular, we say that a function $f\colon \mathbb{R} \to \mathbb{R}$ *has the Darboux property* (or *is a Darboux function*) if it satisfies property (DP). Thus, the intermediate value theorem says that every continuous function $f\colon \mathbb{R} \to \mathbb{R}$ has the Darboux property.

Not every Darboux function, however, is continuous. For example, the function $f(x) = \sin(1/x)$ for $x \neq 0$ and $f(0) = 0$ is a discontinuous Darboux function. The next theorem tells us that there are also everywhere-discontinuous Darboux functions (that is, Darboux functions that are discontinuous at every point). For this we will need one more definition. We say that a function $f\colon \mathbb{R} \to \mathbb{R}$ is *strongly Darboux* if

(SD)   for every $a < b$ and every number y there is an $x \in (a, b)$ such that $f(x) = y$.

Thus f is strongly Darboux if $f[(a, b)] = \mathbb{R}$ for every $a < b$. Clearly, every strongly Darboux function is Darboux and everywhere discontinuous.

Theorem 7.2.1 *There exists a strongly Darboux function* $f\colon \mathbb{R} \to \mathbb{R}$. *In particular, f is Darboux and everywhere discontinuous.*

Proof The set A constructed in Theorem 6.1.1 is the graph of the desired function f.   □

The next theorem is a generalization of Theorem 7.2.1.

Theorem 7.2.2 *Let $\mathcal{G}$ be a family of real functions, $\mathcal{G} \subset \mathbb{R}^{\mathbb{R}}$, with $|\mathcal{G}| \leq \mathfrak{c}$. Then there exists a function $f\colon \mathbb{R} \to \mathbb{R}$ such that $f + g$ is strongly Darboux for every $g \in \mathcal{G}$.*

Proof Let $\mathcal{F} = \{\langle g, I, r\rangle\colon g \in \mathcal{G}\ \&\ r \in \mathbb{R}\ \&\ I = (a, b)$ for some $a < b\}$. Then $|\mathcal{F}| = \mathfrak{c}$, since $\mathfrak{c} = |\mathbb{R}| \leq |\mathcal{F}| \leq |\mathcal{G}| \otimes |\mathbb{R}^3| = \mathfrak{c}$. Let $\{\langle g_\xi, I_\xi, r_\xi\rangle\colon \xi < \mathfrak{c}\}$ be an enumeration of family $\mathcal{F}$. By transfinite induction define a sequence $\{x_\xi\colon \xi < \mathfrak{c}\}$ such that

$$x_\xi \in I_\xi \setminus \{x_\zeta\colon \zeta < \xi\}.$$

Such a choice can be made since $|I_\xi| = |\mathbb{R}| = \mathfrak{c} > |\xi| \geq |\{x_\zeta\colon \zeta < \xi\}|$ for every $\xi < \mathfrak{c}$.

Define $f(x_\xi) = r_\xi - g_\xi(x_\xi)$ and extend it to all of $\mathbb{R}$ arbitrarily. Notice that it has the desired property, since for every nonempty interval (a, b), every $r \in \mathbb{R}$, and every $g \in \mathcal{G}$ there exists a $\xi < \mathfrak{c}$ such that $\langle g_\xi, I_\xi, r_\xi \rangle = \langle g, (a, b), r \rangle$, and so

$$r = r_\xi = f(x_\xi) + g_\xi(x_\xi) \in (f + g_\xi)[I_\xi] = (f + g)[(a, b)]. \qquad \square$$

Notice that Theorem 7.2.2 generalizes Theorem 7.2.1, since we can assume that the function that is identically zero belongs to $\mathcal{G}$. In particular, the function f in Theorem 7.2.2 can be chosen to be strongly Darboux.

Corollary 7.2.3 *Let $\mathcal{G}$ be a family of real functions such that $|\mathcal{G}| \leq \mathfrak{c}$. Then there exists a (strongly) Darboux function $f \colon \mathbb{R} \to \mathbb{R}$ such that $f + g$ is (strongly) Darboux for every $g \in \mathcal{G}$.*

Another way to look at Corollary 7.2.3 is to use the language of cardinal functions as follows. Let $\mathcal{D}$ stand for the family of all Darboux functions from $\mathbb{R}$ to $\mathbb{R}$ and let $\mathcal{A}(\mathcal{D})$ be the minimal cardinality of a family $\mathcal{G}$ for which Corollary 7.2.3 fails, that is,

$$\mathcal{A}(\mathcal{D}) = \min\{|\mathcal{G}| \colon \forall f \in \mathbb{R}^\mathbb{R} \; \exists g \in \mathcal{G} \; (f + g \notin \mathcal{D})\}.$$

Notice that the set $\{|\mathcal{G}| \colon \forall f \in \mathbb{R}^\mathbb{R} \; \exists g \in \mathcal{G} \; (f + g \notin \mathcal{D})\}$ is not empty, since $|\mathbb{R}^\mathbb{R}| = \mathfrak{c}^\mathfrak{c} = 2^\mathfrak{c}$ belongs to this set: For every $f \in \mathbb{R}^\mathbb{R}$ there exists a $g \in \mathcal{G} = \mathbb{R}^\mathbb{R}$ such that $f + g$ is equal to any fixed function h, and such that h can be chosen to be not Darboux. So $\mathcal{A}(\mathcal{D}) \leq 2^\mathfrak{c}$. In this language Corollary 7.2.3 can be stated as follows:

$$\mathfrak{c} < \mathcal{A}(\mathcal{D}) \leq 2^\mathfrak{c}. \tag{7.1}$$

If $2^\mathfrak{c} = \mathfrak{c}^+$, which follows from the generalized continuum hypothesis, then $\mathcal{A}(\mathcal{D}) = 2^\mathfrak{c}$. However, this equation cannot be proved in ZFC alone. It has been proved by K. Ciesielski and A. W. Miller (1994–5) that (7.1) is essentially everything that can be proved in ZFC about $\mathcal{A}(\mathcal{D})$.

In what follows we will consider some other generalizations of Theorem 7.2.1. We start by noticing that the sum of a Darboux function and a continuous function does not have to be Darboux. To see it, modify the proof of Theorem 7.2.2 to get a strongly Darboux function f such that $g(x) = f(x) - x$ is not Darboux, by choosing

$$x_\xi \in I_\xi \setminus (\{x_\zeta \colon \zeta < \xi\} \cup \{r_\xi\})$$

and defining $f(x_\xi) = r_\xi$. Then $f(x_\xi) \neq x_\xi$ and we can extend f to all of $\mathbb{R}$ to have $f(x) \neq x$ for all $x \in \mathbb{R}$. So $g(x) = f(x) - x \neq 0$ for every $x \in \mathbb{R}$ and Proposition 7.2.4, stated next, implies that g is not Darboux.

Proposition 7.2.4 *If f is strongly Darboux, then for every continuous function g,*

$$f + g \text{ is strongly Darboux iff } f + g \text{ is Darboux.}$$

Proof Implication $\Rightarrow$ is obvious. The other implication follows from the fact that on any interval $I = (c, d)$, with $c < d$, the continuous function g is bounded, while $f[I] = \mathbb{R}$. So $f + g$ must be unbounded from below and from above. In particular, for every $y \in \mathbb{R}$ there exist $A < y < B$ and $a, b \in I$ such that $(f + g)(a) = A$ and $(f + g)(b) = B$. Thus there exists an x between a and b with $(f + g)(x) = y$, since $f + g$ is Darboux. Hence $(f + g)[I] = \mathbb{R}$. $\qquad\qquad\qquad\qquad\qquad\qquad\qquad\qquad\qquad\qquad\qquad\square$

The next theorem, due to Kirchheim and Natkaniec (1990–1), generalizes the preceding observation. It concerns continuous nowhere-constant functions, that is, continuous functions that are not constant on any open nonempty interval.

Theorem 7.2.5 *If the continuum hypothesis holds then there is a strongly Darboux function f such that $f + h$ is not Darboux for every continuous nowhere-constant function h.*

Proof Let $\{h_\xi \colon \xi < \mathfrak{c}\}$ be an enumeration of all continuous nowhere-constant functions from $\mathbb{R}$ to $\mathbb{R}$. Notice that for any such function h the set $h^{-1}(r)$ is closed and nowhere dense for every $r \in \mathbb{R}$.

By Proposition 7.2.4 it is enough to find $f \colon \mathbb{R} \to \mathbb{R}$ such that

$$\forall \xi < \mathfrak{c} \, \exists r_\xi \in \mathbb{R} \, \forall x \in \mathbb{R} \, (f(x) + h_\xi(x) \neq r_\xi). \qquad (7.2)$$

For this we will construct sequences $\{x_\xi \in \mathbb{R} \colon \xi < \mathfrak{c}\}$, $\{y_\xi \in \mathbb{R} \colon \xi < \mathfrak{c}\}$, and $\{r_\xi \in \mathbb{R} \colon \xi < \mathfrak{c}\}$ such that the function f defined by

$$f(x_\xi) = y_\xi \qquad (7.3)$$

is strongly Darboux and satisfies condition (7.2). As we know from Theorems 7.2.1 and 7.2.2, it is relatively easy to construct the sequences $\{x_\xi\}$ and $\{y_\xi\}$ such that f defined by (7.3) is strongly Darboux. In previous constructions, however, we did not have to worry about $\{x_\xi \colon \xi < \mathfrak{c}\}$ being equal to $\mathbb{R}$, since any extension of the part of f given by (7.3) was still strongly Darboux. But this time, we also have to take care of condition (7.2), and the extension may fail to have this property. To avoid this problem we will arrange our construction to have

$$\{x_\xi \colon \xi < \mathfrak{c}\} = \mathbb{R}. \qquad (7.4)$$

Now let S and T be subsets of $\mathfrak{c}$ of cardinality $\mathfrak{c}$. Let $\{s_\xi \colon \xi \in S\}$ be an enumeration of $\mathbb{R}$ and let $\{\langle I_\xi, t_\xi \rangle \colon \xi \in T\}$ be an enumeration of the family

$\mathcal{F} = \{\langle I, t \rangle : t \in \mathbb{R} \ \& \ I = (a, b) \text{ for some } a < b\}$. To satisfy (7.4) it is enough to have

(S1) $x_\xi = s_\xi$ for every $\xi \in S$.

To make sure that f is strongly Darboux it is enough to proceed as in Theorems 7.2.1 and 7.2.2; that is, for every $\xi \in T$ choose

$$x_\xi \in I_\xi \setminus \{x_\zeta : \zeta < \xi\} \tag{7.5}$$

and put

(T1) $y_\xi = t_\xi$.

Notice, however, that for $s_\xi \notin I_\xi$ conditions (S1) and (7.5) cannot be satisfied simultaneously. To avoid this problem we will assume that the sets S and T are disjoint. We can also assume that $S \cup T = \mathfrak{c}$.

Property (7.2) will hold if we construct the sequences in such a way that for every $\xi < \mathfrak{c}$

$$\forall \zeta < \xi \ \forall \eta < \xi \ (y_\eta + h_\zeta(x_\eta) \neq r_\zeta). \tag{7.6}$$

To preserve this condition while constructing x_ξ, y_ξ, and r_ξ we will have to choose r_ξ such that $y_\eta + h_\xi(x_\eta) \neq r_\xi$ for every $\eta < \xi$, that is, pick

($\star$) $r_\xi \in \mathbb{R} \setminus \{y_\eta + h_\xi(x_\eta) : \eta < \xi\}$,

and choose x_ξ and y_ξ such that

$$y_\xi + h_\zeta(x_\xi) \neq r_\zeta \quad \text{for every} \quad \zeta \leq \xi. \tag{7.7}$$

Now for $\xi \in T$ we have no freedom in choosing y_ξ since, by (T1), $y_\xi = t_\xi$. But to satisfy (7.7) it is enough to choose $x_\xi \notin \bigcup_{\zeta \leq \xi} h_\zeta^{-1}(r_\zeta - y_\xi)$. Combining this with (7.5) we need only choose

(T2) $x_\xi \in I_\xi \setminus \left(\{x_\zeta : \zeta < \xi\} \cup \bigcup \{h_\zeta^{-1}(r_\zeta - y_\xi) : \zeta \leq \xi\} \right)$ for every $\xi \in T$.

For $\xi \in S$, by (S1), we have $x_\xi = s_\xi$. If $x_\xi = x_\zeta$ for some $\zeta < \xi$ then, by (7.3), we have to define

(S2) $y_\xi = y_\zeta$.

But then conditions (7.6) and (7.7) will be preserved. On the other hand, if $x_\xi \neq x_\zeta$ for every $\zeta < \xi$, in order to maintain (7.7) it is enough to choose

(S3) $y_\xi \in \mathbb{R} \setminus \{r_\eta - h_\eta(x_\xi) : \eta \leq \xi\}$.

The foregoing discussion shows that the function f can be constructed if we can construct the sequences $\{x_\xi \in \mathbb{R} : \xi < \mathfrak{c}\}$, $\{y_\xi \in \mathbb{R} : \xi < \mathfrak{c}\}$, and $\{r_\xi \in \mathbb{R} : \xi < \mathfrak{c}\}$ such that the following conditions hold for every $\xi < \mathfrak{c}$:

$(\star)$ $r_\xi \in \mathbb{R} \setminus \{y_\eta + h_\xi(x_\eta) : \eta < \xi\}$;

(T1) $y_\xi = t_\xi$ for $\xi \in T$;

(T2) $x_\xi \in I_\xi \setminus \left(\{x_\zeta : \zeta < \xi\} \cup \bigcup \{h_\zeta^{-1}(r_\zeta - y_\xi) : \zeta \le \xi\} \right)$ for $\xi \in T$;

(S1) $x_\xi = s_\xi$ for $\xi \in S$;

(S2) $y_\xi = y_\zeta$ for $\xi \in S$ if $x_\xi = x_\zeta$ for some $\zeta < \xi$;

(S3) $y_\xi \in \mathbb{R} \setminus \{r_\zeta - h_\zeta(x_\xi) : \zeta \le \xi\}$ for $\xi \in S$ if $x_\xi \ne x_\zeta$ for all $\zeta < \xi$.

By the recursion theorem it is enough just to argue that such a choice is always possible.

The possibility of the choices in $(\star)$, (T1), (S1), (S2), and (S3) is obvious. The choice of x_ξ in (T2) is possible by the Baire category theorem, since the set $\{x_\zeta : \zeta < \xi\} \cup \bigcup \{h_\zeta^{-1}(r_\zeta - y_\xi) : \zeta \le \xi\}$ is of first category, being an at most countable union of nowhere-dense sets (this is the place where we used the continuum hypothesis!).   □

It is worthwhile to mention that the conclusion of Theorem 7.2.5 is not equivalent to the continuum hypothesis. Essentially the same proof works with a weaker additional set-theoretic axiom, known as the additivity of category, that the union of less than continuum many meager sets is still a meager set (see Theorem 8.2.6). However, it has been recently proved by Steprāns (1993) that the conclusion of Theorem 7.2.5 is independent of ZFC set theory.

EXERCISES

1 Show that every function $f \colon C \to \mathbb{R}$ on a nowhere-dense subset C of $\mathbb{R}$ can be extended to a Darboux function $g \colon \mathbb{R} \to \mathbb{R}$.

Since there are nowhere-dense subsets C of $\mathbb{R}$ of cardinality $\mathfrak{c}$, for example, the Cantor set, conclude that the family $\mathcal{D}$ of Darboux functions has cardinality $2^{\mathfrak{c}}$.

2 Generalize Theorem 7.2.5 to prove the following:

> Let $\mathcal{G}$ be the family of all continuous functions from $\mathbb{R}$ to $\mathbb{R}$ such that for every h in $\mathcal{G}$ there exists a nonempty open interval I such that h is not constant on any nonempty open subinterval of I. If the continuum hypothesis holds then there exists a strongly Darboux function f such that $f + h$ is not Darboux for every $h \in \mathcal{G}$.

3 Show that if h is a continuous function with the property that every nonempty open interval contains a nonempty open subinterval on which h is constant then $f + h$ is strongly Darboux for every strongly Darboux function f.

Remark: There are continuous nonconstant functions with the property just described. For example, if C is the Cantor set then we can define a function $h\colon [0, 1] \to \mathbb{R}$ by $h(x) = \sum_{n=1}^{\infty} x_n/2^{n+1}$, where $\sum_{n=1}^{\infty} x_n/3^n = \max(C \cap [0, x])$ and $x_n \in \{0, 2\}$.

4 Find uncountable sets $X, Y \subset \mathbb{R} \setminus \mathbb{Q}$ such that for every continuous function $f\colon \mathbb{R} \to \mathbb{R}$ if $f[X] \subset Y$ then f is constant.

5 Prove the following theorem of Sierpiński and Zygmund: There exists a function $f\colon \mathbb{R} \to \mathbb{R}$ such that for every continuous function $g\colon \mathbb{R} \to \mathbb{R}$ the set $\{x \in \mathbb{R}\colon f(x) = g(x)\}$ has cardinality less than $\mathfrak{c}$. *Hint:* Use the fact that every continuous function $h\colon X \to \mathbb{R}$, with $X \subset \mathbb{R}$, has a continuous extension $\bar{h}\colon G \to \mathbb{R}$ to a G_δ set $G \subset \mathbb{R}$. (This can be proved by noticing that the set of all $z \in \mathbb{R}$ for which the limit $\lim_{x \to z} h(x)$ does not exist is an F_σ set.)

7.3 Additive functions and Hamel bases

A function $F\colon \mathbb{R} \to \mathbb{R}$ is an *additive function* if $F(x + y) = F(x) + F(y)$ for every $x, y \in \mathbb{R}$. Every linear function $F(x) = ax$ is clearly additive and continuous, and it is quite easy to see that these are the only continuous additive functions. However, there exist discontinuous additive functions. The first example of such a function was constructed by Hamel, with the use of a *Hamel basis*, that is, a linear basis of $\mathbb{R}$ over $\mathbb{Q}$, which exists by Theorem 4.4.1. For its construction we will need some easy facts.

For $B \subset \mathbb{R}$ let $\mathrm{LIN}_{\mathbb{Q}}(B)$ denote the smallest linear subspace of $\mathbb{R}$ over $\mathbb{Q}$ containing B. Notice that $\mathrm{LIN}_{\mathbb{Q}}(B)$ can be obtained by closing set B under the action of the family $\mathcal{F}$ of operations $\langle x, y \rangle \mapsto x + y$ and $x \mapsto qx$ for every $q \in \mathbb{Q}$. Thus $\mathrm{LIN}_{\mathbb{Q}}(B) = \mathrm{cl}_{\mathcal{F}}(B)$, where we are using the notation of Lemma 6.1.6. In particular, Lemma 6.1.6(a) implies that

$$|\mathrm{LIN}_{\mathbb{Q}}(B)| = |B| + \omega \quad \text{for every} \quad B \subset \mathbb{R}. \tag{7.8}$$

From this we immediately conclude that

$$\text{every Hamel basis } H \text{ has cardinality continuum} \tag{7.9}$$

since $\mathrm{LIN}_{\mathbb{Q}}(H) = \mathbb{R}$.

Proposition 7.3.1 *If $F\colon \mathbb{R} \to \mathbb{R}$ is additive then it is linear over $\mathbb{Q}$, that is,*

$$F(px + qy) = pF(x) + qF(y)$$

for every $p, q \in \mathbb{Q}$ and $x, y \in \mathbb{R}$.

Proof Let $F\colon \mathbb{R} \to \mathbb{R}$ be additive. Then $F(0) + F(0) = F(0 + 0) = F(0)$, so

$$F(0) = 0.$$

It is easy to prove by induction on $n \in \omega$ that $F(nx) = nF(x)$ for every $x \in \mathbb{R}$. Since $F(-x) + F(x) = F(-x + x) = F(0) = 0$, we have also that $F(-x) = -F(x)$. Therefore we conclude that

$$F(nx) = nF(x) \text{ for every } x \in \mathbb{R} \text{ and } n \in \mathbb{Z}.$$

In particular, for $0 < m < \omega$ we have $mF\left(\frac{1}{m}x\right) = F\left(m\frac{1}{m}x\right) = F(x)$, that is, $F\left(\frac{1}{m}x\right) = \frac{1}{m}F(x)$. So

$$F(px) = pF(x) \text{ for every } x \in \mathbb{R} \text{ and } p \in \mathbb{Q},$$

and $F(px + qy) = F(px) + F(qy) = pF(x) + qF(y)$ for every $p, q \in \mathbb{Q}$ and $x, y \in \mathbb{R}$. $\square$

Theorem 7.3.2 *If $H \subset \mathbb{R}$ is a Hamel basis then every function $f\colon H \to \mathbb{R}$ can be extended uniquely to an additive function $F\colon \mathbb{R} \to \mathbb{R}$.*

Proof Since every $x \in \mathbb{R}$ has a unique representation $x = q_1 b_1 + \cdots + q_m b_m$ in the basis H (that is, $b_1 < \cdots < b_m$ are from H and $q_1, \ldots, q_m \in \mathbb{Q} \setminus \{0\}$; see Section 4.4), we can define

$$F(x) = \sum_{i=1}^{m} q_i f(b_i). \tag{7.10}$$

This function is clearly additive and extends f. Its uniqueness follows from Proposition 7.3.1. $\square$

Corollary 7.3.3 *There are additive discontinuous functions $F\colon \mathbb{R} \to \mathbb{Q}$.*

Proof Let H be a Hamel basis and let $x \in H$. Define $f\colon H \to \mathbb{R}$ by putting $f(x) = 1$ and $f(y) = 0$ for $y \in H \setminus \{x\}$. Extend f to the additive function F. By (7.10) the range of f is equal to $\mathbb{Q}$. Thus, F cannot be continuous, since it does not have the Darboux property. $\square$

The function from Corollary 7.3.3 has a small (countable) range. It is not difficult to modify the argument from Corollary 7.3.3 to get an additive discontinuous function *onto* $\mathbb{R}$. The next theorem shows that we can even do a lot better than this.

Theorem 7.3.4 *There exists a Hamel basis H and a function $f\colon H \to \mathbb{R}$ such that $f^{-1}(r)$ is a Bernstein set for every $r \in \mathbb{R}$.*

Proof First notice that if $f\colon H \to \mathbb{R}$ is such that

$$f^{-1}(r) \cap P \neq \emptyset \text{ for every } r \in \mathbb{R} \text{ and perfect set } P \subset \mathbb{R}$$

then f has the desired property, since the complement of each $f^{-1}(r)$ contains $f^{-1}(r+1)$ and so it must also intersect every perfect set.

Let $\{\langle P_\xi, r_\xi \rangle \colon \xi < \mathfrak{c}\}$ be an enumeration of $\{P \subset \mathbb{R} \colon P \text{ is perfect}\} \times \mathbb{R}$. We will construct, by induction on $\xi < \mathfrak{c}$, a one-to-one sequence $\{x_\xi \colon \xi < \mathfrak{c}\}$ that is linearly independent over $\mathbb{Q}$ and such that $x_\xi \in P_\xi$ for every $\xi < \mathfrak{c}$.

Notice that this will be enough to construct function f. To see it, extend $\{x_\xi \colon \xi < \mathfrak{c}\}$ to a Hamel basis (see Theorem 4.4.1). Define $f(x_\xi) = r_\xi$ for $\xi < \mathfrak{c}$ and extend f arbitrarily to H. Then for every perfect set $P \subset \mathbb{R}$ and every $r \in \mathbb{R}$ there exists a $\xi < \mathfrak{c}$ such that $\langle P, r \rangle = \langle P_\xi, r_\xi \rangle$, so that $f(x_\xi) = r_\xi = r$ and $x_\xi \in P_\xi = P$. Therefore $f^{-1}(r) \cap P \neq \emptyset$.

To make the sequence $\{x_\xi \colon \xi < \mathfrak{c}\}$ linearly independent over $\mathbb{Q}$ for every $\xi < \mathfrak{c}$ we will choose

(i) $x_\xi \in P_\xi \setminus \mathrm{LIN}_{\mathbb{Q}}(\{x_\zeta \colon \zeta < \xi\})$.

The choice is possible since, by (7.8), $|\mathrm{LIN}_{\mathbb{Q}}(\{x_\zeta \colon \zeta < \xi\})| = |\xi| + \omega < \mathfrak{c}$. Thus, by the recursion theorem, we can find a sequence satisfying (i). To finish the proof it is enough to show that the choice of x_ξs from the complement of $\mathrm{LIN}_{\mathbb{Q}}(\{x_\zeta \colon \zeta < \xi\})$ makes the sequence $\{x_\xi \colon \xi < \mathfrak{c}\}$ linearly independent over $\mathbb{Q}$. This part is left as an exercise. $\square$

Corollary 7.3.5 *There exists a nonmeasurable non-Baire additive function $F\colon \mathbb{R} \to \mathbb{R}$ that is strongly Darboux.*

Proof Let F be an additive extension of f from Theorem 7.3.4. Then $F^{-1}(r)$ is a Bernstein set for every $r \in \mathbb{R}$. $\square$

In fact it can be proved that every discontinuous additive function is neither measurable nor Baire.

Corollary 7.3.6 *There exists a Hamel basis H that is neither measurable nor has the Baire property.*

Proof It is enough to notice that the Hamel basis H from Theorem 7.3.4 is a Bernstein set.

It clearly intersects every perfect set. On the other hand, if $a \in H$, then $a + H = \{a + x \colon x \in H\}$ is disjoint from H. This is the case since otherwise there would exist $x, y \in H$ such that $a + x = y$ and so $\{a, x, y\} \subset H$ would be linearly dependent. Thus $a + H \subset \mathbb{R} \setminus H$. But $a + H$ intersects every perfect set P, since $H \cap (-a+P) \neq \emptyset$ for every perfect set P, the set $-a+P$ being perfect. □

Notice that there are Hamel bases that are measurable and have the Baire property (see Exercise 2).

Corollary 7.3.7 *There exists a Hamel basis H such that $|H \cap P| = \mathfrak{c}$ for every perfect set P.*

Proof It is clear that the basis H from Theorem 7.3.4 has the desired property, since $H \cap P = \bigcup_{r \in \mathbb{R}}(f^{-1}(r) \cap P)$ is the union of continuum many nonempty disjoint sets. □

Next, we will construct an additive function F whose graph is connected as a subset of the plane. The graph of such a function is called a *Jones space*.

Theorem 7.3.8 *There exists a discontinuous additive function $F \colon \mathbb{R} \to \mathbb{R}$ whose graph is connected.*

Proof Let H be the Hamel basis from Corollary 7.3.7 and let $\mathcal{F}$ be the family of all closed subsets P of the plane such that $p[P]$ contains a perfect set, where p is the projection of the plane onto the x-axis. Let $\{P_\xi \colon \xi < \mathfrak{c}\}$ be an enumeration of $\mathcal{F}$. Define, by induction on $\xi < \mathfrak{c}$, a sequence $\{\langle x_\xi, y_\xi \rangle \in \mathbb{R}^2 \colon \xi < \mathfrak{c}\}$ such that

(i) $x_\xi \in (H \cap p[P_\xi]) \setminus \{x_\zeta \colon \zeta < \xi\}$,

(ii) $\langle x_\xi, y_\xi \rangle \in P_\xi$.

The choice as in (i) can be made, since $|H \cap p[P_\xi]| = \mathfrak{c}$.

Now define $f(x_\xi) = y_\xi$ and extend f to H in an arbitrary way. Let F be the additive extension of f. We will show that the graph of F is connected.

To obtain a contradiction assume that the graph of F is not connected. Then there exist disjoint open subsets U and V of $\mathbb{R}^2$ such that $U \cap F \neq \emptyset \neq V \cap F$ and $F \subset U \cup V$. Let $P = \mathbb{R}^2 \setminus (U \cup V)$. We will show that $P \cap F \neq \emptyset$, which will give us the desired contradiction.

If $P \in \mathcal{F}$ then there exists a $\xi < \mathfrak{c}$ such that $P = P_\xi$, and so $\langle x_\xi, y_\xi \rangle \in P_\xi \cap F = P \cap F$. Thus it is enough to prove that $P \in \mathcal{F}$. Clearly, P is closed. So we have to show only that

$$p[P] \text{ contains a perfect subset.}$$

In order to prove it we first show that there exists an $x \in \mathbb{R}$ such that

$$U \cap (\{x\} \times \mathbb{R}) \neq \emptyset \neq V \cap (\{x\} \times \mathbb{R}). \tag{7.11}$$

To obtain a contradiction, assume that (7.11) is false. Then the sets $U_1 = p[U]$ and $V_1 = p[V]$ are disjoint. But they are open in $\mathbb{R}$, since the projection of an open set is open. They are clearly nonempty, since U and V are nonempty. Finally, $U_1 \cup V_1 = p[U \cup V] \supset p[F] = \mathbb{R}$, that is, sets U_1 and V_1 violate the connectedness of $\mathbb{R}$. Condition (7.11) has been proved.

Now let $x_0 \in \mathbb{R}$ be such that $U \cap (\{x_0\} \times \mathbb{R}) \neq \emptyset \neq V \cap (\{x_0\} \times \mathbb{R})$ and let $y_1, y_2 \in \mathbb{R}$ be such that $\langle x_0, y_1 \rangle \in U$ and $\langle x_0, y_2 \rangle \in V$. Then there exists an open interval $I = (a, b)$ containing x_0 such that $I \times \{y_1\} \subset U$ and $I \times \{y_2\} \subset V$. But for every $x \in I$ the interval $J_x = \{x\} \times [y_1, y_2]$ is a connected set intersecting both U and V. Thus J_x must intersect P as well, since it is connected. Therefore $\{x\} = p[J_x \cap P] \subset p[P]$ for every $x \in I$, that is, $p[P] \supset I$. This finishes the proof that $P \in \mathcal{F}$. $\square$

We will finish this section with following theorem of Erdős and Kakutani (1943).

Theorem 7.3.9 *The continuum hypothesis is equivalent to the existence of a countable partition $\{H_n \colon n < \omega\}$ of $\mathbb{R} \setminus \{0\}$ such that every H_n is a Hamel basis.*

In the proof of the theorem we will use the following two lemmas, the first of which is due to Erdős and Hajnal (Erdős 1978–9). They are also very interesting in their own right.

Lemma 7.3.10 *If $f \colon A \times B \to \omega$, where $|A| = \omega_2$ and $|B| = \omega_1$, then for every $n < \omega$ there exist $B_0 \in [B]^n$ and $A_0 \in [A]^{\omega_2}$ such that $f(a_0, b_0) = f(a_1, b_1)$ for every $a_0, a_1 \in A_0$ and $b_0, b_1 \in B_0$.*

Proof Fix $n < \omega$. First we will show that

$$\forall a \in A \; \exists B_a \in [B]^n \; \exists m_a < \omega \; \forall b \in B_a \; (f(a, b) = m_a). \qquad (7.12)$$

To see this, notice that for every $a \in A$ the sets $S_a^m = \{b \in B \colon f(a, b) = m\}$ form a countable partition of an uncountable set B. Thus there is an $m = m_a$ for which the set S_a^m is uncountable and any n-element subset B_a of S_a^m satisfies (7.12).

Now let $F \colon A \to [B]^n \times \omega$ be such that for every $a \in A$ the pair $F(a) = \langle B_a, m_a \rangle$ satisfies (7.12). The set $[B]^n \times \omega$ has cardinality ω_1, so there exists $\langle B_0, m \rangle \in [B]^n \times \omega$ such that $A_0 = F^{-1}(B_0, m)$ has cardinality ω_2. But then, if $a_i \in A_0$ for $i < 2$ then $F(a_i) = \langle B_0, m \rangle$ and

$$f(a_i, b) = m_{a_i} = m$$

for every $b \in B_{a_i} = B_0$. So the sets A_0 and B_0 have the desired property.
$\square$

Notice that Lemma 7.3.10 can also be expressed in the following graph-theoretic language. Let A and B be disjoint sets of cardinality ω_2 and ω_1, respectively, and let $G = \langle V, E \rangle$ be the bipartite graph between A and B, that is, $V = A \cup B$ and $E = \{\{a, b\} : a \in A \ \& \ b \in B\}$. Then for any $n < \omega$ and any coloring $f : E \to \omega$ of the edges of graph G there are $A_0 \in [A]^{\omega_2}$ and $B_0 \in [B]^n$ such that the subgraph $G_0 = \langle A_0 \cup B_0, E \cap \mathcal{P}(A_0 \cup B_0) \rangle$ of G generated by $A_0 \cup B_0$ is monochromatic, that is, the coloring function f is constant on the edges of G_0.

Lemma 7.3.11 *Assume that $\mathfrak{c} > \omega_1$ and let $H \in [\mathbb{R}]^{\omega_2}$. Then for every partition $\{H_n : n < \omega\}$ of $\mathbb{R}$ there exist $n < \omega$ and disjoint sets $A_0 \in [H]^{\omega_2}$ and $B_0 \in [H]^2$ such that*

$$a + b \in H_n \quad \text{for every } a \in A_0 \text{ and } b \in B_0.$$

Proof Choose disjoint $A \in [H]^{\omega_2}$ and $B \in [H]^{\omega_1}$. Define $f : A \times B \to \omega$ by

$$f(a, b) = m \quad \text{if and only if} \quad a + b \in H_m.$$

Then the sets A_0 and B_0 from Lemma 7.3.10 used with f and $n = 2$ have the desired properties. □

Proof of Theorem 7.3.9 $\Leftarrow$: Let H be any Hamel basis and suppose that there exists a partition $\{H_n : n < \omega\}$ of $\mathbb{R} \setminus \{0\}$ into Hamel bases.

To obtain a contradiction assume that $\mathfrak{c} > \omega_1$. Then $|H| = \mathfrak{c} \geq \omega_2$ so we can choose A_0 and B_0 as in Lemma 7.3.11. Take different $a_0, a_1 \in A_0$ and $b_0, b_1 \in B_0$. Then the numbers $x_{ij} = a_i + b_j$ for $i, j \in 2$ are different and belong to the same H_n. However,

$$x_{00} - x_{10} = (a_0 + b_0) - (a_1 + b_0) = (a_0 + b_1) - (a_1 + b_1) = x_{01} - x_{11},$$

contradicting the fact that $x_{00}, x_{10}, x_{01}, x_{11} \in H_n$ are linearly independent over $\mathbb{Q}$.

$\Rightarrow$: Let $\{z_\xi \in \mathbb{R} : \xi < \omega_1\}$ be a Hamel basis and for every $\xi \leq \omega_1$ define $L_\xi = \mathrm{LIN}_{\mathbb{Q}}(\{z_\zeta : \zeta < \xi\})$. Notice that $L_\xi \subset L_\eta$ for $\xi < \eta \leq \omega_1$, that $L_\lambda = \bigcup_{\xi < \lambda} L_\xi$ for limit $\lambda \leq \omega_1$, and that $L_{\omega_1} = \mathbb{R}$ (compare this with Lemma 6.1.6). Thus the sets $\{L_{\xi+1} - L_\xi : \xi < \omega_1\}$ form a partition of $\mathbb{R} \setminus L_0 = \mathbb{R} \setminus \{0\}$. For every $\xi < \omega_1$ choose a bijection $f_\xi : L_{\xi+1} - L_\xi \to \omega$ and let $f = \bigcup_{\xi < \omega_1} f_\xi$. Then $f : \mathbb{R} \setminus \{0\} \to \omega$. Let $H_n = f^{-1}(n)$.

The sets $\{H_n : n < \omega\}$ form a partition of $\mathbb{R} \setminus \{0\}$. To finish the proof it is enough to show that every H_n is a Hamel basis.

So choose $n < \omega$ and let $\{x_\xi\} = (L_{\xi+1} - L_\xi) \cap f^{-1}(n)$ for $\xi < \omega_1$. It is enough to prove that

$$L_\eta = \mathrm{LIN}_{\mathbb{Q}}(\{x_\zeta : \zeta < \eta\}) \tag{7.13}$$

for every $\eta \leq \omega_1$, since then $H_n = \{x_\xi : \xi < \mathfrak{c}\}$ spans $\mathbb{R}$ and $\{x_\xi : \xi < \mathfrak{c}\}$ is linearly independent by Exercise 1. Condition (7.13) can be proved by induction on $\eta \leq \omega_1$. So let $\alpha \leq \omega_1$ be such that (7.13) holds for every $\eta < \alpha$.

If α is a limit ordinal then

$$L_\alpha = \bigcup_{\eta < \alpha} L_\eta = \bigcup_{\eta < \alpha} \mathrm{LIN}_{\mathbb{Q}}(\{x_\zeta : \zeta < \eta\}) = \mathrm{LIN}_{\mathbb{Q}}(\{x_\zeta : \zeta < \alpha\}).$$

If $\alpha = \eta + 1$ then $L_\eta = \mathrm{LIN}_{\mathbb{Q}}(\{x_\zeta : \zeta < \eta\})$ and

$$\{x_\eta\} \cup \mathrm{LIN}_{\mathbb{Q}}(\{x_\zeta : \zeta < \eta\}) = \{x_\eta\} \cup L_\eta \subset L_{\eta+1}$$

so $\mathrm{LIN}_{\mathbb{Q}}(\{x_\zeta : \zeta < \eta + 1\}) \subset L_{\eta+1}$.

To prove the other inclusion, recall that $x_\eta \in L_{\eta+1} - L_\eta$. Therefore $x_\eta = q z_\eta + \sum_{i<m} q_i b_i$ for some $q, q_0, \dots, q_{m-1} \in \mathbb{Q}$ and $b_0, \dots, b_{m-1} \in \{x_\zeta : \zeta < \eta\}$. Moreover, $q \neq 0$ because $x_\eta \notin L_\eta = \mathrm{LIN}_{\mathbb{Q}}(\{x_\zeta : \zeta < \eta\})$. So

$$z_\eta = \frac{1}{q}\left(x_\eta - \sum_{i<m} q_i b_i\right) \in \mathrm{LIN}_{\mathbb{Q}}(\{x_\zeta : \zeta < \eta + 1\}).$$

Now $\{z_\eta\} \cup L_\eta \subset \mathrm{LIN}_{\mathbb{Q}}(\{x_\zeta : \zeta < \eta + 1\})$, and so we can conclude that $L_{\eta+1} \subset \mathrm{LIN}_{\mathbb{Q}}(\{x_\zeta : \zeta < \eta + 1\})$. $\qquad \square$

EXERCISES

1 Complete the proof of Theorem 7.3.4 by showing that if $\{x_\xi \in \mathbb{R} : \xi < \mathfrak{c}\}$ is such that $x_\xi \notin \mathrm{LIN}_{\mathbb{Q}}(\{x_\zeta : \zeta < \xi\})$ for every $\xi < \mathfrak{c}$, then $\{x_\xi \in \mathbb{R} : \xi < \mathfrak{c}\}$ is linearly independent over $\mathbb{Q}$.

2 Let $C \subset \mathbb{R}$ be the Cantor set. Show that there is a Hamel basis contained in C. Since C is nowhere dense and has measure zero, it follows that Hamel bases can be measurable and have the Baire property. *Hint:* Use the fact that $C + C = \{a + b : a, b \in C\}$ contains the unit interval $[0, 1]$.

3 Let $\sim$ be an equivalence relation on $\mathbb{R}$ defined by $x \sim y$ if and only if $x - y \in \mathbb{Q}$. If V is a selector from the family of all equivalence classes of $\sim$ then V is called a *Vitali set*. It is known that any Vitali set is neither measurable nor has the Baire property.

Construct a Vitali set V such that

(a) $V + V = \mathbb{R}$,

(b) $V + V$ is a Bernstein set.

4 Prove that the graph of any additive discontinuous function $F : \mathbb{R} \to \mathbb{R}$ is dense in $\mathbb{R}^2$.

7.4   Symmetrically discontinuous functions

This section is motivated in part by the following generalization of the continuity of real functions. A function $f\colon \mathbb{R} \to \mathbb{R}$ is said to be *symmetrically continuous at the point* $x \in \mathbb{R}$ if

$$\lim_{h \to 0^+} [f(x-h) - f(x+h)] = 0,$$

that is, if for every $\varepsilon > 0$ there exists a $d > 0$ such that

$$(0, d) \subset S_x^\varepsilon,$$

where $S_x^\varepsilon = \{h > 0 \colon |f(x-h) - f(x+h)| < \varepsilon\}$. A function $f\colon \mathbb{R} \to \mathbb{R}$ is *symmetrically continuous* if it is symmetrically continuous at every point $x \in \mathbb{R}$.

Clearly, every continuous function is symmetrically continuous. The converse implication is not true, since there are symmetrically continuous functions that are not continuous in the usual sense. For example, the characteristic function $\chi_{\{0\}}$ of a singleton set $\{0\}$ is symmetrically continuous and discontinuous at 0. However, symmetrically continuous functions cannot behave too badly in this respect: It can be proved that the set of points of discontinuity of a symmetrically continuous function must be of first category and have measure zero.

The study of symmetrically continuous functions is an important subject in real analysis, motivated by Fourier analysis. In this section, however, this notion serves merely as a motivation, since we will study here the functions that are not symmetrically continuous.

It is easy to find a function that is not symmetrically continuous. For example, a characteristic function of any nontrivial interval is symmetrically discontinuous at the endpoints of that interval. In fact, it is also not difficult to construct a function $f\colon \mathbb{R} \to \mathbb{R}$ that is nowhere symmetrically continuous. Such a function must have the property that for every $x \in \mathbb{R}$ there exists an $\varepsilon > 0$ such that

$$\forall d > 0 \ \exists h \in (0, d) \ (|f(x-h) - f(x+h)| \geq \varepsilon), \tag{7.14}$$

or, equivalently, such that

$$(0, d) \not\subset S_x^\varepsilon \quad \text{for every } d > 0. \tag{7.15}$$

To get such an example define f as the characteristic function χ_H of any dense Hamel basis H. Notice that such a basis exists by Theorem 7.3.4. To see that χ_H is nowhere symmetrically continuous, take $x \in \mathbb{R}$, $\varepsilon \in (0, 1)$, and an arbitrary $d > 0$. We have to find $h \in (0, d)$ with the property that $|\chi_H(x - h) - \chi_H(x + h)| \geq \varepsilon$. So let $q_1 b_1 + \cdots + q_n b_n$ be a representation of x in basis H. Then, by the density of H, there exists a $b \in H \cap (x - d, x) \setminus \{b_1, \ldots, b_n\}$. Pick $h = x - b$. Then $h \in (0, d)$ and $\chi_H(x - h) = \chi_H(b) = 1$. On the other hand, $\chi_H(x + h) = 0$ as $x + h = 2x - b = 2q_1 b_1 + \cdots + 2q_n b_n - b \notin H$. Therefore $|\chi_H(x - h) - \chi_H(x + h)| = 1 > \varepsilon$.

Considerations of how badly nowhere–symmetrically continuous functions can behave led several people[1] to to ask whether there exists a function f satisfying (7.14) with the quantifiers $\forall d > 0 \ \exists h \in (0, d)$ replaced by the reversed quantifiers $\exists d > 0 \ \forall h \in (0, d)$, that is, whether there exists a function $f \colon \mathbb{R} \to \mathbb{R}$ such that for every $x \in \mathbb{R}$ there exists an $\varepsilon > 0$ with

$$\exists d > 0 \ \forall 0 < h < d \ (|f(x - h) - f(x + h)| \geq \varepsilon),$$

or, equivalently, such that

$$(0, d) \cap S_x^\varepsilon = \emptyset \quad \text{for some } d > 0.$$

Replacing ε and d by their minima, we can rephrase this problem by asking whether there exists a function $f \colon \mathbb{R} \to \mathbb{R}$ such that

$(\star)$ for every $x \in \mathbb{R}$ there exists a $d > 0$ with

$$(0, d) \cap S_x^d = \emptyset, \tag{7.16}$$

where $S_x^d = \{h > 0 \colon |f(x - h) - f(x + h)| < d\}$.

A function f satisfying $(\star)$ will be called a *uniformly antisymmetric function*.

The existence of uniformly antisymmetric functions can be inferred from the following theorem due to Ciesielski and Larson (1993–4).

Theorem 7.4.1 *There exists a partition $\mathcal{P} = \{P_n \colon n \in \mathbb{N}\}$ of $\mathbb{R}$ such that for every $x \in \mathbb{R}$ the set*

$$S_x = \bigcup_{n \in \mathbb{N}} \{h > 0 \colon x - h, x + h \in P_n\} \tag{7.17}$$

is finite.

Before proving this theorem, we first show how to use it to construct a uniformly antisymmetric function.

Corollary 7.4.2 *There exists a uniformly antisymmetric function $f \colon \mathbb{R} \to \mathbb{N}$.*

---

[1] Evans and Larson in 1984; Kostyrko in 1991.

Proof Let $\mathcal{P} = \{P_n : n \in \mathbb{N}\}$ be a partition satisfying (7.17) and for $x \in \mathbb{R}$ define

$$f(x) = n \quad \text{if and only if} \quad x \in P_n.$$

Then f is uniformly antisymmetric, since for every $x \in \mathbb{R}$ the set

$$S_x^1 = \{h > 0 : |f(x-h) - f(x+h)| < 1\} = \{h > 0 : f(x-h) = f(x+h)\} = S_x$$

is finite, so $d = \min(S_x \cup \{1\})$ satisfies $(\star)$, as $S_x^d = S_x^1$. $\qquad\square$

Proof of Theorem 7.4.1 Let H be a Hamel basis. Then every $x \in \mathbb{R}$ has a unique representation $x = q_1 b_1 + \cdots + q_n b_n$ in the basis H (that is, $b_1 < \cdots < b_n$ are from H and $q_1, \ldots, q_n \in \mathbb{Q} \setminus \{0\}$). Let $B_x = \{b_1, \ldots, b_n\} \subset H$ and define $c_x : B_x \to \mathbb{Q}$ by putting $c_x(b_i) = q_i$ for $1 \le i \le n$. Moreover, extend each c_x to $\hat{c}_x : H \to \mathbb{Q}$ by putting

$$\hat{c}_x(b) = \begin{cases} c_x(b) & \text{for } b \in B_x, \\ 0 & \text{otherwise.} \end{cases}$$

Then, for every $x \in \mathbb{R}$,

$$x = \sum_{b \in B_x} c_x(b)\, b = \sum_{b \in H} \hat{c}_x(b)\, b.$$

We will start the argument by proving that there exists a countable set $D = \{g_n : n < \omega\}$ of functions from H into $\mathbb{Q}$ such that

$$\forall x \in \mathbb{R}\ \exists n < \omega \ (c_x \subset g_n). \tag{7.18}$$

To see it, let $\mathcal{U} = \{(p,q) \cap H : p, q \in \mathbb{Q}\}$ and notice that $\mathcal{U}$ is countable. Define

$$D = \left\{ \sum_{i<n} q_i \chi_{U_i} \in \mathbb{Q}^H : n \in \omega \text{ and } \langle q_i, U_i \rangle \in \mathbb{Q} \times \mathcal{U} \text{ for every } i < n \right\}.$$

Clearly, D is countable, since it is indexed by a countable set $\bigcup_{n<\omega}(\mathbb{Q} \times \mathcal{U})^n$. To see that it satisfies (7.18), take $x \in \mathbb{R}$ and find a family $\{U_b \in \mathcal{U} : b \in B_x\}$ of disjoint sets such that $b \in U_b$ for every $b \in B_x$. Then $g = \sum_{b \in B_x} c_x(b)\, \chi_{U_b}$ belongs to D and $c_x(b) = g(b)$ for every $b \in B_x$. So $c_x \subset g$. Condition (7.18) has been proved.

Now let $\{g_n : n < \omega\}$ be as in (7.18). Define $f : \mathbb{R} \to \omega$ by

$$f(x) = \min\{n < \omega : c_x \subset g_n\}$$

and let $P_n = f^{-1}(n)$ for every $n < \omega$. We will show that the partition $\mathcal{P} = \{P_n : n \in \mathbb{N}\}$ of $\mathbb{R}$ satisfies (7.17).

So fix $x \in \mathbb{R}$ and let $h \in S_x$. We will show that

$$\hat{c}_{x+h}(b) \in \{0, \hat{c}_x(b), 2\hat{c}_x(b)\} \text{ for every } b \in H. \qquad (7.19)$$

This will finish the proof, since $\{0, \hat{c}_x(b), 2\hat{c}_x(b)\} = \{0\}$ for $b \in H \setminus B_x$ and so there are at most $3^{|B_x|} < \omega$ numbers $x + h$ such that $h \in S_x$. Thus S_x is finite.

To see (7.19), notice that $(x + h) + (x - h) = 2x$, so

$$\sum_{b \in H} \hat{c}_{x+h}(b) \, b + \sum_{b \in H} \hat{c}_{x-h}(b) \, b = 2 \sum_{b \in H} \hat{c}_x(b) \, b.$$

In particular,

$$\hat{c}_{x+h}(b) + \hat{c}_{x-h}(b) = 2\hat{c}_x(b) \qquad (7.20)$$

for every $b \in H$.

Now, if $x + h, x - h \in P_n$ then $c_{x+h} \subset g_n$ and $c_{x-h} \subset g_n$. Hence

$$c_{x+h}(b) = g_n(b) = c_{x-h}(b) \text{ for every } b \in B_{x+h} \cap B_{x-h}.$$

Therefore, by (7.20), $\hat{c}_{x+h}(b) = \hat{c}_x(b)$ for every $b \in B_{x+h} \cap B_{x-h}$. But for $b \in H \setminus B_{x+h}$ we have $\hat{c}_{x+h}(b) = 0$, and for every $b \in B_{x+h} \setminus B_{x-h}$ we have $\hat{c}_{x+h}(b) = \hat{c}_{x+h}(b) + \hat{c}_{x-h}(b) = 2\hat{c}_x(b)$. Thus $\hat{c}_{x+h}(b) \in \{0, \hat{c}_x(b), 2\hat{c}_x(b)\}$ for every $b \in H$. This finishes the proof of condition (7.20) and Theorem 7.4.1. $\square$

An interesting open problem is whether or not there exists a uniformly antisymmetric function with finite or bounded range. Some partial results are known in this direction. Komjáth and Shelah (1993–4) proved that there is no function $f \colon \mathbb{R} \to \mathbb{R}$ with finite range such that all the sets S_x from Theorem 7.4.1 are finite. It has also been proved by Ciesielski (1995–6) that the range of any uniformly antisymmetric function must have at least four elements. The next theorem, due to Ciesielski and Larson (1993–4), shows only that the range of a uniformly antisymmetric function must have at least three elements.

Theorem 7.4.3 *If $f \colon \mathbb{R} \to \{0, 1\}$ then f is not uniformly antisymmetric.*

Proof To obtain a contradiction assume that there exists a uniformly antisymmetric function $f \colon \mathbb{R} \to \{0, 1\}$ and for every $x \in \mathbb{R}$ let $n_x \in \mathbb{N}$ be such that for every $h \in (0, 1/n_x)$,

$$|f(x - h) - f(x + h)| \geq 1/n_x.$$

Fix $n \in \mathbb{N}$ such that $L = \{x \in \mathbb{R} \colon n_x = n\}$ is uncountable. Then

$$f(x - h) \neq f(x + h) \text{ for every } x \in L \text{ and } h \in (0, 1/n). \qquad (7.21)$$

Since $\{(\frac{k}{n}, \frac{k+1}{n}] \cap L : k \in \mathbb{Z}\}$ forms a countable partition of L, at least one of these sets must be uncountable. In particular, we can choose $x, y, z \in L$ such that $x < y < z < x+1/n$. Now we would like to find points $a, b, c \in \mathbb{R}$ such that x, y, and z are the midpoints of pairs $\langle a, b \rangle$, $\langle b, c \rangle$, and $\langle a, c \rangle$, respectively, that is, such that $a + b = 2x$, $b + c = 2y$, and $a + c = 2z$. It is easy to see that the points

$$a = x - y + z, \quad b = x + y - z, \quad c = -x + y + z$$

have these properties. Moreover, for $h = z - y \in (0, 1/n)$ we have $a = x+h$ and $b = x - h$. Hence, by (7.21), $f(a) \neq f(b)$. Similarly, $f(b) \neq f(c)$ and $f(c) \neq f(a)$. But this is impossible, since the points a, b, and c are distinct and f attains only two values.   □

We proved Corollary 7.4.2, via Theorem 7.4.1, by showing that for every $x \in \mathbb{R}$ the set $S_x^1 = S_x$ of exceptional points, being finite, is bounded away from 0. What if we allow the sets S_x to be countable? Certainly, such sets do not have to be bounded away from 0. But we can replace the condition of being "bounded away from 0" by the weaker condition of being "almost bounded away from 0," in the sense that $S_x \cap (0, d)$ is countable for some $d > 0$. Can we prove then an analog of Theorem 7.4.1? Surprisingly, the answer depends on the continuum hypothesis, as is proved in the next theorem.

The general form of this theorem is due to Ciesielski and Larson (1993–4). However, implication (i)⇒(ii) was first proved by Sierpiński (1936). The equivalence of (i) and (ii) is also implicitly contained in a work of Freiling (1989–90).

Theorem 7.4.4 *The following conditions are equivalent.*

(i)   *The continuum hypothesis.*

(ii) *There exists a partition $\mathcal{P} = \{A_0, A_1\}$ of $\mathbb{R}$ such that for every $x \in \mathbb{R}$ the set $S_x = \bigcup_{i<2}\{h > 0 : x - h, x + h \in A_i\}$ is at most countable.*

(iii) *There exists a function $f : \mathbb{R} \to \{0, 1\}$ such that for every $x \in \mathbb{R}$ there is a $d > 0$ with the property that $\left| S_x^d \cap (0, d) \right| \leq \omega$.*

Proof (i)⇒(ii): Let $H = \{b_\zeta : \zeta < \omega_1\}$ be a Hamel basis. For $x \in \mathbb{R} \setminus \{0\}$ let $q(x) = q_n$, where $x = q_1 b_{\zeta_1} + \cdots + q_n b_{\zeta_n}$ is the unique representation of x in the basis H, with $\zeta_1 < \cdots < \zeta_n$ and $q_i \neq 0$. Put

$$x \in A_0 \text{ if and only if } q(x) > 0,$$

and $A_1 = \mathbb{R} \setminus A_0$. We will show that the partition $\mathcal{P} = \{A_0, A_1\}$ satisfies (ii).

For $\xi < \omega_1$ let $K_\xi = \mathrm{LIN}_{\mathbb{Q}}(\{b_\zeta : \zeta \leq \xi\})$. Notice that $\mathbb{R} = \bigcup_{\xi < \omega_1} K_\xi$ and that every K_ξ is countable. We will show that if $x \in K_\xi$ then $S_x \subset K_\xi$. This will finish the proof.

But it is easy to see that for $h \in \mathbb{R} \setminus K_\xi$ and $x \in K_\xi$,

$$h \in A_0 \text{ if and only if } x + h \in A_0,$$

since $q(h) = q(x + h)$. However, h and $-h$ cannot belong to the same A_i for $h \neq 0$, so $S_x \subset K_\xi$.

(ii)$\Rightarrow$(iii): For $x \in \mathbb{R}$ define $f(x) = i$ if $x \in A_i$. Then $S_x^d = S_x$ is countable for every $x \in \mathbb{R}$ and $d \in (0, 1)$.

(iii)$\Rightarrow$(i): Let f be as in (iii), and for every $x \in \mathbb{R}$ let $n_x \in \mathbb{N}$ be such that the set

$$C_x = S_x^{1/n_x} \cap (0, 1/n_x) = \{h \in (0, 1/n_x) \colon f(x - h) = f(x + h)\}$$

is countable.

To obtain a contradiction, assume that the continuum hypothesis fails and let B be a linearly independent subset of $\mathbb{R}$ over $\mathbb{Q}$ of cardinality ω_2. Choose $K \subset B$ of cardinality ω_2 such that for some $n \in \mathbb{N}$ we have $n_x = n$ for all $x \in K$. Let U be an open interval of length less than $1/n$ such that the set $L = K \cap U$ has cardinality ω_2. Then, in particular,

$$f(x - h) \neq f(x + h) \text{ for } x \in L \text{ and } h \in (0, 1/n) \setminus C_x, \qquad (7.22)$$

and

$$|x - y| < 1/n \text{ for every } x, y \in L. \qquad (7.23)$$

Define, by transfinite induction, a sequence $\langle t_\xi \in L \colon \xi < \omega_2 \rangle$ such that

$$t_\xi \in L \setminus T_\xi \text{ for every } \xi < \omega_2, \qquad (7.24)$$

where T_ξ is the smallest linear subspace of $\mathbb{R}$ containing $\{t_\zeta \colon \zeta < \xi\}$ and such that

$$C_x \subset T_\xi \text{ for every } x \in T_\xi. \qquad (7.25)$$

Such T_ξ is obtained by applying Lemma 6.1.6(a) to the set $Z = \{t_\zeta \colon \zeta < \xi\}$ and the family $\mathcal{F}$ of operations $x \mapsto C_x$, $\langle x, y \rangle \mapsto x + y$, and $x \mapsto qx$ for every $q \in \mathbb{Q}$. Then T_ξ has cardinality $\leq \omega_1$, so the induction can be done easily.

Now put $x = t_0$, $z = t_{\omega_1}$ and, for $0 < \xi < \omega_1$, consider the numbers $|-x + t_\xi|$. All these numbers are different, so there is $0 < \eta < \omega_1$ such that

$$|-x + t_\eta| \notin C_z. \qquad (7.26)$$

Put $y = t_\eta$ and proceed as in Theorem 7.4.3. Define

$$a = x - y + z, \quad b = x + y - z, \quad c = -x + y + z. \tag{7.27}$$

Then $(a+b)/2 = x$, $(b+c)/2 = y$, and $(c+a)/2 = z$. We will show that

$$f(a) \neq f(b), \quad f(b) \neq f(c), \quad f(c) \neq f(a), \tag{7.28}$$

which will give us the desired contradiction, since the points a, b, and c are distinct and f admits only two values.

To prove (7.28), notice first that (7.28) follows from (7.22) as long as

$$|a - x| < 1/n, \quad |b - y| < 1/n, \quad |c - z| < 1/n \tag{7.29}$$

and

$$|a - x| \notin C_x, \quad |b - y| \notin C_y, \quad |c - z| \notin C_z. \tag{7.30}$$

But (7.29) follows easily from (7.27) and (7.23). Finally, (7.30) can be proved as follows.

$|a - x| \notin C_x$, since otherwise we would have $z - y = a - x \in T_{\omega_1}$ and $z \in y + T_{\omega_1} \subset T_{\omega_1}$, contradicting $z = t_{\omega_1} \notin T_{\omega_1}$.

$|b - y| \notin C_y$, since otherwise we would have $z - x = y - b \in T_{\omega_1}$ and $z \in x + T_{\omega_1} \subset T_{\omega_1}$, contradicting $z = t_{\omega_1} \notin T_{\omega_1}$.

$|c - z| \notin C_z$, since otherwise we would have $|-x + t_\eta| = |-x + y| = |c - z| \in C_z$, contradicting (7.26). $\qquad \square$

To motivate the last theorem of this section we need to reformulate Theorem 7.4.1 in another language. For this we need the following definitions. We say that *partition $\mathcal{P}$ of $\mathbb{R}$ is ω sum free* if for every $x \in \mathbb{R}$ the equation $a + b = x$ has less than ω solutions with a and b being in the same element of the partition, that is, when for every $x \in \mathbb{R}$ the set

$$T_x = \{\langle a, b\rangle \colon a + b = x \text{ and } a, b \in P \text{ for some } P \in \mathcal{P}\}$$

is finite. Similarly, we say that *partition $\mathcal{P}$ of $\mathbb{R}$ is ω difference free* if the set

$$D_x = \{\langle a, b\rangle \colon a - b = x \text{ and } a, b \in P \text{ for some } P \in \mathcal{P}\}$$

is finite for every $x \in \mathbb{R}$, $x \neq 0$.

In this language Theorem 7.4.1 reads as follows.

Corollary 7.4.5 *There exists a countable partition $\mathcal{P}$ of $\mathbb{R}$ that is ω sum free.*

Proof It is enough to notice that $T_{2x} = \{\langle x + h, x - h\rangle \colon |h| \in S_x \cup \{0\}\}$, where S_x is from Theorem 7.4.1. $\qquad \square$

Can we prove the same results about ω-difference-free partitions of $\mathbb{R}$? The answer is given by the next theorem, due to Ciesielski (1996).

Theorem 7.4.6 *The continuum hypothesis is equivalent to the existence of a countable partition $\mathcal{P}$ of $\mathbb{R}$ that is ω difference free.*

Proof The proof of the theorem is similar to that of Theorem 7.3.9.

$\Leftarrow$: To obtain a contradiction assume that $\mathfrak{c} \geq \omega_2$ and that there exists a countable partition $\mathcal{P}$ of $\mathbb{R}$ that is ω difference free.

Let H be any Hamel basis. Since $|H| = \mathfrak{c} \geq \omega_2$ by Lemma 7.3.11 we can choose disjoint sets $A_0 \in [H]^{\omega_2}$ and $B_0 \in [H]^2$ and $P \in \mathcal{P}$ such that

$$a + b \in P \text{ for every } a \in A_0 \text{ and } b \in B_0.$$

Take different $b_0, b_1 \in B_0$. Then

$$(a + b_0) - (a + b_1) = b_0 - b_1$$

for every $a \in A_0$, that is, $D_{b_0 - b_1}$ contains all pairs $\langle a + b_0, a + b_1 \rangle$. This contradicts the assumption that D_x is finite for $x = b_0 - b_1 \neq 0$.

$\Rightarrow$: Represent $\mathbb{R}$ as the union of an increasing sequence $\langle V_\alpha : \alpha < \omega_1 \rangle$ of countable linear subspaces V_α of $\mathbb{R}$ over $\mathbb{Q}$ such that $V_\lambda = \bigcup_{\alpha < \lambda} V_\alpha$ for every limit ordinal $\lambda < \omega_1$. Such a sequence exists by Lemma 6.1.6(b) applied to the family $\mathcal{F}$ of operations $\langle x, y \rangle \mapsto x + y$ and $x \mapsto qx$ for every $q \in \mathbb{Q}$. For convenience we will also assume that $V_0 = \emptyset$.

Thus $\{V_{\alpha+1} \setminus V_\alpha : \alpha < \omega_1\}$ is a partition of $\mathbb{R}$ into countable sets. For $\alpha < \omega_1$ let $\{p_n^\alpha : n < \omega\}$ be an enumeration of $V_{\alpha+1} \setminus V_\alpha$. By induction on $\alpha < \omega_1$ we will define one-to-one functions $f : V_{\alpha+1} \setminus V_\alpha \to \omega$ such that the following inductive condition holds[2]

$$f(p_n^\alpha) \in \omega \setminus \{f(p) : p \in V_\alpha \ \& \ p = p_n^\alpha \pm p_j^\alpha \text{ for some } j \leq n\}.$$

We will show that the partition $\mathcal{P} = \{f^{-1}(n) : n < \omega\}$ of $\mathbb{R}$ is ω difference free.

So choose an arbitrary $x = p_n^\xi \neq 0$ and consider the pairs $\langle a, b \rangle$ satisfying $a - b = x$ with a and b being from the same element of $\mathcal{P}$, that is, such that $f(a) = f(b)$. It is enough to show that $\{a, b\} \cap \{p_j^\xi : j \leq n\} \neq \emptyset$.

Let $a = p_m^\alpha$ and $b = p_k^\beta$. Then $p_m^\alpha - p_k^\beta = p_n^\xi$. Notice that $\delta = \max\{\xi, \alpha, \beta\}$ must be equal to at least two of ξ, α, and β, since otherwise the number p with the index δ would belong to V_δ. Moreover, $\alpha \neq \beta$, since otherwise $f(p_m^\alpha) = f(a) = f(b) = f(p_k^\alpha)$, contradicting the fact that f is one-to-one on $V_{\alpha+1} \setminus V_\alpha$. We are left with two cases:

If $\xi = \alpha > \beta$ then $p_m^\xi - p_n^\xi = a - x = b = p_k^\beta \in V_\xi$. So $f(p_m^\xi) = f(a) = f(b) = f(p_m^\xi - p_n^\xi)$ implies that $m < n$ and $a = p_m^\xi \in \{p_j^\xi : j \leq n\}$.

---

[2] We define here f separately on each set from the partition $\{V_{\alpha+1} \setminus V_\alpha : \alpha < \omega_1\}$ of $\mathbb{R}$. Formally, we should be using a different symbol, such as f_α, for such a part of f, and define f as the union of all f_αs. However, this would obscure the clarity of this notation.

If $\xi = \beta > \alpha$ then $p_k^\xi + p_n^\xi = b + x = a = p_m^\alpha \in V_\xi$. So $f(p_k^\xi) = f(b) = f(a) = f(p_k^\xi + p_n^\xi)$ implies that $k < n$ and $b = p_k^\xi \in \{p_j^\xi : j \le n\}$. $\square$

EXERCISES

1 Prove that if f is uniformly antisymmetric then it does not have the Baire property. *Hint:* First prove the following fact due to Kuratowski:

> For every function $f \colon \mathbb{R} \to \mathbb{R}$ with the Baire property there exists a first-category set S such that $f|_{\mathbb{R} \setminus S} \colon \mathbb{R} \setminus S \to \mathbb{R}$ is continuous.

2 (Project) Consider the following classes of subsets of $\mathbb{R}$:

- $\mathcal{B} = \{B \subset \mathbb{R} \colon B \text{ is a Bernstein set}\}$;

- $\mathcal{H} = \{H \subset \mathbb{R} \colon H \text{ is a Hamel basis}\}$;

- $\mathcal{V} = \{V \subset \mathbb{R} \colon V \text{ is a Vitali set}\}$;

- $\mathcal{T} = \{T \subset \mathbb{R} \colon B \text{ is a transcendental basis of } \mathbb{R} \text{ over } \mathbb{Q}\}$.

Moreover, if $\mathcal{F}$ is any of the preceding families let

- $\mathcal{F}^* = \{F \subset \mathbb{R} \colon F + F \in \mathcal{F}\}$.

Find the complete intersection/subset relations among the classes $\mathcal{B}$, $\mathcal{H}$, $\mathcal{V}$, $\mathcal{T}$, $\mathcal{B}^*$, $\mathcal{H}^*$, $\mathcal{V}^*$, and $\mathcal{T}^*$.

Part IV

When induction is too short

Chapter 8

Martin's axiom

8.1 Rasiowa–Sikorski lemma

The previous chapter was devoted to constructing objects by transfinite induction. A typical scheme for such constructions was a diagonalization argument like the following. To find a subset S of a set X concerning a family $\mathcal{P} = \{P_\alpha \colon \alpha < \kappa\}$ we chose $S = \{x_\xi \in X \colon \xi < \kappa\}$ by picking each x_ξ to take care of a set P_ξ. But what can we do if the cardinal κ is too big compared to the freedom of choice of the x_ξs; for example, if the set X has cardinality less than κ?

There is no absolute answer to this question. In some cases you can do nothing. For example, if you try to construct a subset S of ω different from every set from the family $\mathcal{P}(\omega) = \{B_\xi \colon \xi < \mathfrak{c}\}$, then you are obviously condemned to failure. The inductive construction does not work, since you would have to take care of continuum many conditions, having the freedom to choose only countably many points for S.

In some other cases you can reduce a family $\mathcal{P}$ to the appropriate size. This was done, for example, in Theorem 6.3.7 (on the existence of Bernstein sets) in which we constructed a nonmeasurable subset B of $\mathbb{R}^n$: The natural family $\mathcal{P} = \mathcal{L}$ of cardinality $2^\mathfrak{c}$ was replaced by the family $\mathcal{P}_0 = \{P_\xi \colon \xi < \mathfrak{c}\}$ of all perfect subsets of $\mathbb{R}^n$. The difficult part of the proof, that is, Theorem 6.3.6, was to show that the family $\mathcal{P}_0$ does the job.

Yet in other cases, such as Theorems 7.3.9, 7.4.4, and 7.4.6, the induction could be performed in only ω_1 steps while we had to take care of a family $\mathcal{P}$ of cardinality $\mathfrak{c}$. These cases resulted in us assuming that $\mathfrak{c} = \omega_1$. In other words, in order to rescue a diagonal argument, we had to assume an additional set-theoretic assumption, the continuum hypothesis. In fact, in Theorems 7.3.9, 7.4.4, and 7.4.6 we also proved that this assumption was necessary.

The continuum hypothesis was also assumed in the proof of Theorem 7.2.5. In this theorem, however, we did not show that the continuum hypothesis is a necessary assumption, though we stated that the theorem cannot be proved in ZFC alone. The reason for this is that Theorem 7.2.5 can also be proved under weaker set-theoretic assumptions, which can be true even if the continuum hypothesis fails.

In remainder of this text we will be mainly interested in studying different kinds of set-theoretic axioms that will allow us to solve the problem of having a "too short induction." For this we will need some more definitions and notations.

Consider a partially ordered set $\langle \mathbb{P}, \leq \rangle$. A subset $D \subset \mathbb{P}$ is said to be *dense* in $\mathbb{P}$ provided for every $p \in \mathbb{P}$ there exists a $d \in D$ such that $d \leq p$.

Examples 1. If $X \neq \emptyset$ and $\langle \mathbb{P}, \leq \rangle = \langle \mathcal{P}(X), \subset \rangle$ then $D = \{\emptyset\}$ is dense in $\mathbb{P}$. Notice that $\emptyset$ is the smallest element of $\mathbb{P}$.

In fact, $D = \{m\}$ is dense in $\langle \mathbb{P}, \leq \rangle$ if and only if m is the smallest element of $\mathbb{P}$. Also, if $\langle \mathbb{P}, \leq \rangle$ has the smallest element m, then $D \subset \mathbb{P}$ is dense in $\langle \mathbb{P}, \leq \rangle$ if and only if $m \in D$. Dense sets such as these are too easy to describe to be of much interest. To avoid them, we will usually study partially ordered sets without a smallest element.

2. If $\langle \mathbb{P}, \leq \rangle = \langle \mathbb{R}, \leq \rangle$ then the set $D = (-\infty, 0)$ is dense in $\langle \mathbb{P}, \leq \rangle$. Notice that the word "dense" is used here in a different sense than it is usually used for the linearly ordered sets (see (8.1)). However, there will be very little chance to confuse these two notions of density, since we will usually use our new definition of density for partially ordered sets that are not linearly ordered.

3. Let $\langle X, \tau \rangle$ be any topological space. If $\langle \mathbb{P}, \leq \rangle = \langle \tau \setminus \{\emptyset\}, \subset \rangle$ then any base $\mathcal{B}$ in X is dense in $\mathbb{P}$.

A subset F of a partially ordered set $\langle \mathbb{P}, \leq \rangle$ is a *filter* in $\mathbb{P}$ if

(F1)   for every $p, q \in F$ there is an $r \in F$ such that $r \leq p$ and $r \leq q$, and

(F2)   if $q \in F$ and $p \in \mathbb{P}$ are such that $q \leq p$ then $p \in F$.

Note that a simple induction argument shows that condition (F1) is equivalent to the following stronger condition.

(F1')   For every finite subset F_0 of F there exists an $r \in F$ such that $r \leq p$ for every $p \in F_0$.

Examples 1. For any chain F in a partially ordered set $\langle \mathbb{P}, \leq \rangle$ the family

$$F^\star = \{p \in \mathbb{P} \colon \exists q \in F(q \leq p)\}$$

is a filter in $\mathbb{P}$.

2. Let $\mathcal{G}$ be any family of subsets of a nonempty set X that is closed under finite intersections and let $\langle \mathbb{P}, \leq \rangle = \langle \mathcal{G}, \subset \rangle$. If $x \in X$ then the family $F_x = \{Y \in \mathcal{G} \colon x \in Y\}$ is a filter in $\mathbb{P}$. A filter F in such $\langle \mathbb{P}, \leq \rangle$ is called a *principal filter* if it is of this form, that is, if $F = F_x$ for some $x \in X$.

3. Let $\mathcal{G}$ be a nonempty family of subsets of an infinite set X and let $\omega \leq \kappa \leq |X|$. If $\mathcal{G}$ is closed under finite unions then $\mathcal{G} \cap [X]^{<\kappa}$ is a filter in $\langle \mathbb{P}, \leq \rangle = \langle \mathcal{G}, \supset \rangle$. If $\mathcal{G}$ is closed under finite intersections then $F_\kappa = \{Y \in \mathcal{G} \colon |X \setminus Y| < \kappa\}$ is a filter in $\langle \mathbb{P}, \leq \rangle = \langle \mathcal{G}, \subset \rangle$.

4. Let $\langle \mathbb{P}, \leq \rangle = \langle \mathcal{P}(X), \subset \rangle$. Then $F \subset \mathbb{P}$ is a filter in $\mathbb{P}$ if (1) $A \cap B \in F$ provided $A, B \in F$; and (2) if $A \subset B \subset X$ and $A \in F$ then $B \in F$. (Compare this with Exercise 3 in Section 4.4.)

Before we formulate our next example we introduce the following notation. For nonempty sets X and Y we will use the symbol $\text{Func}(X, Y)$ to denote the family of all partial functions from X into Y, that is,

$$\text{Func}(X, Y) = \bigcup \{Y^D \colon D \subset X\}.$$

Also, for an infinite cardinal number κ we put

$$\text{Func}_\kappa(X, Y) = \bigcup \{Y^D \colon D \in [X]^{<\kappa}\} = \{s \in \text{Func}(X, Y) \colon |s| < \kappa\}.$$

5. Let $\mathcal{F} \subset \text{Func}(X, Y)$ be such that for every $g_0, g_1 \in \mathcal{F}$

$$\text{if } g_0 \cup g_1 \in \text{Func}(X, Y) \text{ then } g_0 \cup g_1 \in \mathcal{F}.$$

(In particular, $\text{Func}_\kappa(X, Y)$ satisfies this condition.) Then for every $f \in \text{Func}(X, Y)$ the set $G_f = \{g \in \mathcal{F} \colon g \subset f\}$ is a filter in $\langle \mathbb{P}, \leq \rangle = \langle \mathcal{F}, \supset \rangle$. To see (F1) let $g_0, g_1 \in G_f$. Then $g_0 \cup g_1 \subset f$, so $h = g_0 \cup g_1 \in \text{Func}(X, Y)$. Thus $h \in \mathcal{F}$ and, clearly, $h \leq g_0$ and $h \leq g_1$. To see (F2) notice that $g \in G_f$, $h \in \mathcal{F}$, and $g \leq h$ imply that $h \subset g \subset f$, so $h \in G_f$.

In what follows we will very often use the following fact. Its main part can be viewed as a kind of converse of the previous Example 5.

Proposition 8.1.1 *Let $\langle \mathbb{P}, \leq \rangle = \langle \mathcal{F}, \supset \rangle$ for some $\mathcal{F} \subset \mathrm{Func}(X,Y)$. If $F \subset \mathbb{P}$ is a filter in $\mathbb{P}$ then $f = \bigcup F$ is a function and $F \subset \{g \in \mathcal{F} : g \subset f\}$. Moreover,*

(a) *if F intersects every set $D_x = \{s \in \mathbb{P} : x \in \mathrm{dom}(s)\}$ for $x \in X$ then $\mathrm{dom}(f) = X$;*

(b) *if F intersects every set $R_y = \{s \in \mathbb{P} : y \in \mathrm{range}(s)\}$ for $y \in Y$ then $\mathrm{range}(f) = Y$.*

Proof Let $F \subset \mathbb{P}$ be a filter in $\mathbb{P}$ and let $g_0, g_1 \in F$. To prove that f is a function it is enough to show that $g_0(x) = g_1(x)$ for every $x \in \mathrm{dom}(g_0) \cap \mathrm{dom}(g_1)$. So let $x \in \mathrm{dom}(g_0) \cap \mathrm{dom}(g_1)$ and let $h \in F$ be such that $h \leq g_0$ and $h \leq g_1$. Then $h \supset g_0$ and $h \supset g_1$. In particular, $x \in \mathrm{dom}(h)$ and $g_0(x) = h(x) = g_1(x)$.

The inclusion $F \subset \{g \in \mathcal{F} : g \subset f\}$ is obvious.

To see (a) let $x \in X$. Then there exists an $s \in F \cap D_x$. But $s \subset f$, so $x \in \mathrm{dom}(s) \subset \mathrm{dom}(f)$.

Condition (b) is proved similarly. $\qquad\qquad\qquad\qquad\qquad\qquad\square$

In what follows we will often use filters in partially ordered sets to construct functions in a manner similar to that of Proposition 8.1.1. Usually we will be interested in the entire functions, and part (a) of Proposition 8.1.1 suggests how to achieve this goal. This and the next theorem lead to the following definition.

Let $\langle \mathbb{P}, \leq \rangle$ be a partially ordered set and let $\mathcal{D}$ be a family of dense subsets of $\mathbb{P}$. We say that a filter F in $\mathbb{P}$ is $\mathcal{D}$-*generic* if

$$F \cap D \neq \emptyset \quad \text{for all } D \in \mathcal{D}.$$

The partial orders used in the context of $\mathcal{D}$-generic filters will often be called *forcings*. Also, if $\langle \mathbb{P}, \leq \rangle$ is a forcing then elements of $\mathbb{P}$ will sometimes be referred to as *conditions*. For conditions $p, q \in \mathbb{P}$ we say that p *is stronger than* q provided $p \leq q$.

In this terminology, if all sets D_x are dense in $\mathbb{P}$ and $\{D_x : x \in X\} \subset \mathcal{D}$ then for every $\mathcal{D}$-generic filter F in $\mathbb{P}$ the domain of $f = \bigcup F$ is equal to X, that is, $f\colon X \to Y$. Similarly, if all sets R_y are dense in $\mathbb{P}$ and $\{R_y : y \in Y\} \subset \mathcal{D}$ then $\mathrm{range}(\bigcup F) = Y$ for every $\mathcal{D}$-generic filter F in $\mathbb{P}$.

The next theorem, due to Rasiowa and Sikorski, shows that there are some interesting $\mathcal{D}$-generic filters.

Theorem 8.1.2 (Rasiowa–Sikorski lemma) *Let $\langle \mathbb{P}, \leq \rangle$ be a partially ordered set and $p \in \mathbb{P}$. If $\mathcal{D}$ is a countable family of dense subsets of $\mathbb{P}$ then there exists a $\mathcal{D}$-generic filter F in $\mathbb{P}$ such that $p \in F$.*

Proof Let $\mathcal{D} = \{D_n : n < \omega\}$. We define a sequence $\langle p_n : n < \omega \rangle$ by induction on $n < \omega$. We start by picking $p_0 \in D_0$ such that $p_0 \leq p$. We continue by choosing $p_{n+1} \in D_{n+1}$ such that $p_{n+1} \leq p_n$.

Now let $E = \{p_n : n < \omega\}$ and put $F = E^\star = \{p \in \mathbb{P} : \exists q \in E(q \leq p)\}$. Then F is a filter in $\mathbb{P}$ intersecting every $D \in \mathcal{D}$. $\qquad \square$

The Rasiowa–Sikorski lemma is one of the most fundamental facts that will be used in the remaining sections. Its importance, however, does not come from its power. Its proof is too simple for this. It is the language of generic filters it employs that makes it so useful. In particular, it motivates the different generalizations described in the next sections, which are consistent with ZFC and can be used for our problem of a "too short induction."

In most of the applications of the Rasiowa–Sikorski lemma and its generalizations the intuition behind the proofs comes from an attempt at proving the theorem by (transfinite) induction. More precisely, a partial order $\mathbb{P}$ used to construct an object will usually be built on the basis of an attempted inductive construction of the object. That is, conditions (elements of $\mathbb{P}$) will be chosen as a "description of the current stage of induction." The inductive steps will be related to the dense subsets of $\mathbb{P}$ in the sense that the density of a particular set $D_x = \{p \in \mathbb{P} : \varphi(p, x)\}$ will be equivalent to the fact that at an arbitrary stage q of the inductive construction we can make the next inductive step by extending the condition q to p having the property $\varphi(p, x)$. In particular, the family $\mathcal{D}$ of dense subsets of $\mathbb{P}$ will always represent the set of all inductive conditions of which we have to take care, and a $\mathcal{D}$-generic filter in $\mathbb{P}$ will be an "oracle" that "takes care of all our problems," and from which we will recover the desired object.

Evidently, if the number $|\mathcal{D}|$ of conditions we have to take care of is not more than the number of steps in our induction, then usually the (transfinite) induction will be powerful enough to construct the object, and the language of forcing will be redundant. In particular, this will be the case for all the applications of the Rasiowa–Sikorski lemma presented in the rest of this section. These applications, however, are presented here to see the use of the generic-filters technique in the simplest situations. Moreover, in the next sections the same theorems will be either generalized or used for some motivation.

To state the next theorem let us recall that, for a linearly ordered set $\langle X, \leq \rangle$, a subset D of X is *dense in X* if

for every $x, y \in X$ with $x < y$ there is a $d \in D$ such that $x < d < y$. (8.1)

A linearly ordered set $\langle X, \leq \rangle$ is said to be *dense* if it is dense in itself, that is, if X is dense in X.

Theorem 8.1.3 *Any two countable dense linearly ordered sets, neither of which has a first or a last element, are order isomorphic.*

Proof Let $\langle X, \leq \rangle$ and $\langle Y, \preceq \rangle$ be two linearly ordered sets as in the theorem. The inductive proof of the theorem may go as follows. Enumerate X and Y as $X = \{x_n \colon n < \omega\}$ and $Y = \{y_n \colon n < \omega\}$. Construct, by induction on $n < \omega$, a sequence $h_0 \subset h_1 \subset h_2 \subset \cdots$ of functions such that each h_n is an order isomorphism between $X_n \in [X]^{<\omega}$ and $Y_n \in [Y]^{<\omega}$, where $x_n \in X_n$ and $y_n \in Y_n$ for every $n < \omega$. Then $h = \bigcup_{n < \omega} h_n$ is an order isomorphism between X and Y. The difficult part of the proof is to extend h_n to h_{n+1} while maintaining the condition $x_{n+1} \in X_{n+1}$ and $y_{n+1} \in Y_{n+1}$.

To translate this proof into the language of partially ordered sets, let $\mathbb{P}$ be the set of all possible functions $h_n \colon X_n \to Y_n$ from our inductive construction, that is, the set of all finite partial isomorphisms from X to Y:

$$\mathbb{P} = \{h \in \mathrm{Func}_\omega(X, Y) \colon h \text{ is strictly increasing}\}.$$

Consider $\mathbb{P}$ to be ordered by reverse inclusion, that is, $\langle \mathbb{P}, \leq \rangle = \langle \mathbb{P}, \supset \rangle$. For $x \in X$ and $y \in Y$ put

$$D_x = \{h \in \mathbb{P} \colon x \in \mathrm{dom}(h)\} \quad \text{and} \quad R_y = \{h \in \mathbb{P} \colon y \in \mathrm{range}(h)\}.$$

These sets are the counterparts of the conditions "$x_n \in X_n$" and "$y_n \in Y_n$." We will show that

$$\text{the sets } D_x \text{ and } R_y \text{ are dense in } \mathbb{P}, \tag{8.2}$$

which is a translation of the fact that the inductive step of our inductive construction can always be made.

Before we prove (8.2), let us show how it implies the theorem. Let $\mathcal{D} = \{D_x \colon x \in X\} \cup \{R_y \colon y \in Y\}$. Then $\mathcal{D}$ is countable since the sets X and Y are countable. Thus, by the Rasiowa–Sikorski lemma, there exists a $\mathcal{D}$-generic filter F in $\mathbb{P}$. Then, by Proposition 8.1.1, $f = \bigcup F$ is a function from X onto Y. It remains only to show that f is strictly increasing.

To see this, let $x_0, x_1 \in X$ with $x_0 < x_1$. Then there are $g_0, g_1 \in F$ such that $x_0 \in \mathrm{dom}(g_0)$ and $x_1 \in \mathrm{dom}(g_1)$. Let $g \in F$ be such that $g \leq g_0$ and $g \leq g_1$. Then $g_0 \cup g_1 \subset g \subset f$. So $f(x_0) = g(x_0) \prec g(x_1) = f(x_1)$, since g is strictly increasing. Thus f is strictly increasing. We have proved that $f \colon X \to Y$ is an order isomorphism.

To finish the proof it is enough to show (8.2). So let $x \in X$. To prove that D_x is dense, let $g \in \mathbb{P}$. We have to find $h \in D_x$ such that $h \leq g$, that is, a function $h \in \mathbb{P}$ such that $h \supset g$ and $x \in \mathrm{dom}(h)$. If $x \in \mathrm{dom}(g)$ then $h = g$ works. So assume that $x \notin \mathrm{dom}(g)$. Let $\mathrm{dom}(g) = \{x_0, x_1, \ldots, x_n\}$ with $x_0 < x_1 < \cdots < x_n$ and let $y_i = g(x_i)$ for $i \leq n$. Then $y_0 \prec y_1 \prec \cdots \prec y_n$, since g is strictly increasing. We will define h as $g \cup \{\langle x, y \rangle\}$, where $y \in Y$ is chosen in such a way that h is strictly increasing. To do this, it is enough to pick $y \in Y \setminus \{y_0, y_1, \ldots, y_n\}$ such that the following holds for every $i \leq n$:

$$x < x_i \iff y \prec y_i. \tag{8.3}$$

If $x < x_0$, pick $y \prec y_0$. Such y exists, since Y does not have a smallest element. If $x > x_n$, pick $y \succ y_n$, which exists since Y does not have a largest element. So assume that $x_i < x < x_{i+1}$ for some $i < n$. Then choose $y \in Y$ such that $y_i \prec y \prec y_{i+1}$, which exists since Y is dense. It is easy to see that such y satisfies (8.3).

We have proved that each D_x is dense in $\mathbb{P}$. The proof that every R_y is dense in $\mathbb{P}$ is almost identical, and is left as an exercise. $\square$

Evidently $\langle \mathbb{Q}, \leq \rangle$ is an example of a dense linearly ordered set with neither a first nor a last element. Thus Theorem 8.1.3 says that any countable linearly ordered dense set without a first or a last element is isomorphic to $\langle \mathbb{Q}, \leq \rangle$. The order type of this class is usually denoted by the letter η.

In fact, it can be proved that $\langle \mathbb{Q}, \leq \rangle$ is universal for the class of all countable linearly ordered sets, in the sense that every countable linearly ordered set is isomorphic to $\langle S, \leq \rangle$ for some $S \subset \mathbb{Q}$ (see Exercise 1).

Clearly, $\langle \mathbb{R}, \leq \rangle$ is not isomorphic to $\langle \mathbb{Q}, \leq \rangle$. The order type of $\langle \mathbb{R}, \leq \rangle$ is usually denoted by λ. To give a characterization of the order type λ similar to that of Theorem 8.1.3 we will need the following definition. A linearly ordered set $\langle X, \leq \rangle$ is said to be *complete* if every subset of X that is bounded from above has a least upper bound. That is, if the set $B(S) = \{b \in X : \forall x \in S \ (x \leq b)\}$ is not empty for some $S \subset X$ then $B(S)$ has a least element, denoted by $\sup S$. It is a fundamental fact that $\langle \mathbb{R}, \leq \rangle$ is complete.

Theorem 8.1.4 *Any two complete linearly ordered sets both having countable dense subsets, and having neither a least nor a largest element, are order isomorphic.*

Proof Let $\langle X, \leq \rangle$ and $\langle Y, \preceq \rangle$ be linearly ordered sets as described in the theorem. Let X_0 and Y_0 be countable dense subsets of X and Y, respectively. It is easy to see that X_0 and Y_0 satisfy the assumptions of Theorem 8.1.3.

Let $f_0 \colon X_0 \to Y_0$ be an order isomorphism between $\langle X_0, \leq \rangle$ and $\langle Y_0, \preceq \rangle$. Define $f \colon X \to Y$ by $f(x) = \sup\{f_0(x_0) \colon x_0 \in X_0 \ \& \ x_0 \leq x\}$. It is not difficult to prove that f is an order isomorphism. The details are left as an exercise. $\square$

The next theorem will show that we can also use the Rasiowa–Sikorski lemma in proofs by transfinite induction. To formulate it we need the following definition. For $f, g \colon \omega \to \omega$ we define

$$f <^* g \iff f(n) < g(n) \text{ for all but finitely many } n < \omega. \tag{8.4}$$

It is easy to see that the relation $<^*$ is transitive on ω^ω.

Theorem 8.1.5 *If $G \subset \omega^\omega$ has cardinality $\leq \omega_1$ then there exists a $<^*$-increasing sequence $\langle f_\xi \colon \xi < \omega_1 \rangle$ such that for every $g \in G$ there exists a $\xi < \omega_1$ with $g <^* f_\xi$.*

Before we prove it, notice first that it immediately implies the following corollary. For its formulation we need the following definition. A sequence $\langle f_\xi \in \omega^\omega \colon \xi < \kappa \rangle$ is called a *scale* in ω^ω if it is $<^*$-increasing and if for every $g \in \omega^\omega$ there exists a $\xi < \kappa$ such that $g <^* f_\xi$.

Corollary 8.1.6 *If the continuum hypothesis holds then there exists a scale $\langle f_\xi \in \omega^\omega \colon \xi < \mathfrak{c} \rangle$.*

In the next section we will show that Corollary 8.1.6 can also be proved when the continuum hypothesis is false.

The proof of Theorem 8.1.5 will be based on two lemmas. The first of these lemmas is not essential for the proof, but it serves as a good approximation for the second one.

Lemma 8.1.7 *If $G \subset \omega^\omega$ has cardinality $\leq \omega$ then there exists an $f \in \omega^\omega$ such that for every $g \in G$*

$$g(n) < f(n) \quad \text{for infinitely many } n < \omega.$$

Proof An inductive construction of such f can be done as follows. Let $G \times \omega = \{\langle g_n, k_n \rangle \colon n < \omega\}$. Construct a sequence $f_0 \subset f_1 \subset f_2 \subset \cdots$ of functions such that each $f_n \in \mathrm{Func}_\omega(\omega, \omega)$ and for every $n < \omega$ there exists an $m \in \mathrm{dom}(f_n)$ with $m \geq k_n$ such that $f_n(m) > g_n(m)$. Then $\hat{f} = \bigcup_{n < \omega} f_n \in \mathrm{Func}(\omega, \omega)$ and any extension f of $\hat{f}$ to ω has the desired property, since for every $g \in G$ and $k < \omega$ there exists an $m \geq k$ with $f(m) > g(m)$.

To translate this proof into the language of partially ordered sets, let $\mathbb{P}$ be the set of all possible functions f_n as before, that is, take

$$\mathbb{P} = \mathrm{Func}_\omega(\omega, \omega),$$

and order it by reverse inclusion $\supset$. For $g \in \omega^\omega$ and $k < \omega$ define

$$D_g^k = \{s \in \mathbb{P} \colon \exists m \geq k \ (m \in \mathrm{dom}(s) \ \& \ s(m) > g(m))\}.$$

These sets are the counterparts of the inductive condition "$f_n(m) > g_n(m)$ for some $m \geq k_n$."

Notice that the sets D_g^k are dense in $\mathbb{P}$, since for every $t \in \mathbb{P}$ there exists an $m \in \omega \setminus (k \cup \mathrm{dom}(t))$ and for $s = t \cup \{\langle m, g(m) + 1 \rangle\}$ we have $s \in D_g^k$ and $s \leq t$.

Let $\mathcal{D} = \{D_g^k \colon k < \omega \ \& \ g \in G\}$ and let F be a $\mathcal{D}$-generic filter in $\mathbb{P}$. Then, by Proposition 8.1.1, $\bigcup F$ is a function. Let $f \colon \omega \to \omega$ be any extension of $\bigcup F$. Then f has the desired property, since for every $g \in G$ and $k < \omega$ there are $s \in F \cap D_g^k$ and $m \geq k$ such that $f(m) = s(m) > g(m)$. $\square$

The lemma that we really need for the proof of Theorem 8.1.5 is the following.

Lemma 8.1.8 *If $G \subset \omega^\omega$ has cardinality $\leq \omega$ then there exists an $f \in \omega^\omega$ such that*
$$g <^* f \quad \text{for every } g \in G.$$

Proof An inductive construction of such f can be done as follows. Let $G = \{g_n \colon n < \omega\}$ and $\omega = \{x_n \colon n < \omega\}$. Construct a sequence $f_0 \subset f_1 \subset f_2 \subset \cdots$ of functions such that $f_n \in \text{Func}_\omega(\omega, \omega)$ and $x_n \in \text{dom}(f_n)$ for every $n < \omega$. Moreover, at the inductive step $n < \omega$ choose f_n such that $f_n(x) > g_i(x)$ for all functions g_i looked at so far and all numbers x that have not yet been considered up to this point of the induction, that is, such that

$$f_n(x) > g_i(x) \quad \text{for all } x \in \text{dom}(f_n) \setminus \text{dom}(f_{n-1}) \text{ and } i < n. \tag{8.5}$$

Then $f = \bigcup_{n<\omega} f_n \colon \omega \to \omega$ and $g_i <^* f$ for every $i < \omega$.

One problem in finding an appropriate forcing for this construction is that the "current stage of induction" includes information not only on "functions f_n constructed so far" but also on "all functions g_i considered so far." The first part of this information is included in functions from $\mathbb{P} = \text{Func}_\omega(\omega, \omega)$, as in Lemma 8.1.7. However, we need the second piece of information as well, which is to be coded by finite subsets of G. Thus we define the partially ordered set $\mathbb{P}^* = \mathbb{P} \times [G]^{<\omega}$. Moreover, the partial order on $\mathbb{P}^*$, in order to describe the extension as in (8.5), must be defined as follows. For $\langle s, A \rangle, \langle t, B \rangle \in \mathbb{P}^*$ we define $\langle s, A \rangle \leq \langle t, B \rangle$ provided $s \supset t$, $A \supset B$, and

$$s(n) > g(n) \quad \text{for all } n \in \text{dom}(s) \setminus \text{dom}(t) \text{ and } g \in B. \tag{8.6}$$

Now the inductive conditions "$x_n \in \text{dom}(f_n)$" and "every element g of G is taken care of at some stage of the induction" are coded by the following subsets of $\mathbb{P}^*$:

$$D_n = \{\langle s, A \rangle \in \mathbb{P}^* \colon n \in \text{dom}(s)\} \quad \text{and} \quad E_g = \{\langle s, A \rangle \in \mathbb{P}^* \colon g \in A\},$$

where $g \in G \subset \omega^\omega$ and $n < \omega$.

The sets E_g are dense in $\mathbb{P}^*$, since for every $\langle s, A \rangle \in \mathbb{P}^*$ we can pick $\langle s, A \cup \{g\} \rangle \in E_g$ and clearly $\langle s, A \cup \{g\} \rangle \leq \langle s, A \rangle$, as condition (8.6) is then satisfied vacuously.

To see that a set D_n is dense in $\mathbb{P}^*$ take $p = \langle t, A \rangle \in \mathbb{P}^*$. We have to find $q \in D_n$ such that $q \leq p$. If $n \in \mathrm{dom}(t)$ then $p \in D_n$ and $q = p$ works. So assume that $n \notin \mathrm{dom}(t)$. Define $q = \langle s, A \rangle \in D_n$ with $s = t \cup \{\langle n, k \rangle\}$ and $k = \sup\{g(n) + 1 \colon g \in A\}$. Then $q \leq p$, since clearly $s \supset t$, $A \supset A$, and (8.6) holds, as $s(n) = k > g(n)$ for all $g \in A$.

Now let $\mathcal{D} = \{D_n \colon n < \omega\} \cup \{E_g \colon g \in G\}$. Then $\mathcal{D}$ is a countable family of dense subsets of $\mathbb{P}^*$ so, by the Rasiowa–Sikorski lemma, we can find a $\mathcal{D}$-generic filter F in $\mathbb{P}^*$. Notice that $F_0 = \{s \colon \langle s, A \rangle \in F\}$ is a filter in $\mathbb{P}$. Thus, by Proposition 8.1.1, $f = \bigcup F_0$ is a function. Notice also that $D_n \cap F \neq \emptyset$ implies $n \in \mathrm{dom}(f)$. Therefore f maps ω into itself.

To finish the proof it is enough to show that $g <^* f$ for every $g \in G$. So let $g \in G$. Then there exists $\langle t, B \rangle \in F \cap E_g$. We will prove that

$$f(n) > g(n) \ \text{ for every } \ n \in \omega \setminus \mathrm{dom}(t).$$

This is the case since for every $n \in \omega \setminus \mathrm{dom}(t)$ there is $\langle t', B' \rangle \in F \cap D_n$. Take $\langle s, A \rangle \in F$ with $\langle s, A \rangle \leq \langle t, B \rangle$ and $\langle s, A \rangle \leq \langle t', B' \rangle$. Then $n \in \mathrm{dom}(t') \subset \mathrm{dom}(s)$ so $n \in \mathrm{dom}(s) \setminus \mathrm{dom}(t)$ and $g \in B$. Therefore condition (8.6) for $\langle s, A \rangle \leq \langle t, B \rangle$ implies that $f(n) = s(n) > g(n)$.   □

Proof of Theorem 8.1.5 Let $G = \{g_\xi \colon \xi < \omega_1\}$. Define $\langle f_\xi \colon \xi < \omega_1 \rangle$ by induction on $\xi < \omega_1$ by choosing f_ξ as the function f from Lemma 8.1.8 applied to $G = \{f_\zeta \colon \zeta < \xi\} \cup \{g_\zeta \colon \zeta < \xi\}$. It is easy to see that this sequence has the desired property.   □

EXERCISES

1 Prove that for every countable linearly ordered set $\langle X, \preceq \rangle$ there exists a strictly increasing function $f \colon X \to \mathbb{Q}$. (Such a function f is said to be an *order embedding* of X into $\mathbb{Q}$, and it establishes an order isomorphism between $\langle X, \preceq \rangle$ and $\langle f[X], \leq \rangle$.)

2 Prove that $\langle \mathbb{R}, \leq \rangle$ is complete as a linearly ordered set (use Dedekind's definition of real numbers).

3 Complete the details of the proof of Theorem 8.1.4 by showing that

(a)   the sets X_0 and Y_0 are dense in themselves and have neither least nor largest elements;

(b)   the function $f \colon X \to Y$ is an order isomorphism.

4 A subset D of a linearly ordered set $\langle X, \leq \rangle$ is *weakly dense* if

for every $x, y \in X$ with $x < y$ there is a $d \in D$ such that $x \leq d \leq y$.

Prove that for every linearly ordered set $\langle X, \leq \rangle$ containing a countable weakly dense subset there exists a strictly increasing function $f \colon X \to \mathbb{R}$.

5 Prove that the relation $\leq$ defined on the set $\mathbb{P}^*$ as in Lemma 8.1.8 is indeed a partial-order relation.

6 For $f, g \colon \omega \to \omega$ we define

$$ f \leq^* g \Leftrightarrow f(n) \leq g(n) \quad \text{for all but finitely many } n < \omega. $$

Show that $\leq^*$ is a preorder relation on ω^ω (see Exercise 4 from Section 2.4). The equivalence relation generated by $\leq^*$ is usually denoted by $=^*$. Show also that the partial-order relation $\preceq^*$ induced by $\leq^*$ on the family of all equivalence classes of $=^*$ is a linear-order relation.

8.2 Martin's axiom

In this section we would like to introduce an axiom that says that in a large number of situations the statement "for countably many" can be replaced by "for less than continuum many," even when the continuum hypothesis fails.

In particular, we would like to be able to make such a replacement in the Rasiowa–Sikorski lemma, which leads us to the following statement.

($\star$)   Let $\langle \mathbb{P}, \leq \rangle$ be a partially ordered set. If $\mathcal{D}$ is a family of dense subsets of $\mathbb{P}$ such that $|\mathcal{D}| < \mathfrak{c}$, then there exists a $\mathcal{D}$-generic filter F in $\mathbb{P}$.

Clearly ($\star$) is implied by the continuum hypothesis. However, it is false under the negation of the continuum hypothesis. To see it, consider the partially ordered set $\langle \mathbb{P}, \leq \rangle = \langle \mathrm{Func}_\omega(\omega, \omega_1), \supset \rangle$ and, for $\xi < \omega_1$, the sets $R_\xi = \{s \in \mathbb{P} \colon \xi \in \mathrm{range}(s)\}$. Each set R_ξ is dense, since for every $s \in \mathbb{P}$ we have $t = s \cup \{\langle n, \xi \rangle\} \in R_\xi$ and $t \leq s$ for any $n \in \omega \setminus \mathrm{dom}(s)$. But $\mathcal{D} = \{R_\xi \colon \xi < \omega_1\}$ has cardinality $\leq \omega_1 < \mathfrak{c}$. So ($\star$) implies the existence of a $\mathcal{D}$-generic filter F in $\mathbb{P}$, and, by Proposition 8.1.1, $\bigcup F$ is a function from a subset of ω *onto* ω_1. This is clearly impossible.

Thus, in order to find a ($\star$)-like axiom that is consistent with the negation of the continuum hypothesis we will have to restrict the class of partially ordered sets allowed in its statement. In particular, we will have to exclude forcings such as $\langle \mathrm{Func}_\omega(\omega, \omega_1), \supset \rangle$. To define such a class we need some new definitions.

Let $\langle \mathbb{P}, \leq \rangle$ be a partially ordered set.

- $x, y \in \mathbb{P}$ are *comparable* if either $x \leq y$ or $y \leq x$. Thus a chain in $\mathbb{P}$ is a subset of $\mathbb{P}$ of pairwise-comparable elements.

- $x, y \in \mathbb{P}$ are *compatible (in $\mathbb{P}$)* if there exists a $z \in \mathbb{P}$ such that $z \leq x$ and $z \leq y$. In particular, condition (F1) from the definition of a filter says that any two elements of a filter F are compatible in F.

- $x, y \in \mathbb{P}$ are *incompatible* if they are not compatible.

- A subset A of $\mathbb{P}$ is an *antichain (in $\mathbb{P}$)* if every two distinct elements of A are incompatible. An antichain is *maximal* if it is not a proper subset of any other antichain. An elementary application of the Hausdorff maximal principle shows that every antichain in $\mathbb{P}$ is contained in some maximal antichain.

To illustrate these notions consider $\langle \mathbb{P}, \leq \rangle = \langle \operatorname{Func}_\omega(X, Y), \supset \rangle$. Then $s, t \in \mathbb{P}$ are compatible if and only if $s \cup t \in \operatorname{Func}_\omega(X, Y)$. Therefore elements $s, t \in \mathbb{P}$ are incompatible if there exists an $x \in \operatorname{dom}(s) \cap \operatorname{dom}(t)$ such that $s(x) \neq t(x)$. For any nonempty $D \in [X]^{<\omega}$ the set $A = Y^D$ is a maximal antichain. On the other hand, if $C \subset \mathbb{P}$ is a family of functions with pairwise-disjoint domains, then any two elements of C are compatible.

The forcings that will be used in our $(\star)$-like axiom are defined in terms of antichains in the following way.

- A partially ordered set $\langle \mathbb{P}, \leq \rangle$ is *ccc* (or satisfies the *countable chain condition*) if every *antichain* of $\mathbb{P}$ is at most countable.[1]

Clearly, every countable partially ordered set is ccc. In particular, the forcing $\langle \mathbb{P}, \leq \rangle = \langle \operatorname{Func}_\omega(\omega, \omega), \supset \rangle$ from Lemma 8.1.7 is ccc. Notice also that the forcing

$$\mathbb{P}^\star = \mathbb{P} \times [G]^{<\omega} \text{ from Lemma 8.1.8 is ccc} \qquad (8.7)$$

for an arbitrary $G \subset \omega^\omega$, including $G = \omega^\omega$. To see this, take an uncountable subset $A = \{p_\xi \in \mathbb{P}^\star : \xi < \omega_1\}$ of $\mathbb{P}^\star$ with $p_\xi = \langle s_\xi, A_\xi \rangle$ and notice that there must be a $\zeta < \xi < \omega_1$ such that $s_\zeta = s_\xi$. But then, for $p = \langle s_\zeta, A_\zeta \cup A_\xi \rangle \in \mathbb{P}^\star$ we have $p \leq p_\zeta$ and $p \leq p_\xi$, condition (8.6) being satisfied vacuously. So A is not an antichain.

On the other hand, the forcing $\langle \mathbb{P}, \leq \rangle = \langle \operatorname{Func}_\omega(\omega, \omega_1), \supset \rangle$ is not ccc, since $\{\{\langle 0, \xi \rangle\} \in \mathbb{P} : \xi < \omega_1\}$ is an uncountable antichain in $\mathbb{P}$. Thus,

---

[1] The name "countable chain condition" is certainly misleading. A more appropriate name would be "countable antichain condition" or "cac." However, the tradition of this name is very strong and outweighs reason. This tradition can be explained by the fact that every partially ordered set can be canonically embedded into a complete Boolean algebra, and for such algebras the maximal sizes of chains and antichains are equal.

restricting $(\star)$ to the ccc forcings removes the immediate threat of the previously described contradiction with ¬CH, the negation of the continuum hypothesis. In fact such a restriction also removes all possibility of any contradiction with ¬CH. More precisely, consider the following axiom, known as Martin's axiom and usually abbreviated by MA.

Martin's axiom Let $\langle \mathbb{P}, \leq \rangle$ be a ccc partially ordered set. If $\mathcal{D}$ is a family of dense subsets of $\mathbb{P}$ such that $|\mathcal{D}| < \mathfrak{c}$, then there exists a $\mathcal{D}$-generic filter F in $\mathbb{P}$.

Clearly CH implies MA. But MA is also consistent with ZFC and ¬CH, as stated by the next theorem.

Theorem 8.2.1 *Martin's axiom plus the negation of the continuum hypothesis MA+¬CH is consistent with ZFC set theory.*

The proof of Theorem 8.2.1 will be postponed until Section 9.5 where we will prove that MA is consistent with $\mathfrak{c} = \omega_2$. However, the same proof can be used to prove the consistency of MA with $\mathfrak{c} = \kappa$ for most regular cardinals κ.

In the remainder of this section we will see several consequences of MA. First we will see the following generalization of Corollary 8.1.6.

Theorem 8.2.2 *If MA holds then there exists a scale $\langle f_\xi \in \omega^\omega : \xi < \mathfrak{c} \rangle$.*

Proof The argument is essentially identical to that for Theorem 8.1.5. First notice that the following generalization of Lemma 8.1.8 is implied by Martin's axiom.

(I)   If $G \subset \omega^\omega$ has cardinality $< \mathfrak{c}$ then there exists an $f \in \omega^\omega$ such that $g <^\star f$ for every $g \in G$.

To see why, consider the forcing $\mathbb{P}^\star = \mathrm{Func}_\omega(\omega, \omega) \times [G]^{<\omega}$ defined as in Lemma 8.1.8, that is, ordered by

$$\langle s, A \rangle \leq \langle t, B \rangle \;\Leftrightarrow\; s \supset t \;\&\; A \supset B$$
$$\&\; s(n) > g(n) \text{ for all } n \in \mathrm{dom}(s) \setminus \mathrm{dom}(t) \text{ and } g \in B,$$

and its dense subsets

$$D_n = \{\langle s, A \rangle \in \mathbb{P}^\star : n \in \mathrm{dom}(s)\} \quad \text{and} \quad E_g = \{\langle s, A \rangle \in \mathbb{P}^\star : g \in A\}$$

for $g \in \omega^\omega$ and $n < \omega$. Then $\mathcal{D} = \{D_n : n < \omega\} \cup \{E_g : g \in G\}$ has cardinality $|G| + \omega < \mathfrak{c}$. Since by (8.7) $\mathbb{P}^\star$ is ccc, MA implies that there exists a $\mathcal{D}$-generic filter F in $\mathbb{P}^\star$. Now we argue as in Lemma 8.1.8. We define $F_0 = \{s : \langle s, A \rangle \in F\}$ and notice that $f = \bigcup F_0$ is a function. Then

$D_n \cap F \neq \emptyset$ implies that $n \in \text{dom}(f)$, that is, that f maps ω into ω. Finally, we notice that $E_g \cap F \neq \emptyset$ implies $g <^* f$.

To prove the theorem from (I) enumerate ω^ω as $\{g_\xi : \xi < \mathfrak{c}\}$ and define the scale $\langle f_\xi : \xi < \mathfrak{c} \rangle$ by induction on $\xi < \mathfrak{c}$ by choosing g_ξ as a function f from (I) applied to $G = \{f_\zeta : \zeta < \xi\} \cup \{g_\zeta : \zeta < \xi\}$. □

Another application of Martin's axiom is stated in the next theorem.

Theorem 8.2.3 *Assume MA. If $X \in [\mathbb{R}]^{<\mathfrak{c}}$ then every subset Y of X is a G_δ subset of X, that is, there exists a G_δ set $G \subset \mathbb{R}$ such that $G \cap X = Y$.*

Proof Let $X \in [\mathbb{R}]^{<\mathfrak{c}}$ and fix $Y \subset X$. We will show that Y is G_δ in X.

Let $\mathcal{B} = \{B_n : n < \omega\}$ be a countable base for $\mathbb{R}$. First notice that it is enough to find a set $\hat{A} \subset \omega$ such that for every $x \in X$

$$x \in Y \quad \Leftrightarrow \quad x \in B_n \text{ for infinitely many } n \text{ from } \hat{A}. \tag{8.8}$$

To see why, define for every $k < \omega$ an open set $G_k = \bigcup\{B_n : n \in \hat{A} \,\&\, n > k\}$ and put $G = \bigcap_{k<\omega} G_k$. Then G is a G_δ set and, by (8.8), for every $x \in X$ we have

$$x \in Y \quad \Leftrightarrow \quad x \in G_k \text{ for all } k < \omega.$$

Thus $G \cap X = Y$.

For countable X we could prove (8.8) in the following way. Let $X \setminus Y = \{z_n : n < \omega\}$ and $Y \times \omega = \{\langle y_n, k_n \rangle : n < \omega\}$. Then construct $\hat{A}$ as a union of an increasing sequence $\langle A_n : n < \omega \rangle$ of finite subsets of ω such that for every $n < \omega$ there is an $m \in A_n$ such that $m > k_n$ and $y_n \in B_m$, and such that $z_i \notin B_m$ for all new ms from A_n and all points z_i considered so far, that is, such that

$$z_i \notin B_m \text{ for all } m \in A_n \setminus A_{n-1} \text{ and } i \leq n. \tag{8.9}$$

We will transform this idea into a forcing similarly as in Lemma 8.1.8. We define the partially ordered set $\langle \mathbb{P}, \leq \rangle$ by putting $\mathbb{P} = [\omega]^{<\omega} \times [X \setminus Y]^{<\omega}$ and define $\leq$ to take care of (8.9). That is, for $\langle A_1, C_1 \rangle, \langle A_0, C_0 \rangle \in \mathbb{P}$ we define $\langle A_1, C_1 \rangle \leq \langle A_0, C_0 \rangle$ provided $A_1 \supset A_0$, $C_1 \supset C_0$, and

$$c \notin B_m \text{ for all } m \in A_1 \setminus A_0 \text{ and } c \in C_0. \tag{8.10}$$

Now inductive conditions are coded by the following subsets of $\mathbb{P}$:

$$D_y^k = \{\langle A, C \rangle \in \mathbb{P} : \exists m \in A \, (m \geq k \,\&\, y \in B_m)\}$$

and

$$E_z = \{\langle A, C \rangle \in \mathbb{P} : z \in C\},$$

where $y \in Y$, $k < \omega$, and $z \in X \setminus Y$. We will use Martin's axiom to find a $\mathcal{D}$-generic filter for

$$\mathcal{D} = \{D_y^k \colon y \in Y \ \& \ k < \omega\} \cup \{E_z \colon z \in X \setminus Y\}.$$

To use Martin's axiom, we have to check whether its assumptions are satisfied.

Clearly $|\mathcal{D}| \le |X| + \omega < \mathfrak{c}$.

To see that $\mathbb{P}$ is ccc consider an uncountable subset $\{\langle A_\xi, C_\xi \rangle \colon \xi < \omega_1\}$ of $\mathbb{P}$. Since $[\omega]^{<\omega}$ is countable, there are $A \in [\omega]^{<\omega}$ and $\zeta < \xi < \omega_1$ such that $A_\zeta = A_\xi = A$. Then $\langle A_\zeta, C_\zeta \rangle = \langle A, C_\zeta \rangle$ and $\langle A_\xi, C_\xi \rangle = \langle A, C_\xi \rangle$ are compatible, since $\langle A, C_\zeta \cup C_\xi \rangle \in \mathbb{P}$ extends them both, as condition (8.10) is satisfied vacuously.

To see that each set E_z is dense in $\mathbb{P}$ take $\langle A, C \rangle \in \mathbb{P}$ and notice that $\langle A, C \cup \{z\} \rangle \in E_z$ extends $\langle A, C \rangle$.

Finally, to see that a set D_y^k is dense in $\mathbb{P}$, take $\langle A, C \rangle \in \mathbb{P}$. Notice that there exist infinitely many basic open sets B_m such that

$$y \in B_m \text{ and } C \cap B_m = \emptyset. \tag{8.11}$$

Take $m > k$ satisfying (8.11), and notice that $\langle A \cup \{m\}, C \rangle \in D_y^k$ extends $\langle A, C \rangle$.

Now apply Martin's axiom to find a $\mathcal{D}$-generic filter F in $\mathbb{P}$, and define $\hat{A} = \bigcup\{A \colon \langle A, C \rangle \in F\}$. We will show that $\hat{A}$ satisfies (8.8). So let $x \in X$.

If $x \in Y$ then for every $k < \omega$ there exists $\langle A, C \rangle \in F \cap D_x^k$. In particular, there exists an $m \in A \subset \hat{A}$ with $m > k$ such that $x \in B_m$. So $x \in B_m$ for infinitely many m from $\hat{A}$.

If $x \in X \setminus Y$ then there exists $\langle A_0, C_0 \rangle \in F \cap E_x$. In particular, $x \in C_0$. It is enough to prove that $x \notin B_m$ for every $m \in \hat{A} \setminus A_0$. So take $m \in \hat{A} \setminus A_0$. By the definition of $\hat{A}$ there exists $\langle A, C \rangle \in F$ such that $m \in A$. But, by the definition of a filter, there exists $\langle A_1, C_1 \rangle \in F$ extending $\langle A, C \rangle$ and $\langle A_0, C_0 \rangle$. Now $\langle A_1, C_1 \rangle \le \langle A_0, C_0 \rangle$, $m \in A \subset A_1$, $m \notin A_0$, and $x \in C_0$. Hence, by (8.10), $x \notin B_m$. $\qquad \square$

Corollary 8.2.4 *If MA holds then* $2^\omega = 2^\kappa$ *for every infinite cardinal* $\kappa < \mathfrak{c}$.

Proof Let $\kappa < \mathfrak{c}$ be an infinite cardinal number. Then clearly $2^\omega \le 2^\kappa$. To see the other inequality take $X \in [\mathbb{R}]^\kappa$. Then, by Theorem 8.2.3,

$$2^\kappa = |\mathcal{P}(X)| = |\{B \cap X \colon B \in \mathcal{B}or\}| \le |\mathcal{B}or| = 2^\omega. \qquad \square$$

Corollary 8.2.5 *If MA holds then $\mathfrak{c}$ is a regular cardinal.*

Proof Notice that by Theorem 5.3.8

$$2^{\mathrm{cf}(\mathfrak{c})} = (2^\omega)^{\mathrm{cf}(\mathfrak{c})} = \mathfrak{c}^{\mathrm{cf}(\mathfrak{c})} > \mathfrak{c}.$$

Hence, by Corollary 8.2.4, $\mathrm{cf}(\mathfrak{c}) \geq \mathfrak{c}$.   □

The next theorem tells us that MA implies the continuum additivity of category, which has been mentioned in the remark after Theorem 7.2.5. In particular, it implies that Theorem 7.2.5 can be proved when CH is replaced with MA.

Theorem 8.2.6 *If MA holds then a union of less than continuum many meager subsets of $\mathbb{R}^n$ is meager in $\mathbb{R}^n$, that is,*

$$\bigcup \mathcal{F} \in \mathcal{M} \quad \text{for every} \quad \mathcal{F} \in [\mathcal{M}]^{<\mathfrak{c}}.$$

Proof The idea of the proof is very similar to that of Theorem 8.2.3.

Let $\mathcal{F} \in [\mathcal{M}]^{<\mathfrak{c}}$. Since every F from $\mathcal{F}$ is a countable union of nowhere-dense sets, $\bigcup \mathcal{F}$ is a union of $|\mathcal{F}| + \omega < \mathfrak{c}$ nowhere-dense sets. Thus we may assume that every set in $\mathcal{F}$ is nowhere dense.

Let $\mathcal{B} = \{B_j \neq \emptyset \colon j < \omega\}$ be a countable base for $\mathbb{R}^n$. Notice that it is enough to find a set $\hat{A} \subset \omega$ such that

$$\{m \in \hat{A} \colon B_m \cap F \neq \emptyset\} \text{ is finite for every } F \in \mathcal{F} \qquad (8.12)$$

and

$$\{m \in \hat{A} \colon B_m \subset B_j\} \text{ is infinite for every } j < \omega. \qquad (8.13)$$

To see why, define $U_k = \bigcup\{B_m \colon m \in \hat{A} \ \& \ m > k\}$ for every $k < \omega$. Notice that by (8.13) the sets U_k are dense and open. Therefore the sets $\mathbb{R}^n \setminus U_k$ are nowhere dense. But condition (8.12) implies that $\bigcup \mathcal{F} \subset \bigcup_{k<\omega}(\mathbb{R}^n \setminus U_k) \in \mathcal{M}$, since for every $F \in \mathcal{F}$ there exists a $k < \omega$ such that

$$U_k \cap F = \bigcup\{B_m \cap F \colon m \in \hat{A} \ \& \ m > k\} = \emptyset.$$

To prove the existence of such a set $\hat{A}$ consider the partial order $\mathbb{P} = [\omega]^{<\omega} \times [\mathcal{F}]^{<\omega}$, where we put $\langle A_1, \mathcal{C}_1 \rangle \leq \langle A_0, \mathcal{C}_0 \rangle$ provided $A_1 \supset A_0$, $\mathcal{C}_1 \supset \mathcal{C}_0$, and

$$B_m \cap F = \emptyset \text{ for all } m \in A_1 \setminus A_0 \text{ and } F \in \mathcal{C}_0. \qquad (8.14)$$

We will define $\hat{A}$ as the union of all A such that $\langle A, C \rangle$ belongs to an appropriate generic filter in $\mathbb{P}$. In particular, in the condition $\langle A, C \rangle \in \mathbb{P}$

the set A approximates $\hat{A}$. The sets F from $\mathcal{C}$ represent "elements of $\mathcal{F}$ looked at so far" in our "induction" and (8.14) guarantees that the basic open sets indexed by "new elements of $\hat{A}$" will not intersect these "old" Fs. This will take care of (8.12).

The forcing $\mathbb{P}$ is ccc since $[\omega]^{<\omega}$ is countable and any conditions $\langle A, \mathcal{C}_0 \rangle$ and $\langle A, \mathcal{C}_1 \rangle$ are compatible, having $\langle A, C_0 \cup C_1 \rangle$ as a common extension.

Conditions (8.13) and (8.12) are related to the following dense subsets of $\mathbb{P}$:

$$D_j^k = \{\langle A, \mathcal{C} \rangle \in \mathbb{P} \colon \exists m \in A \ (m \geq k \ \& \ B_m \subset B_j)\}$$

and

$$E_F = \{\langle A, \mathcal{C} \rangle \in \mathbb{P} \colon F \in \mathcal{C}\},$$

where $k, j < \omega$ and $F \in \mathcal{F}$. The sets E_F are dense, since $\langle A, \mathcal{C} \cup \{F\} \rangle \in E_F$ extends $\langle A, \mathcal{C} \rangle$ for every $\langle A, \mathcal{C} \rangle \in \mathbb{P}$. To see that the sets D_j^k are dense take $\langle A, \mathcal{C} \rangle \in \mathbb{P}$. Since $\bigcup \mathcal{C}$ is nowhere dense, there exist infinitely many basic open sets B_m such that

$$B_m \subset B_j \setminus \bigcup \mathcal{C}. \tag{8.15}$$

Take $m > k$ satisfying (8.15) and notice that $\langle A \cup \{m\}, \mathcal{C} \rangle \in D_j^k$ extends $\langle A, \mathcal{C} \rangle$.

Hence, by Martin's axiom, there exists a $\mathcal{D}$-generic filter $\hat{F}$ in $\mathbb{P}$, where

$$\mathcal{D} = \{D_j^k \colon k, j < \omega\} \cup \{E_F \colon F \in \mathcal{F}\}.$$

Define $\hat{A} = \bigcup \{A \colon \langle A, \mathcal{C} \rangle \in \hat{F}\}$. It is enough to show that $\hat{A}$ satisfies properties (8.13) and (8.12).

To see (8.13) take $j < \omega$. It is enough to prove that for every $k < \omega$ there exists an $m \in \hat{A}$ with $m > k$ and $B_m \subset B_j$. So fix $k < \omega$ and pick $\langle A, \mathcal{C} \rangle \in \hat{F} \cap D_j^k$. Then, by the definition of D_j^k, there exists an $m \in A \subset \hat{A}$ with $m > k$ such that $B_m \subset B_j$.

To see (8.12) take $F \in \mathcal{F}$. Then there exists $\langle A_0, \mathcal{C}_0 \rangle \in \hat{F} \cap E_F$. In particular, $F \in \mathcal{C}_0$. It is enough to prove that $B_m \cap F = \emptyset$ for every $m \in \hat{A} \setminus A_0$. Take $m \in \hat{A} \setminus A_0$. By the definition of $\hat{A}$ there exists $\langle A, \mathcal{C} \rangle \in \hat{F}$ such that $m \in A$. So there exists $\langle A_1, \mathcal{C}_1 \rangle \in \hat{F}$ extending $\langle A, \mathcal{C} \rangle$ and $\langle A_0, \mathcal{C}_0 \rangle$. Now $\langle A_1, \mathcal{C}_1 \rangle \leq \langle A_0, \mathcal{C}_0 \rangle$, $m \in A \subset A_1$, $m \notin A_0$, and $F \in \mathcal{C}_0$. Hence, by (8.14), $B_m \cap F = \emptyset$. $\square$

An analog of Theorem 8.2.6 for the ideal $\mathcal{N}$ of measure-zero subsets of $\mathbb{R}^n$ is also true.

Theorem 8.2.7 *If MA holds then a union of less than continuum many null subsets of $\mathbb{R}^n$ is null in $\mathbb{R}^n$, that is,*

$$\bigcup \mathcal{F} \in \mathcal{N} \ \text{for every} \ \mathcal{F} \in [\mathcal{N}]^{<\mathfrak{c}}.$$

Proof To make the argument simpler, we will prove the theorem only for $\mathbb{R}$. The proof of the general case is essentially the same.

Let $\mathcal{B}_0$ be the family of all open intervals in $\mathbb{R}$ and let $l(I)$ stand for the length of $I \in \mathcal{B}_0$. Recall that $S \in \mathcal{N}$ if for every $\varepsilon > 0$ there exists a sequence $\langle I_k \in \mathcal{B}_0 : k < \omega \rangle$ such that $S \subset \bigcup_{k<\omega} I_k$ and $\sum_{k<\omega} l(I_k) < \varepsilon$.

Now, if $\mathcal{B} = \{I_n : n < \omega\}$ is the family of all intervals with rational endpoints, then the family $\mathcal{B}_0$ can be replaced by $\mathcal{B}$. To see it, take $S \in \mathcal{N}$ and let $\varepsilon > 0$. Then there exists a sequence $\langle J_k \in \mathcal{B}_0 : k < \omega \rangle$ such that $S \subset \bigcup_{k<\omega} J_k$ and $\sum_{k<\omega} l(J_k) < \varepsilon/2$. But for every $k < \omega$ there exists an $I_k \in \mathcal{B}_0$ such that $J_k \subset I_k$ and $l(I_k) < l(J_k) + \varepsilon/2^{k+2}$. So $S \subset \bigcup_{k<\omega} J_k \subset \bigcup_{k<\omega} I_k$ and $\sum_{k<\omega} l(I_k) \leq \sum_{k<\omega} [l(J_k) + \varepsilon/2^{k+2}] < \varepsilon$.

Let $\mathcal{F} \in [\mathcal{N}]^{<\mathfrak{c}}$ and fix $\varepsilon > 0$. We will find an $\hat{A} \subset \omega$ such that

$$\bigcup \mathcal{F} \subset \bigcup_{n \in \hat{A}} I_n \quad \text{and} \quad \sum_{n \in \hat{A}} l(I_n) \leq \varepsilon. \tag{8.16}$$

Define $\mathbb{P} = \{A \subset \omega : \sum_{n \in A} l(I_n) < \varepsilon\}$ and order it by reverse inclusion: $A_1 \leq A_0 \Leftrightarrow A_1 \supset A_0$. To see that $\mathbb{P}$ is ccc let $\mathcal{A} \subset \mathbb{P}$ be uncountable. We have to find different $A, A' \in \mathcal{A}$ that are compatible, that is, such that $\sum_{n \in A \cup A'} l(I_n) < \varepsilon$.

So for every $A \in \mathcal{A}$ let $m_A < \omega$ be such that $\sum_{n \in A} l(I_n) + 1/m_A < \varepsilon$. Since $\mathcal{A}$ is uncountable, there exists an uncountable subset $\mathcal{A}'$ of $\mathcal{A}$ and an $m < \omega$ such that $m_A = m$ for every $A \in \mathcal{A}'$, that is,

$$\sum_{n \in A} l(I_n) + \frac{1}{m} < \varepsilon \quad \text{for every } A \in \mathcal{A}'. \tag{8.17}$$

Next, for every $A \in \mathcal{A}'$ choose $k_A < \omega$ such that $\sum_{n \in A \setminus k_A} l(I_n) < 1/m$. Since $\mathcal{A}'$ is uncountable, there exists an uncountable subset $\mathcal{A}''$ of $\mathcal{A}'$ and a $k < \omega$ such that $k_A = k$ for every $A \in \mathcal{A}''$, that is,

$$\sum_{n \in A \setminus k} l(I_n) < \frac{1}{m} \quad \text{for every } A \in \mathcal{A}''. \tag{8.18}$$

Now we can find two different $A, A' \in \mathcal{A}''$ such that $A \cap k = A' \cap k$. For such A and A' we have, in particular, $A \cup A' = A \cup (A' \setminus k)$. So, by (8.18) and (8.17),

$$\sum_{n \in A \cup A'} l(I_n) \leq \sum_{n \in A} l(I_n) + \sum_{n \in A' \setminus k} l(I_n) < \sum_{n \in A} l(I_n) + \frac{1}{m} < \varepsilon.$$

Thus A and A' are compatible and $\mathbb{P}$ is ccc.

Now for every $F \in \mathcal{F}$ let $D_F = \{A \in \mathbb{P} : F \subset \bigcup_{n \in A} I_n\}$. Notice that every D_F is dense in $\mathbb{P}$. To see this, let $A \in \mathbb{P}$ and put $\delta = \varepsilon - \sum_{n \in A} l(I_n) >$

0. Since $F \in \mathcal{N}$ we can find a $B \subset \omega$ such that $F \subset \bigcup_{n \in B} I_n$ and $\sum_{n \in B} l(I_n) < \delta$. But then $A \cup B \in D_F$ since

$$\sum_{n \in A \cup B} l(I_n) \leq \sum_{n \in A} l(I_n) + \sum_{n \in B} l(I_n) < \sum_{n \in A} l(I_n) + \delta = \varepsilon$$

and $F \subset \bigcup_{n \in A \cup B} I_n$. Since $A \cup B \leq A$, the set D_F is dense.

Let $\mathcal{D} = \{D_F \colon F \in \mathcal{F}\}$. Since $\mathbb{P}$ is ccc and $|\mathcal{D}| \leq |\mathcal{F}| < \mathfrak{c}$, by Martin's axiom there exists a $\mathcal{D}$-generic filter $\hat{F}$ in $\mathbb{P}$. Let $\hat{A} = \bigcup \hat{F}$. We will show that $\hat{A}$ satisfies (8.16).

For every $F \in \mathcal{F}$ there exists an $A \in \hat{F} \cap D_F$. So $F \subset \bigcup_{n \in A} I_n \subset \bigcup_{n \in \hat{A}} I_n$ since $A \subset \hat{A}$. Then $\bigcup \mathcal{F} \subset \bigcup_{n \in \hat{A}} I_n$.

To see $\sum_{n \in \hat{A}} l(I_n) \leq \varepsilon$ it is enough to show that $\sum_{n \in \hat{A} \cap k} l(I_n) < \varepsilon$ for every $k < \omega$. So fix $k < \omega$. Now for every $n \in \hat{A} \cap k$ there exists an $A_n \in \hat{F}$ such that $n \in A_n$. Using condition (F1′) of the definition of a filter we can find an $A \in \hat{F}$ such that $A \leq A_n$ for every $n \in \hat{A} \cap k$. But then $\hat{A} \cap k \subset A$ and

$$\sum_{n \in \hat{A} \cap k} l(I_n) \leq \sum_{n \in A} l(I_n) < \varepsilon.$$

This finishes the proof. □

All previous applications of Martin's axiom could be deduced as well from the continuum hypothesis, but in the remaining part of this section we will show that this will not always be the case, by discussing the consequences of MA+¬CH, which do not follow from CH. Similar results can also be found in the next section.

In the next theorem we will use the following terminology. A subset A of a partially ordered set $\langle \mathbb{P}, \leq \rangle$ is *compatible* if for every finite subset A_0 of A there exists a $p \in \mathbb{P}$ such that $p \leq q$ for all $q \in A_0$ (compare this with condition (F1′) from the definition of a filter).

Theorem 8.2.8 *Assume MA+¬CH and let $\langle \mathbb{P}, \leq \rangle$ be a ccc partially ordered set. If $A \subset \mathbb{P}$ is uncountable, then there exists an uncountable compatible subset $\hat{A}$ of A.*

Proof Let $A \subset \mathbb{P}$ be uncountable. Without loss of generality we may assume that $|A| = \omega_1$.

The inductive approach to the proof is to construct a strictly increasing sequence $\langle A_\xi \colon \xi < \omega_1 \rangle$ of compatible subsets of A such that $\bigcup_{\xi < \omega_1} A_\xi$ is an uncountable compatible subset of A. Although this idea is basically correct, it has a fundamental flaw. If you start with $A_0 \in [A]^{\leq \omega}$ such that A_0 is compatible with at most countably many elements of A, then this construction must fail. Fortunately, there are only countably many sets

$A_0 \in [A]^{\leq \omega}$ that may cause such a problem, and after removing them from A we will be able to follow the idea just described.

So for $p \in \mathbb{P}$ let $A_p = \{q \in A : q$ is compatible with $p\}$. We will show that the set

$$A' = \{q \in A : \exists p \leq q \, (|A_p| \leq \omega)\} \quad \text{is at most countable.} \quad (8.19)$$

Indeed, consider the family $\mathcal{F} = \{C \subset B : C$ is an antichain in $\mathbb{P}\}$, where $B = \{p \in \mathbb{P} : |A_p| \leq \omega\}$. It is easy to see that $\mathcal{F}$ satisfies the assumptions of the Hausdorff maximal principle. Thus we can choose a maximal element C_0 of $\mathcal{F}$. Now C_0 is clearly an antichain in $\mathbb{P}$. So C_0 is at most countable, since $\mathbb{P}$ is ccc. It is enough to prove that

$$A' \subset \bigcup_{r \in C_0} A_r,$$

since the set $\bigcup_{r \in C_0} A_r$ is countable, being a countable union of countable sets. To see the inclusion, take $q \in A'$. Then there exists a $p \leq q$ such that $p \in B$. By the maximality of C_0 there exists an $r \in C_0$ such that r is compatible with p, that is, $s \leq r$ and $s \leq p \leq q$ for some $s \in \mathbb{P}$. In particular, r and q are compatible, so $q \in A_r$. Condition (8.19) has been proved.

Now the set $A^\star = A \setminus A'$ is uncountable. Moreover, if $p \leq q$ for some $q \in A^\star$ then A_p must be uncountable. In particular,

$$A_p^\star = \{q \in A^\star : q \text{ is compatible with } p\} \quad \text{is uncountable} \quad (8.20)$$

for every $p \in \mathbb{P}$ such that $p \leq q$ for some $q \in A^\star$.

Coming back to the idea of an inductive proof, we can see that we could have a hard time extending an infinite set A_ξ to a compatible set $A_{\xi+1}$ properly containing A_ξ. This problem will be solved with the help of MA.

For this, consider a partially ordered set

$$\mathbb{P}^\star = \{F \in [A^\star]^{<\omega} : F \text{ is compatible in } \mathbb{P}\}$$

ordered by reverse inclusion $\supset$. Notice that the forcing $\mathbb{P}^\star$ is ccc. To see why, let $\{F_\xi \in \mathbb{P}^\star : \xi < \omega_1\}$. We will find $\zeta < \xi < \omega_1$ such that F_ζ and F_ξ are compatible, that is, that $F_\zeta \cup F_\xi \in \mathbb{P}^\star$. But for every $\xi < \omega_1$ there exists a $p_\xi \in \mathbb{P}$ with $p_\xi \leq p$ for every $p \in F_\xi$. Moreover, the set $\{p_\xi : \xi < \omega_1\}$ cannot be an antichain, since $\mathbb{P}$ is ccc. Thus there are compatible p_ζ and p_ξ for some $\zeta < \xi < \omega_1$. Now, if $q \in \mathbb{P}$ is such that $q \leq p_\zeta$ and $q \leq p_\xi$ then $q \leq p$ for every $p \in F_\zeta \cup F_\xi$. Therefore $F_\zeta \cup F_\xi \in \mathbb{P}^\star$ and $\mathbb{P}^\star$ is ccc.

Let $\langle q_\xi : \xi < \omega_1 \rangle$ be a one-to-one enumeration of $A^\star$. For $\alpha < \omega_1$ define

$$D_\alpha = \{F \in \mathbb{P}^\star : \exists \xi > \alpha \, (q_\xi \in F)\}$$

and notice that the sets D_α are dense in $\mathbb{P}^\star$. This is so since for every $F \in \mathbb{P}^\star$ there exists a $p \in \mathbb{P}$ such that $p \leq q$ for every $q \in F$, and, by (8.20), p is compatible with uncountably many q_ξ. In particular, there exists a $\xi > \alpha$ such that q_ξ is compatible with p, and it is easy to see that $F \cup \{q_\xi\} \in D_\alpha$ extends F.

Let $\mathcal{D} = \{D_\alpha : \alpha < \omega_1\}$. Then $|\mathcal{D}| \leq \omega_1 < \mathfrak{c}$ so, by MA, there exists a $\mathcal{D}$-generic filter $\hat{F}$ in $\mathbb{P}^\star$. We will show that $\hat{A} = \bigcup \hat{F}$ is an uncountable compatible subset of A.

Clearly, $\hat{A} \subset A$. To see that $\hat{A}$ is uncountable, notice that for every $\alpha < \omega_1$ there exists a $\xi > \alpha$ such that $q_\xi \in \hat{A}$, since $\hat{F} \cap D_\alpha \neq \emptyset$. To finish the proof, it is enough to show that $\hat{A}$ is compatible.

So let $F = \{r_0, \dots, r_n\} \subset \hat{A}$. For every $i \leq n$ there exists an $F_i \in \hat{F}$ such that $r_i \in F_i$. By (F1') we can find an $E \in \hat{F}$ such that $E \leq F_i$ for all $i \leq n$. In particular, $E \supset \{r_0, \dots, r_n\} = F$, that is, F is compatible. $\square$

By definition, a forcing $\mathbb{P}$ is ccc if every uncountable set $\{p_\alpha : \alpha \in A\}$ contains two different compatible elements. Equivalently, $\mathbb{P}$ is ccc if for every sequence $\langle p_\alpha : \alpha \in A \rangle$ (not necessarily one-to-one) with uncountable index set A there are different indices $\alpha, \beta \in A$ such that p_α and p_β are compatible. In this language Theorem 8.2.8 can be restated as follows.

Corollary 8.2.9 *Assume $MA+\neg CH$ and let $\langle \mathbb{P}, \leq \rangle$ be a ccc partially ordered set. If $\langle p_\alpha : \alpha \in A \rangle$ is an uncountable sequence of elements of $\mathbb{P}$ then there exists an uncountable subset $\hat{A}$ of A such that $\{p_\alpha : \alpha \in \hat{A}\}$ is compatible in $\mathbb{P}$.*

To state the next corollary, we need the following important definition. Let $\langle \mathbb{P}_0, \leq_0 \rangle$ and $\langle \mathbb{P}_1, \leq_1 \rangle$ be partially ordered sets. Their *product* $\langle \mathbb{P}, \leq \rangle$ is defined by $\mathbb{P} = \mathbb{P}_0 \times \mathbb{P}_1$ and

$$\langle p_0, p_1 \rangle \leq \langle q_0, q_1 \rangle \Leftrightarrow p_0 \leq_0 q_0 \ \& \ p_1 \leq_1 q_1.$$

Corollary 8.2.10 *If $MA+\neg CH$ holds then the product of two ccc forcings is ccc.*

Proof Let $\langle \mathbb{P}, \leq \rangle$ be a product of ccc forcings $\langle \mathbb{P}_0, \leq_0 \rangle$ and $\langle \mathbb{P}_1, \leq_1 \rangle$ and let $\langle \langle p_\alpha, q_\alpha \rangle : \alpha \in A \rangle$ be an uncountable sequence of elements of $\mathbb{P}$. By Corollary 8.2.9 used for the forcing $\mathbb{P}_0$ and a sequence $\langle p_\alpha : \alpha \in A \rangle$, we can find an uncountable subset $\hat{A}$ of A such that the set $\{p_\alpha : \alpha \in \hat{A}\}$ is compatible in $\mathbb{P}_0$. Then there are different $\alpha, \beta \in \hat{A}$ such that q_α and q_β are compatible in $\mathbb{P}_1$, since $\mathbb{P}_1$ is ccc. It is easy to see that $\langle p_\alpha, q_\alpha \rangle$ and $\langle p_\beta, q_\beta \rangle$ are compatible in $\mathbb{P}$. $\square$

The last theorem of this section shows that Corollary 8.2.10 is false under CH. In its proof we will use the following lemma, which is the main combinatorial tool needed to prove that different kinds of forcing (built with finite sets) are ccc.

Lemma 8.2.11 (Δ-system lemma) *If $\mathcal{A}$ is an uncountable family of finite sets then there exists an uncountable subfamily $\mathcal{A}_0$ of $\mathcal{A}$ and a finite set A such that $X \cap Y = A$ for every distinct $X, Y \in \mathcal{A}_0$.*

Proof Since $|\mathcal{A}| > \omega$ we may assume that there is an $n < \omega$ such that each element of $\mathcal{A}$ has exactly n elements. The proof is by induction on n.

By our assumption we must have $n > 0$. (Otherwise $\mathcal{A} \subset \{\emptyset\}$ has cardinality $\leq 1 < \omega$.) If $n = 1$ then elements of $\mathcal{A}$ must be pairwise disjoint and the theorem holds with $A = \emptyset$ and $\mathcal{A}_0 = \mathcal{A}$. So assume that $n > 1$ and that the theorem holds for $(n-1)$-element sets. Consider two cases.

Case 1: There exists an $A_0 \in \mathcal{A}$ such that A_0 intersects uncountably many $A \in \mathcal{A}$. Then there exists an $a \in A_0$ such that a belongs to uncountably many $A \in \mathcal{A}$. In particular,

$$\mathcal{B} = \{A \setminus \{a\} : A \in \mathcal{A} \ \& \ a \in A\}$$

is an uncountable family of sets of size $n-1$, and by the inductive hypothesis, we can find an uncountable $\mathcal{B}_0 \subset \mathcal{B}$ and a finite set B such that $C \cap D = B$ for every distinct $C, D \in \mathcal{B}_0$. But then $\mathcal{A}_0 = \{C \cup \{a\} : C \in \mathcal{B}_0\} \subset \mathcal{A}$ is uncountable and $X \cap Y = B \cup \{a\}$ for every distinct $X, Y \in \mathcal{A}_0$.

Case 2: For every $A \in \mathcal{A}$ the set

$$S_A = \{B \in \mathcal{A} : A \cap B \neq \emptyset\}$$

is at most countable. Construct, by transfinite induction, a one-to-one sequence $\langle A_\xi : \xi < \omega_1 \rangle$ of pairwise-disjoint subsets of $\mathcal{A}$. This can be done since for every $\xi < \omega_1$ the set

$$\{B \in \mathcal{A} : B \cap A_\zeta \neq \emptyset \text{ for some } \zeta < \xi\} = \bigcup_{\zeta < \xi} S_{A_\zeta}$$

is at most countable. Then $\mathcal{A}_0 = \{A_\xi : \xi < \omega_1\}$ and $A = \emptyset$ satisfy the desired requirements. $\square$

In the next theorem, due to Galvin (1980), for sets A and B we will use the notation

$$A \otimes B = \{\{a, b\} : a \in A \ \& \ b \in B\}.$$

Theorem 8.2.12 *If CH holds then there are two ccc forcings such that their product is not ccc.*

Proof The forcings will be constructed as follows. We will construct a coloring function $f\colon [\omega_1]^2 \to 2$. Then, for $i < 2$, we put $K_i = f^{-1}(i)$ and define

$$\mathbb{P}_i = \{F \in [\omega_1]^{<\omega} \colon [F]^2 \subset K_i\}$$

ordered by reverse inclusion $\supset$.

First we notice that the forcing $\mathbb{P}_0 \times \mathbb{P}_1$ is not ccc, since the family $\{\langle\{\xi\}, \{\xi\}\rangle \in \mathbb{P}_0 \times \mathbb{P}_1 \colon \xi < \omega_1\}$ forms an antichain. This is the case because compatibility of $\langle\{\zeta\}, \{\zeta\}\rangle$ and $\langle\{\xi\}, \{\xi\}\rangle$ implies $\langle\{\zeta, \xi\}, \{\zeta, \xi\}\rangle \in \mathbb{P}_0 \times \mathbb{P}_1$, and this is impossible since $\mathbb{P}_0 \cap \mathbb{P}_1 = [\omega_1]^{\leq 1}$ is disjoint from $[\omega_1]^2$.

Thus it is enough to construct an f such that the forcings $\mathbb{P}_i$ are ccc. We will first translate the ccc property of $\mathbb{P}_i$ into a condition that will be appropriate for the inductive construction of f.

So fix $i < 2$ and let $\langle F_\xi \in \mathbb{P}_i \colon \xi < \omega_1\rangle$. To prove that $\mathbb{P}_i$ is ccc we will have to find $\zeta < \xi < \omega_1$ such that F_ζ and F_ξ are compatible, that is, that

$$[F_\zeta \cup F_\xi]^2 \subset K_i.$$

First notice that by the Δ-system lemma we can assume, choosing a subsequence, if necessary, that for some $F \in [\omega_1]^{<\omega}$

$$F_\zeta \cap F_\xi = F \quad \text{for all } \zeta < \xi < \omega_1.$$

But $[F_\zeta \cup F_\xi]^2 = [F_\zeta]^2 \cup [F_\xi]^2 \cup (F_\zeta \setminus F) \otimes (F_\xi \setminus F)$ and $[F_\zeta]^2 \cup [F_\xi]^2 \subset K_i$. Thus we must find $\zeta < \xi < \omega_1$ such that

$$(F_\zeta \setminus F) \otimes (F_\xi \setminus F) \subset K_i.$$

Replacing F_ξ with $F_\xi \setminus F$, we notice that it is enough to prove that for every sequence $\langle F_\xi \in \mathbb{P}_i \colon \xi < \omega_1\rangle$ of pairwise-disjoint nonempty sets there are $\zeta < \xi < \omega_1$ such that

$$F_\zeta \otimes F_\xi \subset K_i.$$

Moreover, for every $\alpha < \omega_1$ such that $\bigcup_{n<\omega} F_n \subset \alpha$ there exists a $\xi < \omega_1$ with $F = F_\xi \subset \omega_1 \setminus \alpha$. Thus we can reduce our task by showing that for every sequence $\langle F_n \in \mathbb{P}_i \colon n < \omega\rangle$ of pairwise-disjoint nonempty sets there exists an $\alpha < \omega_1$ with $\bigcup_{n<\omega} F_n \subset \alpha$ such that for every $F \in [\omega_1 \setminus \alpha]^{<\omega}$

$$\exists n < \omega \, (F_n \otimes F \subset K_i),$$

that is, that

$$\exists n < \omega \, \forall \beta \in F_n \, \forall \gamma \in F \, (f(\{\beta, \gamma\}) = i). \tag{8.21}$$

Now, by induction on $\xi < \omega_1$, we will construct an increasing sequence of partial functions $f\colon [\xi]^2 \to 2$ such that the entire function $f\colon [\omega_1]^2 \to 2$ will satisfy condition (8.21).[2] For this, let $\langle\langle F_n^\xi\rangle_{n<\omega}\colon \xi < \omega_1\rangle$ be an enumeration of all sequences $\langle F_n\rangle_{n<\omega}$ of pairwise-disjoint nonempty finite subsets of ω_1 such that each sequence appears in the list ω_1 times. Such an enumeration can be chosen by CH, since the family of all such sequences has cardinality $\leq \left|([\omega_1]^{<\omega})^\omega\right| = \mathfrak{c}$. The construction will be done while maintaining the following inductive condition for every $\xi < \omega_1$:

(I_ξ) For every $i < 2$, $\alpha < \xi$, and $F \in [\xi \setminus \alpha]^{<\omega}$, if $\bigcup_{n<\omega} F_n^\alpha \subset \alpha$ then the set

$$\mathcal{E}(i, \alpha, F) = \{F_n^\alpha : \forall \beta \in F_n^\alpha \; \forall \gamma \in F \; (f(\{\beta, \gamma\}) = i)\} \qquad (8.22)$$

is infinite.

Notice that this will finish the proof, since then f will satisfy (8.21) for every $i < 2$ and every sequence $\langle F_n\rangle_{n<\omega}$ of pairwise-disjoint nonempty finite subsets of ω_1. To see this, choose $\alpha < \omega_1$ such that $\bigcup_{n<\omega} F_n \subset \alpha$ and $\langle F_n^\alpha\rangle_{n<\omega} = \langle F_n\rangle_{n<\omega}$, and for every $F \in [\omega_1 \setminus \alpha]^{<\omega}$ find a $\xi < \omega_1$ such that $F \in [\xi \setminus \alpha]^{<\omega}$. Then, by (I_ξ), there exists an $n < \omega$ such that $F_n^\alpha \in \mathcal{E}(i, \alpha, F)$. This n satisfies (8.21).

To make an inductive step, let $\eta < \omega_1$ be such that the construction is already made for all $\xi < \eta$.

If η is a limit ordinal, then $f\colon [\eta]^2 \to 2$ is already constructed and it is easy to see that f satisfies (I_η). So assume that $\eta = \xi + 1$. We have to extend f to $\{\{\beta, \xi\}\colon \beta < \xi\}$ while maintaining (I_η). So let

$$\mathcal{F} = \left\{\langle\mathcal{E}(i, \alpha, F), i\rangle\colon i < 2, \; \alpha \leq \xi, \; F \in [\xi \setminus \alpha]^{<\omega}, \; \bigcup_{n<\omega} F_n^\alpha \subset \alpha\right\}.$$

Clearly $\mathcal{F}$ is countable, being indexed by a countable set. Note also that if $\langle\mathcal{E}(i, \alpha, F), i\rangle \in \mathcal{F}$ then $\mathcal{E}(i, \alpha, F)$ is infinite. For $\alpha < \xi$ this follows directly from the inductive assumption (I_ξ). But if $\alpha = \xi$ then $F = \emptyset$ and $\mathcal{E}(i, \alpha, F) = \mathcal{E}(i, \alpha, \emptyset) = \{F_n^\alpha : n < \omega\}$ is infinite as well.

Let $\langle\langle\mathcal{E}_m, i_m\rangle\colon m < \omega\rangle$ be an enumeration of $\mathcal{F}$ with each pair appearing infinitely many times. Since each $\mathcal{E}_m$ is an infinite family of pairwise-disjoint finite sets, we can construct by induction on $m < \omega$ a sequence $\langle E_m : n < \omega\rangle$ of pairwise-disjoint sets such that $E_m \in \mathcal{E}_m$ for every $m < \omega$. Define $f(\{\beta, \xi\}) = i_m$ for every $\beta \in E_m$ and $m < \omega$, and extend it arbitrarily to $[\eta]^2$. It is enough to show that f satisfies condition (I_η).

So let $i < 2$, $\alpha < \eta = \xi + 1$, and $\hat{F} \in [\eta \setminus \alpha]^{<\omega}$ be such that $\bigcup_{n<\omega} F_n^\alpha \subset \alpha$. If $\hat{F} = \emptyset$ then $\mathcal{E}(i, \alpha, \hat{F}) = \{F_n^\alpha : n < \omega\}$ is infinite, and (8.22) holds. So

---

[2] Formally we are defining an increasing sequence of functions $f_\xi\colon [\xi]^2 \to 2$, aiming for f to be their union. But such an additional index would only obscure a clear idea.

assume that $\hat{F} \neq \emptyset$. Now, if $\xi \notin \hat{F}$ then (8.22) holds for $\hat{F}$ by (I_ξ), since $\emptyset \neq \hat{F} \subset \xi \setminus \alpha$ implies $\alpha < \xi$. So assume that $\xi \in \hat{F}$ and let $F = \hat{F} \setminus \{\xi\}$. Then $\langle \mathcal{E}(i, \alpha, F), i \rangle \in \mathcal{F}$ and $\langle \mathcal{E}(i, \alpha, F), i \rangle = \langle \mathcal{E}_m, i_m \rangle$ for infinitely many $m < \omega$. In particular, for every such m we have

$$f(\{\beta, \xi\}) = i_m = i \tag{8.23}$$

for every $\beta \in E_m$. We claim that every such E_m belongs to $\mathcal{E}(i, \alpha, \hat{F})$. This will finish the proof, since all sets E_m are different. But for every $\beta \in E_m$

$$f(\{\beta, \gamma\}) = i$$

holds for $\gamma \in F = \hat{F} \setminus \{\xi\}$, since $E_m \in \mathcal{E}_m = \mathcal{E}(i, \alpha, F)$, and for $\gamma = \xi$ by (8.23). This finishes the proof. $\qquad\square$

EXERCISES

1 Let X and Y be nonempty sets. Show that the forcing $\langle \mathrm{Func}_\omega(X, Y), \supset \rangle$ is ccc if and only if $|Y| \leq \omega$. *Hint:* Use the Δ-system lemma.

2 A subset Z of $\mathbb{R}$ has *strong measure zero* if for every sequence $\langle \varepsilon_n \colon n < \omega \rangle$ of positive numbers there exists a sequence $\langle J_n \colon n < \omega \rangle$ of open intervals such that each J_n has length less than ε_n and $Z \subset \bigcup_{n<\omega} J_n$. Assuming MA show that every $Z \in [\mathbb{R}]^{<\mathfrak{c}}$ has strong measure zero. *Hint:* Let $\mathcal{B}$ be the family of all open intervals with rational endpoints. For every $\langle \varepsilon_n \colon n < \omega \rangle$ of positive numbers use forcing

$$\mathbb{P} = \{J \in \mathcal{B}^n \colon n < \omega \ \& \ \text{for every } k < n \text{ the length of } J(k) \text{ is less than } \varepsilon_k\}$$

ordered by reverse inclusion $\supset$.

3 (Due to Solovay) Infinite sets A and B are said to be *almost disjoint* provided $A \cap B$ is finite. Show that MA implies the following fact: If $\mathcal{A}$ is a family of almost-disjoint subsets of ω, $|\mathcal{A}| < \mathfrak{c}$, and $\mathcal{C} \subset \mathcal{A}$, then there exists a set $S \subset \omega$ such that $S \cap C$ is finite for every $C \in \mathcal{C}$ and $S \cap A$ is infinite for $A \in \mathcal{A} \setminus \mathcal{C}$. *Hint:* Use a forcing $\mathbb{P} = [\omega]^{<\omega} \times [\mathcal{C}]^{<\omega}$ ordered by $\langle s, \mathcal{E} \rangle \leq \langle s', \mathcal{E}' \rangle$ if and only if $s \supset s'$, $\mathcal{E} \supset \mathcal{E}'$, and $(s \setminus s') \cap \bigcup \mathcal{E}' = \emptyset$.

4 We say that a set A is *almost contained* in B and write $A \subset^* B$ if $A \setminus B$ is finite. Let $\mathcal{B} \subset [\omega]^\omega$ be such that $\bigcap \mathcal{B}_0$ is infinite for every finite $\mathcal{B}_0 \subset \mathcal{B}$. If MA holds and $|\mathcal{B}| < \mathfrak{c}$ show that there exists an $A \in [\omega]^\omega$ such that $A \subset^* B$ for every $B \in \mathcal{B}$.

5 Assume MA and let $\mathcal{B}$ be a family of almost-disjoint subsets of ω such that $|\mathcal{B}| < \mathfrak{c}$. Show that for every $\mathcal{A} \in [\mathcal{B}]^{\leq \omega}$ there exists a $d \subset \omega$ such that $a \subset^\star d$ for every $a \in \mathcal{A}$ and $d \cap b$ is finite for every $b \in \mathcal{B} \setminus \mathcal{A}$. *Hint:* Let $\mathcal{A}^\star = \{a \setminus n \colon n < \omega \ \& \ a \in \mathcal{A}\}$ and define $\mathbb{P} = [\mathcal{A}^\star]^{<\omega} \times [\mathcal{B} \setminus \mathcal{A}]^{<\omega}$, ordered by $\langle A, B \rangle \leq \langle A', B' \rangle$ if and only if $A \supset A'$, $B \supset B'$, and $\bigcup(A \setminus A') \cap \bigcup B' = \emptyset$. Use the forcing $\mathbb{P}$ to define the set d as the union of all sets $\bigcup A$ with $\langle A, B \rangle$ from an appropriate generic filter in $\mathbb{P}$.

6 (Challenging) Assume MA and let $\kappa < \mathfrak{c}$ be an uncountable regular cardinal. Show that for every family $\mathcal{A}$ of countable subsets of κ such that $|\mathcal{A}| < \mathfrak{c}$ there exists a $B \in [\kappa]^\kappa$ such that $A \cap B$ is finite for every $A \in \mathcal{A}$. *Hint:* Use a forcing similar to that from Exercise 3. Use the Δ-system lemma to prove that it is ccc.

7 (Challenging) Generalize Exercise 6 as follows. Assume MA and let $\kappa < \mathfrak{c}$ be an uncountable regular cardinal. Show that for every family $\mathcal{A}$ of countable subsets of κ such that $|\mathcal{A}| < \mathfrak{c}$ there exists a countable cover $\{P_n \colon n < \omega\}$ of κ such that $A \cap P_n$ is finite for every $A \in \mathcal{A}$ and $n < \omega$. *Hint:* If $\langle \mathbb{P}, \preceq \rangle$ is the partially ordered set used in Exercise 6, use the forcing $\mathbb{P}^\star = \mathrm{Func}_\omega(\omega, \mathbb{P})$ ordered by $s \leq t$ if $\mathrm{dom}(s) \supset \mathrm{dom}(t)$ and $s(n) \preceq t(n)$ for every $n \in \mathrm{dom}(t)$.

8.3   Suslin hypothesis and diamond principle

Let $\langle X, \leq \rangle$ be a linear order. For $a, b \in X$, an open interval with endpoints a and b is defined in a natural way as $(a, b) = \{x \in X \colon a < x < b\}$. A linearly ordered set X is *ccc* provided every family of pairwise-disjoint open intervals in X is at most countable or, equivalently, when the partial order $\langle \tau \setminus \{\emptyset\}, \subset \rangle$ is ccc, where τ is an *order topology* on X, that is, the topology generated by the family of all open intervals in X.

In this terminology, Theorem 6.2.1(i) says that $\langle \mathbb{R}, \leq \rangle$ is ccc.

Now let us turn our attention to Theorem 8.1.4. It says that a linearly ordered set $\langle X, \leq \rangle$ is isomorphic to $\langle \mathbb{R}, \leq \rangle$ if and only if it has the following properties.

(a) X has neither a first nor a last element, and is dense in itself.

(b) X is complete.

(c) X contains a countable dense subset.

In 1920 Suslin asked whether, in the preceding characterization, condition (c) can be replaced by the weaker condition

(c$'$) X is ccc.

The *Suslin hypothesis SH* is the statement that such a replacement indeed can be made, that is, that any ccc linearly ordered set $\langle X, \leq \rangle$ satisfying (a) and (b) contains a countable dense subset.

In what follows we will use the term *Suslin line* for a ccc linearly ordered set $\langle X, \leq \rangle$ satisfying (a) that does not contain a countable dense subset. Thus a complete Suslin line is a counterexample for the Suslin hypothesis. In Exercise 1 we sketch the proof that the existence of a Suslin line implies the existence of a complete Suslin line. Thus the Suslin hypothesis is equivalent to the nonexistence of a Suslin line.[3]

In what follows we will show that the Suslin hypothesis is independent of the ZFC axioms. We start with the following theorem.

Theorem 8.3.1 *MA+¬CH implies the Suslin hypothesis.*

Proof Assume MA+¬CH and, to obtain a contradiction, that there exists a Suslin line $\langle X, \leq \rangle$. Then the partially ordered set $\langle \mathbb{P}, \leq \rangle = \langle \mathcal{J}, \subset \rangle$ is ccc, where $\mathcal{J}$ is the family of all nonempty open intervals in $\langle X, \leq \rangle$. We will show that the product forcing $\mathbb{P} \times \mathbb{P}$ is not ccc, contradicting Corollary 8.2.10.

For this, we construct by induction a sequence $\langle \langle a_\xi, b_\xi, c_\xi \rangle \in X^3 : \xi < \omega_1 \rangle$ such that for every $\xi < \omega_1$

(i) $a_\xi < b_\xi < c_\xi$, and

(ii) $(a_\xi, c_\xi) \cap \{b_\zeta : \zeta < \xi\} = \emptyset$.

Such a construction can be easily made, since for every $\xi < \omega_1$ the countable set $\{b_\zeta : \zeta < \xi\}$ cannot be dense, so there must exist $a_\xi < c_\xi$ in X such that $(a_\xi, c_\xi) \cap \{b_\zeta : \zeta < \xi\} = \emptyset$. Thus such an element b_ξ can be chosen, since X is dense in itself.

Now let $U_\xi = \langle (a_\xi, b_\xi), (b_\xi, c_\xi) \rangle \in \mathbb{P} \times \mathbb{P}$. It is enough to show that U_ζ and U_ξ are incompatible in $\mathbb{P} \times \mathbb{P}$ for every $\zeta < \xi < \omega_1$. This is the case since by (ii) either $b_\zeta \leq a_\xi$, in which case $(a_\zeta, b_\zeta) \cap (a_\xi, b_\xi) = \emptyset$, or $b_\zeta \geq c_\xi$, in which case $(b_\zeta, c_\zeta) \cap (b_\xi, c_\xi) = \emptyset$. $\quad\square$

The remainder of this section is devoted to the proof that the existence of a Suslin line is consistent with ZFC.

Notice that in the proof of Theorem 8.3.1 we used MA+¬CH only to conclude that the product of ccc forcings is ccc, from which we deduced

---

[3] In the literature it is probably more common to find the term "Suslin line" used for a ccc linearly ordered set that, considered as a topological space with the order topology, does not have a countable dense subset. However, every such ordering can be extended to a Suslin line in the sense defined here (see Exercise 2). Thus, independently of which meaning of Suslin line is used, the Suslin hypothesis is equivalent to the nonexistence of a Suslin line.

that there is no Suslin line. Moreover, from Theorem 8.2.12 we know that
CH implies the existence of ccc forcings whose product is not ccc. Can we
generalize this argument to construct a Suslin line $\langle X, \leq \rangle$ under CH?

In what follows we will describe some difficulties that await any attempted inductive construction of a Suslin line. Then we will show the
way to overcome these problems. In particular, the next lemma will be
used to achieve both of these goals. It will illustrate the aforementioned
difficulties, and will be used in our construction of a Suslin line. To state
the lemma, we need the following notation, which will be used for the
remainder of this section.

For $\alpha \leq \omega_1$ the symbol λ_α will denote the αth limit ordinal number
in $\omega_1 + 1$. Thus $\{\lambda_\alpha \colon \alpha \leq \omega_1\}$ is a one-to-one increasing enumeration of
the set of all limit ordinals in $\omega_1 + 1$. In particular, $\lambda_0 = 0$, $\lambda_1 = \omega$,
$\lambda_2 = \omega + \omega, \dots$, and

$$\lambda_{\omega_1} = \omega_1 \ \text{ and } \ \lambda_\alpha = \bigcup_{\beta < \alpha} \lambda_\beta \ \text{ for every limit ordinal } \ \alpha \leq \omega_1.$$

Also, for $\alpha < \omega_1$ let $\mathbb{Q}_\alpha = \{\lambda_\alpha + n \colon n < \omega\}$ and let $\leq_\alpha$ be a linear-order
relation on $\mathbb{Q}_\alpha$ giving it the order type of $\langle \mathbb{Q}, \leq \rangle$. Notice that $\lambda_{\alpha+1} =
\lambda_\alpha \cup \mathbb{Q}_\alpha$ for every $\alpha < \omega_1$ and that

$$\alpha \leq \lambda_\alpha = \bigcup \{\mathbb{Q}_\zeta \colon \zeta < \alpha\} \ \text{ for every } \ \alpha \leq \omega_1.$$

Lemma 8.3.2 *Let* $\alpha \leq \omega_1$. *If* $\langle S_\beta \subset \lambda_\beta \colon \beta < \alpha \rangle$ *and* $\langle \preceq_\beta \subset \lambda_\beta \times \lambda_\beta \colon \beta \leq \alpha \rangle$
are such that for every $\beta < \gamma \leq \alpha$

(1) $\preceq_\beta$ *is a linear-order relation on* λ_β;

(2) S_β *is a proper initial segment of* $\langle \lambda_\beta, \preceq_\beta \rangle$;

(3) $\preceq_\gamma = \bigcup_{\delta < \gamma} (\preceq_\delta)$ *if* γ *is a limit ordinal;*

(4)   *if* $\gamma = \xi + 1$ *then the relation* $\preceq_\gamma$ *on* $\lambda_\gamma = \lambda_\xi \cup \mathbb{Q}_\xi$ *is defined from* $\preceq_\xi$
      by "sticking" the entire $\langle \mathbb{Q}_\xi, \leq_\xi \rangle$ *between* S_ξ *and* $\lambda_\xi \setminus S_\xi$; *that is,* $\preceq_\gamma$
      extends $\preceq_\xi \cup \leq_\xi$ *and is defined for every* $\langle x, q \rangle \in \lambda_\xi \times \mathbb{Q}_\xi$ *by:* $x \preceq_\gamma q$
      if $x \in S_\xi$, *and* $q \preceq_\gamma x$ *if* $x \in \lambda_\xi \setminus S_\xi$;

then for every $\gamma \leq \alpha$

(I) $\langle \lambda_\gamma, \preceq_\gamma \rangle$ *is a linearly ordered set that is dense in itself and has neither
      a largest nor a smallest element.*

Moreover, if $\alpha = \omega_1$ *then* $\langle \omega_1, \preceq_{\omega_1} \rangle$ *does not have a countable dense subset.*

Proof Condition (I) is easily proved by induction on $\alpha \leq \omega_1$. Its proof is left as an exercise.

To see that $\langle \omega_1, \preceq_{\omega_1} \rangle$ does not have a countable dense subset take a countable set $D \subset \omega_1$. Then there is a $\xi < \omega_1$ such that $D \subset \lambda_\xi$. Choose $a, b \in \mathbb{Q}_\xi$ such that $a \prec_\xi b$. Then, by (4), there is no $d \in D$ such that $a \prec_{\omega_1} d \prec_{\omega_1} b$. So D is not dense. $\qquad\square$

The linear order $\langle \omega_1, \preceq_{\omega_1} \rangle$ constructed as in Lemma 8.3.2 has already all the properties of a Suslin line, except possibly that of being ccc. Thus the question is, Can we make it ccc?

Clearly, we have the freedom of choice of initial segments S_α. Such a choice can "kill" a potential uncountable antichain $\langle (a_\zeta, b_\zeta) \colon \zeta < \omega_1 \rangle$ by making sure that, starting from some $\alpha < \omega_1$, all $\mathbb{Q}_\xi$ for $\xi \geq \alpha$ will be placed inside some (a_ζ, b_ζ) for $\zeta < \alpha$. This would imply that there are only countably many points of ω_1 outside $\bigcup_{\zeta < \alpha} (a_\zeta, b_\zeta)$, making it impossible for $\langle (a_\zeta, b_\zeta) \colon \zeta < \omega_1 \rangle$ to be an antichain. There is, however, a problem in carrying out such a construction using only CH, since we have an induction of only length ω_1, and we have to "kill" 2^{ω_1} potential antichains $\langle (a_\zeta, b_\zeta) \colon \zeta < \omega_1 \rangle$. Thus once more we face the problem of having a "too short induction." Moreover, Theorem 8.3.1 shows that Martin's axiom can't help us this time. Someone may still have a hope that, through some trick, we can reduce the number of steps necessary to rescue this construction. Indeed, this is what we will do. However, such a "trick" can't be found with the help of CH alone, as it is known that the existence of a Suslin line cannot be concluded just from the continuum hypothesis. Thus we will need another "magic" axiom, which will show us the way out of our dilemma. To formulate it, we need some new definitions.

A subset C of ω_1 is *closed* if for every $S \subset C$ its union $\bigcup S$ is in $C \cup \{\omega_1\}$ or, equivalently, when every limit ordinal $\lambda < \omega_1$ is in C provided $C \cap \lambda$ is unbounded in λ. (In fact, being closed in ω_1 is equivalent to being closed in the order topology of ω_1.) For example, a nonzero ordinal number $\alpha < \omega_1$ is closed in ω_1 (as a subset) if and only if it is a successor ordinal.

In what follows we will be primarily interested in closed *unbounded* subsets of ω_1 (see Section 5.3). Notice that these sets are quite big, as follows from the next proposition.

Proposition 8.3.3 *If C and D are closed, unbounded subsets of ω_1 then $C \cap D$ is also closed and unbounded.*

Proof Clearly, $C \cap D$ is closed, since for every $S \subset C \cap D$ its union $\bigcup S$ is both in $C \cup \{\omega_1\}$ and in $D \cup \{\omega_1\}$.

To see that $C \cap D$ is unbounded in ω_1, pick $\alpha < \omega_1$ and define by induction the sequences $\langle \gamma_n \in C \colon n < \omega \rangle$ and $\langle \delta_n \in D \colon n < \omega \rangle$ such that $\alpha < \gamma_n < \delta_n < \gamma_{n+1}$ for every $n < \omega$. It can be done since sets C and

D are unbounded in ω_1. Then $\beta = \bigcup_{n<\omega} \gamma_n = \bigcup_{n<\omega} \delta_n < \omega_1$ belongs to $C \cap D$ and is greater than α. $\qquad\square$

One of the most important facts we will use about the closed unbounded sets is the following proposition.

Proposition 8.3.4 *Let $\mathcal{F}$ be a countable family of functions of the form $f \colon \omega_1^n \to \omega_1$ or $f \colon \omega_1^n \to [\omega_1]^{\leq \omega}$ with $n < \omega$. If*

$$D = \{\alpha < \omega_1 \colon \alpha \text{ is closed under the action of } \mathcal{F}\}$$

then D is closed and unbounded.

Proof It is easy to see that D is closed (compare the proof of Lemma 6.1.6). To see that it is unbounded pick $\beta < \omega_1$ and let $i \colon \omega_1 \to [\omega_1]^{\leq \omega}$ be given by $i(\alpha) = \alpha = \{\beta \colon \beta < \alpha\} \in [\omega_1]^{\leq \omega}$. Then, by Lemma 6.1.6(a) used with $Z = \beta$, there is a countable subset Y of ω_1 that is closed under the action of $\mathcal{F} \cup \{i\}$. In particular, closure under the action of $\{i\}$ implies that Y is an ordinal number. So Y is a countable ordinal number containing β and belonging to D. $\qquad\square$

A subset S of ω_1 is *stationary* if $S \cap C \neq \emptyset$ for every closed unbounded set $C \subset \omega_1$. The theory of stationary sets is one of the most beautiful and useful parts of set theory. However, due to lack of space, we will not be able to develop it here in full. Some traces of its power can be found in the rest of this section and in the exercises.

The additional axiom we will need is called the *diamond principle* and is usually denoted by the symbol $\Diamond$. It reads as follows.

Diamond principle $\Diamond$ There exists a sequence $\langle A_\alpha \subset \omega_1 \colon \alpha < \omega_1 \rangle$, known as a $\Diamond$-*sequence*, such that for every $A \subset \omega_1$ the set

$$\{\alpha < \omega_1 \colon A \cap \alpha = A_\alpha\}$$

is stationary.

Notice that

Proposition 8.3.5 $\Diamond$ *implies CH.*

Proof Let $A \subset \omega$. Since $C = \omega_1 \setminus \omega$ is closed and unbounded, there exists an $\alpha \in C$ such that $A_\alpha = A \cap \alpha = A \cap \omega = A$. Thus $\{A_\alpha \cap \omega \colon \alpha < \omega_1\} = \mathcal{P}(\omega)$. $\qquad\square$

In the next chapter we will show that $\Diamond$ is consistent with the ZFC axioms. It is also true that it does not follow from CH. Thus we have the following implications, none of which can be reversed:

$$\Diamond \Rightarrow CH \Rightarrow MA.$$

As we have just seen a $\Diamond$-sequence lists all subsets of ω. But it also captures a lot of information regarding uncountable subsets of ω_1. This will be enough to overcome the obstacles to the construction of a Suslin line described earlier. For the construction we need the following easy lemma.

Lemma 8.3.6 *If $\Diamond$ holds then there exists a sequence $\langle B_\alpha \subset \alpha \times \alpha \colon \alpha < \omega_1 \rangle$ such that for every $B \subset \omega_1 \times \omega_1$ the set*

$$\{\alpha < \omega_1 \colon B \cap (\alpha \times \alpha) = B_\alpha\}$$

is stationary.

Proof Let $f_0, f_1 \colon \omega_1 \to \omega_1$ be such that $f \colon \omega_1 \to \omega_1^2$ given by $f(\xi) = \langle f_0(\xi), f_1(\xi) \rangle$ is a bijection and let $B_\alpha = f[A_\alpha] \cap (\alpha \times \alpha)$. Notice that, by Proposition 8.3.4 used with the family $\mathcal{F} = \{f_0, f_1, f^{-1}\}$, the set $D = \{\alpha \colon f[\alpha] = \alpha \times \alpha\}$ is closed and unbounded.

To see that the sequence $\langle B_\alpha \colon \alpha < \omega_1 \rangle$ has the desired properties, take $B \subset \omega_1 \times \omega_1$ and let $C \subset \omega_1$ be closed and unbounded. We must find an $\alpha \in C$ such that $B_\alpha = B \cap (\alpha \times \alpha)$.

Since $C \cap D$ is closed and unbounded and $\{\alpha < \omega_1 \colon f^{-1}(B) \cap \alpha = A_\alpha\}$ is stationary, there is an $\alpha \in C \cap D$ such that $A_\alpha = f^{-1}(B) \cap \alpha$. But $\alpha \in D$ implies that

$$A_\alpha = f^{-1}(B) \cap \alpha = f^{-1}(B) \cap f^{-1}(\alpha \times \alpha) = f^{-1}(B \cap (\alpha \times \alpha)).$$

Hence $B_\alpha = f[A_\alpha] \cap (\alpha \times \alpha) = B \cap (\alpha \times \alpha)$. $\qquad\square$

Theorem 8.3.7 *If $\Diamond$ holds then there exists a Suslin line $\langle X, \preceq \rangle$. In particular, the Suslin hypothesis fails.*

Proof We will construct $\langle X, \preceq \rangle = \langle \omega_1, \preceq_{\omega_1} \rangle$ as in Lemma 8.3.2, following the idea described after its proof. The sequence $\langle B_\alpha \subset \alpha \times \alpha \colon \alpha < \omega_1 \rangle$ from Lemma 8.3.6 will be used as an "oracle" that will tell us how to choose initial segments S_α so as to "kill" *all* potential uncountable antichains.

Thus we will construct inductively the sequences $\langle S_\alpha \subset \lambda_\alpha \colon \alpha < \omega_1 \rangle$ and $\langle \preceq_\alpha \subset \lambda_\alpha \times \lambda_\alpha \colon \alpha \leq \omega_1 \rangle$ such that for every $\beta < \alpha \leq \omega_1$

(1) $\preceq_\alpha$ is a linear-order relation on λ_α;

(2) S_β is a proper initial segment of $\langle \lambda_\beta, \preceq_\beta \rangle$;

(3) $\preceq_\alpha = \bigcup_{\beta < \alpha} (\preceq_\beta)$ if α is a limit ordinal;

(4)   if $\alpha = \xi + 1$ then $\preceq_\alpha$ extends $\preceq_\xi \cup \leq_\xi$ and for every $\langle x, q \rangle \in \lambda_\xi \times \mathbb{Q}_\xi$ is defined by: $x \preceq_\alpha q$ if $x \in S_\xi$, and $q \preceq_\alpha x$ if $x \in \lambda_\xi \setminus S_\xi$.

Note that by Lemma 8.3.2 we know that such a construction can be made and that $\langle X, \preceq \rangle$ obtained this way is linearly ordered, is dense in itself, has neither a largest nor a smallest element, and does not have a countable dense subset. So it is enough to show that we can choose sets S_α such that $\langle X, \preceq \rangle$ will be ccc.

So assume that for some $\alpha < \omega_1$ the sequences $\langle S_\beta \subset \lambda_\beta \colon \beta < \alpha \rangle$ and $\langle \preceq_\beta \subset \lambda_\beta \times \lambda_\beta \colon \beta \leq \alpha \rangle$ are already constructed. We must construct the proper initial segment S_α of λ_α. It will be done as follows.

For $a, b \in \lambda_\alpha$ let $(a, b)_\alpha$ be an open interval in $\langle \lambda_\alpha, \preceq_\alpha \rangle$ with the endpoints a and b, that is,

$$(a, b)_\alpha = \{ x \in \lambda_\alpha \colon a \prec_\alpha x \prec_\alpha b \},$$

and for $\zeta \leq \alpha$ let

$$\mathcal{C}_\zeta^\alpha = \{ (a, b)_\alpha \colon \langle a, b \rangle \in B_\zeta \}.$$

Notice that $B_\zeta \subset \zeta \times \zeta \subset \lambda_\alpha \times \lambda_\alpha$, so $\mathcal{C}_\zeta^\alpha$ is well defined. Define the family $\mathcal{F}_\alpha$ as

$$\{ \mathcal{C}_\zeta^\alpha \colon \zeta \leq \alpha \text{ and } \mathcal{C}_\zeta^\alpha \text{ is a maximal family of disjoint intervals in } \langle \lambda_\alpha, \preceq_\alpha \rangle \}$$

and let $\{ \mathcal{E}_n \colon n < \omega \} = \mathcal{F}_\alpha \cup \{ \hat{\mathcal{C}} \}$, where $\hat{\mathcal{C}}$ is an arbitrary maximal family of pairwise-disjoint intervals in $\langle \lambda_\alpha, \preceq \rangle$. (We add $\hat{\mathcal{C}}$ to this list just to avoid a problem when $\mathcal{F}_\alpha = \emptyset$.)

The idea of the proof is that the elements of $\bigcup_{\alpha < \omega_1} \mathcal{F}_\alpha$ will be approximations of all the potential families of pairwise-disjoint intervals in $\langle X, \preceq \rangle$. We will make sure that every family $\mathcal{C} \in \mathcal{F}_\alpha$ remains maximal after step α by "sticking" every $\mathbb{Q}_\beta$ for $\alpha \leq \beta < \omega_1$ inside some interval $I \in \mathcal{C}$, thus adding no new elements of X outside $\bigcup \mathcal{C}$.

So define, by induction on $n < \omega$, a sequence $\langle \langle a_n, b_n \rangle \in \lambda_\alpha \times \lambda_\alpha \colon n < \omega \rangle$ such that for every $n < \omega$

(A) $a \prec_\alpha a_n \prec_\alpha b_n \prec_\alpha b$ for some $(a, b)_\alpha \in \mathcal{E}_n$; and

(B) $a_{n-1} \prec_\alpha a_n \prec_\alpha b_n \prec_\alpha b_{n-1}$ for $n > 0$.

Such a construction can be made since, by Lemma 8.3.2, $\langle \lambda_\alpha, \preceq_\alpha \rangle$ is dense in itself and, by the maximality of $\mathcal{E}_n$, the interval $(a_{n-1}, b_{n-1})_\alpha$ must intersect some $(a, b)_\alpha \in \mathcal{E}_n$. We define $S_\alpha = \{ x \in \lambda_\alpha \colon \exists n < \omega \ (x \preceq_\alpha a_n) \}$. This finishes the construction.

Now notice that if $\mathcal{C}_\zeta^\alpha \in \mathcal{F}_\alpha$ for some $\zeta \leq \alpha < \omega_1$, then there are $n < \omega$ and $a, b \in \lambda_\alpha$ such that $a \prec_\alpha a_n \prec_\alpha b_n \prec_\alpha b$, $(a, b)_\alpha \in \mathcal{C}_\zeta^\alpha$, and $a_n \in S_\alpha$ but $b_n \notin S_\alpha$. In particular, $\mathbb{Q}_\alpha \subset (a_n, b_n)_{\alpha+1} \subset \bigcup \mathcal{C}_\zeta^{\alpha+1}$ and $\mathcal{C}_\zeta^{\alpha+1}$ remains a maximal family of pairwise-disjoint intervals in $\langle \lambda_{\alpha+1}, \preceq_{\alpha+1} \rangle$. So $\mathcal{C}_\zeta^{\alpha+1} \in \mathcal{F}_{\alpha+1}$.

Using this, we can easily prove by induction on $\beta < \omega_1$ that for every $\alpha < \omega_1$

if $C_\alpha^\alpha \in \mathcal{F}_\alpha$ and $\alpha < \beta < \omega_1$ then $\displaystyle\bigcup_{\alpha \le \gamma < \beta} \mathbb{Q}_\gamma \subset \bigcup C_\alpha^\beta$ and $C_\alpha^\beta \in \mathcal{F}_\beta$.

This, in particular, implies that $\bigcup_{\alpha \le \gamma < \omega_1} \mathbb{Q}_\gamma \subset \bigcup C_\alpha^{\omega_1}$ provided $C_\alpha^\alpha \in \mathcal{F}_\alpha$, that is, that

$C_\alpha^{\omega_1}$ is a maximal family of disjoint intervals in $\langle X, \preceq \rangle$, if $C_\alpha^\alpha \in \mathcal{F}_\alpha$. (8.24)

To see that $\langle X, \preceq \rangle$ is ccc assume, to obtain a contradiction, that there is an uncountable family $\mathcal{A}_0$ of nonempty pairwise-disjoint intervals in $\langle X, \preceq \rangle$. Using Zorn's lemma we can extend $\mathcal{A}_0$ to a maximal family $\mathcal{A}$ of nonempty pairwise-disjoint intervals in $\langle X, \preceq \rangle$. Put $B = \{\langle a, b \rangle \colon (a, b)_{\omega_1} \in \mathcal{A}\}$. The maximality of $\mathcal{A}$ means that for every $a, b \in X$, with $a \prec b$, there exists a nonempty interval $(l, r)_{\omega_1} = (l(a, b), r(a, b))_{\omega_1} \in \mathcal{A}$ intersecting $(a, b)_{\omega_1}$. Let $l, r \colon \omega_1^2 \to \omega_1$ be functions having the foregoing property and let $\Lambda \colon \omega_1 \to \omega_1$ be given by $\Lambda(\xi) = \lambda_\xi + 1$. Put

$$D = \{\alpha < \omega_1 \colon \alpha \text{ is closed under the action of } \{\Lambda, l, r\}\}.$$

By Proposition 8.3.4, D is closed and unbounded in ω_1. Moreover, for every $\alpha \in D$,

$$\lambda_\alpha = \alpha, \tag{8.25}$$

and

$$\mathcal{D}_\alpha = \{(l, r)_\alpha \colon \langle l, r \rangle \in B \cap (\alpha \times \alpha)\} \tag{8.26}$$

is a maximal family of pairwise-disjoint intervals in $\langle \lambda_\alpha, \preceq_\alpha \rangle = \langle \alpha, \preceq_\alpha \rangle$.

Condition (8.25) follows from the fact that α is closed under the action of Λ. To see why, notice first that α is a limit ordinal, since $\xi + 1 \le \lambda_\xi + 1 = \Lambda(\xi) < \alpha$ for every $\xi < \alpha$. Now $\lambda_\alpha = \bigcup_{\xi < \alpha} \lambda_\xi \le \alpha \le \lambda_\alpha$.

Condition (8.26) follows from the closure of α under the actions of l and r, since for every $a, b \in \alpha = \lambda_\alpha$, with $a \prec_\alpha b$, there are $l = l(a, b) < \alpha$ and $r = r(a, b) < \alpha$ with $(l, r)_{\omega_1} \in \mathcal{A}$ intersecting $(a, b)_{\omega_1}$. This clearly implies that $(l, r)_\alpha \in \mathcal{D}_\alpha$ intersects $(a, b)_\alpha$, making $\mathcal{D}_\alpha$ maximal in $\langle \lambda_\alpha, \preceq_\alpha \rangle = \langle \alpha, \preceq_\alpha \rangle$.

By the choice of the sequence $\langle B_\alpha \subset \alpha \times \alpha \colon \alpha < \omega_1 \rangle$ there exists an $\alpha \in D$ such that $B \cap (\alpha \times \alpha) = B_\alpha$. Hence, by (8.25) and (8.26),

$$C_\alpha^\alpha = \{(l, r)_\alpha \colon \langle l, r \rangle \in B_\alpha\} = \{(l, r)_\alpha \colon \langle l, r \rangle \in B \cap (\alpha \times \alpha)\} = \mathcal{D}_\alpha$$

is a maximal family of pairwise-disjoint intervals in $\langle \alpha, \preceq_\alpha \rangle = \langle \lambda_\alpha, \preceq_\alpha \rangle$. Hence, by the definition of $\mathcal{F}_\alpha$, $C_\alpha^\alpha \in \mathcal{F}_\alpha$. So, by (8.24), $C_\alpha^{\omega_1}$ is a maximal antichain in $\langle X, \preceq \rangle$, contradicting the fact that $C_\alpha^{\omega_1}$ is a proper subset of an uncountable antichain $\mathcal{A}$. Thus we have proved that $\langle X, \preceq \rangle$ is ccc. $\qquad\square$

EXERCISES

1 Let $\langle X, \leq \rangle$ be a Suslin line and let X^* be the family of all proper nonempty initial segments of X without last elements. Prove that $\langle X^*, \subset \rangle$ is a complete Suslin line.

2 Let $\langle X, \leq \rangle$ be a ccc linearly ordered set and assume that X, considered as a topological space with the order topology, does not have a countable dense subset. Show that there exists an $X_0 \subset X$ such that $\langle X_0, \leq \rangle$ is a Suslin line and for every $a, b \in X_0$, with $a < b$, there is no countable dense subset of $(a, b) \cap X_0$. *Hint:* Define an equivalence relation $\sim$ on X by putting $x \sim y$ if and only if the interval between them contains a countable subset that is dense in it with respect to the order topology. Choose X_0 to be any selector from the family of all equivalence classes of $\sim$, from which the least and the greatest elements are removed, if they exist.

3 Complete the proof of Lemma 8.3.2 by proving condition (I).

4 Prove that if sets $C_n \subset \omega_1$ are closed and unbounded then $\bigcap_{n < \omega} C_n$ is closed and unbounded.

5 Let $\{C_\alpha \subset \omega_1 : \alpha < \omega_1\}$ be a family of closed unbounded sets. Prove that the set

$$D = \{\gamma < \omega_1 : \gamma \in C_\alpha \text{ for all } \alpha < \gamma\},$$

known as the *diagonal intersection* of the sets C_α, is closed and unbounded.

6 Prove the following theorem, known as the pressing-down lemma or Fodor's theorem:

> Let $S \subset \omega_1$ be a stationary set and let $f: S \to \omega_1$ be such that $f(\gamma) < \gamma$ for every $\gamma \in S$ (such a function is called a *regressive function*). Then there exists an $\alpha < \omega_1$ such that $f^{-1}(\{\alpha\})$ is stationary.

Hint: Otherwise for every α there exists a closed and unbounded set C_α disjoint from $f^{-1}(\{\alpha\})$. Consider the diagonal intersection D of the sets C_α and show that it is disjoint from S.

7 A *tree* is a partially ordered set $\langle T, \le \rangle$ with a smallest element such that the relation $<$ is well founded, that is, that every initial segment of T is well ordered. A *Suslin tree* is an uncountable tree that, considered as a forcing with the reversed order $\langle T, \le \rangle$, has neither uncountable chains nor uncountable antichains. Show the existence of a Suslin line implies the existence of a Suslin tree. *Hint:* Start with a Suslin line $\langle X_0, \le \rangle$ as in Exercise 2. Define, by induction on $\xi < \omega_1$, a sequence $\langle T_\xi : \xi < \omega_1 \rangle$ such that $T_0 = X_0$ and each T_ξ for $\xi > 0$ is a maximal family of pairwise-disjoint intervals in X_0, none of which contain any endpoint of any interval from $\bigcup_{\zeta < \xi} T_\zeta$. Then $T = \bigcup_{\xi < \omega_1} T_\xi$ ordered by reverse inclusion $\supset$ is a Suslin tree.

8 Show that the existence of a Suslin tree $\langle T, \le \rangle$ implies the existence of a Suslin line. (Thus the existence of a Suslin line is equivalent to the existence of a Suslin tree.) *Hint:* Let $\preceq$ be any linear order on T and let X be the family of all strictly increasing functions f from an ordinal number α into $\langle T, \le \rangle$ such that $f[\alpha]$ is a maximal chain in T (such an f is called a *branch* of T). For different $f, g \in X$ define $f < g$ if $f(\beta) \prec g(\beta)$, where β is the smallest ordinal for which $f(\beta) \ne g(\beta)$. Show that $\langle X, \le \rangle$ is a ccc linearly ordered set such that X, considered as a topological space with the order topology, does not have a countable dense subset. Then use Exercise 2.

9 For a tree $\langle T, \le \rangle$ and an ordinal number α, the *α-level* of T is the set of all $t \in T$ for which the initial segment generated by t has order type α. (Thus any level of T is an antichain.) An *Aronszajn tree* is an uncountable tree with no uncountable chains and all levels countable. (Thus every Suslin tree is an Aronszajn tree.) Show, in ZFC, that there exists an Aronszajn tree. *Hint:* For $f, g \colon \alpha \to \omega$ write $f =^\star g$ if the set $\{ \beta < \alpha \colon f(\beta) \ne g(\beta) \}$ is finite. Define, by induction on $\alpha < \omega_1$, a sequence of one-to-one functions $\langle s_\alpha \in \omega^\alpha \colon \alpha < \omega_1 \rangle$ such that $\omega \setminus \mathrm{range}(s_\alpha)$ is infinite and $s_\alpha|_\beta =^\star s_\beta$ for every $\beta < \alpha$. Define $T = \{ s \in \omega^\alpha \colon \alpha < \omega_1 \ \& \ s =^\star s_\alpha \}$ and relation $\le$ as reverse inclusion $\supset$.

Chapter 9

Forcing

In this chapter we will describe a technique for proving that some set-theoretic statements are independent of the ZFC axioms. This technique is known as the *forcing method*. We will not prove here all the theorems needed to justify this method. (Sketches of some of the missing proofs are included in Appendix B. The complete proofs can be found, for example, in Kunen (1980).) Instead, we will describe only the essentials for its use and concentrate on its applications.[1]

9.1 Elements of logic and other forcing preliminaries

We will start here with some definitions, which will serve as technical tools to develop the forcing method.

A set M is said to be *transitive* if $x \subset M$ for every $x \in M$, that is, if $a \in x$ and $x \in M$ imply that $a \in M$.

Lemma 9.1.1 *For every set x there exists a smallest transitive set* $\mathrm{trcl}(x)$ *such that $x \subset \mathrm{trcl}(x)$.*

Proof Define, by induction on $n < \omega$,

$$U_0(x) = x \quad \text{and} \quad U_{n+1}(x) = \bigcup U_n(x) \quad \text{for } n < \omega.$$

---

[1] The material included in Appendix B is not essential for the applicability of the forcing method and can be completely skipped. However, those interested in reading it should consider waiting at least until the end of this section, since the material included here should make Appendix B easier to follow.

Then $\mathrm{trcl}(x) = \bigcup_{n<\omega} U_n(x)$ is the desired set. It is transitive, since for every $y \in \mathrm{trcl}(x)$ there is an $n < \omega$ such that $y \in U_n(x)$. Then $y \subset \bigcup U_n(x) = U_{n+1}(x) \subset \mathrm{trcl}(x)$.

To prove the minimality of $\mathrm{trcl}(x)$ it is enough to take a transitive set M with $x \subset M$ and show, by induction on $n < \omega$, that $U_n \subset M$ for every $n < \omega$. The details are left as an exercise. $\qquad\square$

The set $\mathrm{trcl}(x)$ from Lemma 9.1.1 is called the *transitive closure* of x.

For an ordinal number α define inductively a sequence $\langle R(\beta) \colon \beta \leq \alpha \rangle$ by putting

(a) $R(0) = \emptyset$,

(b) $R(\beta + 1) = \mathcal{P}(R(\beta))$ for $\beta < \alpha$, and

(c) $R(\lambda) = \bigcup_{\beta < \lambda} R(\beta)$ for every limit ordinal $\lambda \leq \alpha$.

Note that the sequence $\langle R(\beta) \colon \beta \leq \alpha \rangle$ is increasing, that is, that

$$R(\beta) \subset R(\gamma) \text{ for every ordinal numbers } \gamma < \beta \leq \alpha. \qquad (9.1)$$

An easy inductive proof of this fact is left as an exercise.

Lemma 9.1.2 *For every set x there exists an ordinal number α such that $x \in R(\alpha)$.*

Proof First notice that for any set x,

($\star$) if for every $y \in x$ there exists an ordinal number α such that $y \in R(\alpha)$, then $x \in R(\alpha_0)$ for some ordinal α_0.

To see this, let r be a function such that $r(y) = \min\{\beta \colon y \in R(\beta + 1)\}$ for every $y \in x$. Such a function exists by the axiom of replacement. Let $\gamma = \bigcup_{y \in x} r(y)$. Then γ is an ordinal number and $x \subset R(\gamma)$. So $x \in \mathcal{P}(R(\gamma)) = R(\gamma + 1)$.

Now, to obtain a contradiction, assume that there exists an x such that $x \notin R(\alpha)$ for every ordinal number α. So, by ($\star$), there exists a $y \in x$ such that $y \notin R(\alpha)$ for every ordinal number α. In particular, the set

$$A = \{y \in \mathrm{trcl}(x) \colon y \notin R(\alpha) \text{ for every ordinal number } \alpha\}$$

is not empty. So, by the axiom of regularity, A contains an $\in$-minimal element, that is, there exists an $x_0 \in A$ such that $A \cap x_0 = \emptyset$. But $x_0 \in A \subset \mathrm{trcl}(x)$. Hence $x_0 \subset \mathrm{trcl}(x)$, since $\mathrm{trcl}(x)$ is transitive. Therefore x_0 satisfies the assumption of ($\star$) so $x_0 \in R(\alpha_0)$ for some ordinal number α_0. But this contradicts $x_0 \in A$. $\qquad\square$

For a set x define the *rank* of x by

$$\mathrm{rank}(x) = \min\{\beta\colon x \in R(\beta+1)\}.$$

Notice that by Lemma 9.1.2 $\mathrm{rank}(x)$ is defined for every set x.

Lemma 9.1.3

(a) $\mathrm{rank}(x) < \alpha$ *if and only if* $x \in R(\alpha)$.

(b)   *If* $x \in y$ *then* $\mathrm{rank}(x) < \mathrm{rank}(y)$.

Proof (a) This is obvious by (9.1) and the definition of $\mathrm{rank}(x)$.

(b) If $\beta = \mathrm{rank}(y)$ then $y \in R(\beta+1) = \mathcal{P}(R(\beta))$. So $x \in y \subset R(\beta)$. Therefore $\mathrm{rank}(x) < \beta = \mathrm{rank}(y)$. $\square$

Before we describe the forcing method, we will reexamine some notions discussed in Sections 1.1 and 1.2. This should give us a better understanding of what we mean when we say that a property is "independent of ZFC."

In what follows the term "formula" will always be understood as a formula of the language of set theory, that is, the language described in Section 1.2. In particular, the set of all formulas is defined *by induction on their length* from only two basic kinds of formula: "$x \in y$" and "$x = y$," where symbols x and y represent variables. A more complicated formula can be built from the less complicated formulas φ and ψ only by connecting them with logical connectors, $\varphi \& \psi$, $\varphi \lor \psi$, $\varphi \rightarrow \psi$, $\varphi \leftrightarrow \psi$, or by preceding one of them with the negation, $\neg \varphi$, or a quantifier, $\exists x \varphi$, $\forall x \varphi$. For example, "$x = y \ \& \ \exists x \ \forall y \ (x \in y)$" is a correct formula, whereas expressions such as "$x \in y \in z$" or "$z = t \lor c$" are not. (Although "$x \in y \in z$" could be interpreted as "$x \in y \ \& \ y \in z$," the latter being a correct formula.)

A *variable* in a formula is any symbol that represents a set. More precisely, it is any symbol that is neither of the following: "$\in$," "$=$," "$\neg$," "$\&$," "$\lor$," "$\rightarrow$," "$\leftrightarrow$," "$\exists$," "$\forall$," "$($," or "$)$." In a given formula not all variables must be alike. For example, in the formula φ defined as "$x = y \ \& \ \exists z \ (z = z)$" the variables x and y are parameters and φ might be either true or false, depending on what value we associate with x and y. On the other hand, the variable z lies within the scope of the quantifier "$\exists$" and is not a parameter.

The variables of a formula φ that are within the scope of a quantifier are said to be *bound* in φ. The parameters of a formula will be referred to as *free variables*. Notice that the same symbol in the same formula may be used both as a bound variable and as a free variable. For example, this is the case in the formula "$\exists x \ (x \in y) \ \& \ x = y$," where the first x is bound, while the second plays the role of a free variable. (Notice the importance of the distribution of parentheses.)

Formulas without free variables are called *sentences*. Since all variables in a sentence are bound, no interpretation of free variables is necessary to decide whether it is true or false. For this reason only sentences will be used as axioms. On the other hand, if $x_1, \ldots, x_n$ are free variables of a formula φ, we often write $\varphi(x_1, \ldots, x_n)$ instead of φ to emphasize its dependence on $x_1, \ldots, x_n$. In particular, we will usually write $\exists x \varphi(x)$ and $\forall x \varphi(x)$ in place of $\exists x \varphi$ and $\forall x \varphi$.

By a *theory* we will mean any set of sentences. The sentences belonging to a theory T will be treated as its axioms.[2] In particular, ZFC is a theory, and it consists of infinitely many axioms since each of the scheme axioms (comprehension or replacement) stands for infinitely many axioms.

We say that a sentence φ is a *consequence of a theory T* (or that φ can be *proved in a theory T*) and write it as $T \vdash \varphi$ if there is a formal proof of φ using as axioms only sentences from T. It is also equivalent to the fact that there exists a finite set $T_0 = \{\psi_0, \ldots, \psi_n\}$ of axioms from T such that $(\psi_0 \& \cdots \& \psi_n) \rightarrow \varphi$ is a consequence of axioms of logic. (See Appendix B for more details.)

A theory T is said to be *inconsistent* if there is a sentence φ such that $T \vdash (\varphi \& \neg \varphi)$, that is, if it leads to a contradiction. Equivalently, T is inconsistent if every sentence ψ is a consequence of T. (The equivalence follows from the fact that the implication $(\varphi \& \neg \varphi) \rightarrow \psi$ is true for all formulas φ and ψ.) Conversely, we say that a theory T is *consistent* and write $\mathrm{Con}(T)$ if T is not inconsistent, that is, if T does not imply a contradiction. A sentence "ψ" is *consistent with theory T* if $T + $ "ψ" is consistent.

Evidently, from the point of view that theories should carry useful information, only consistent ones are interesting. In particular, we will be assuming here that ZFC is consistent. (Recall that by the Gödel's second incompleteness theorem, Theorem 1.1.1, there is no hope of proving the consistency of ZFC in itself as long as it is really consistent.)

The relation between the notions of "being a consequence of a theory" and "being consistent with a theory" is best captured by the following fact, whose easy proof is left as an exercise:

$$T \nvdash \varphi \quad \text{if and only if} \quad \mathrm{Con}(T + \text{``}\neg\varphi\text{''}). \tag{9.2}$$

Recall also that a sentence φ is *independent of theory T* if neither $T \vdash \varphi$ nor $T \vdash \neg\varphi$ or, equivalently, when both theories $T + $ "φ" and $T + $ "$\neg\varphi$" are consistent. Thus Gödel's first incompleteness theorem, Theorem 1.1.2, implies that there are sentences of set theory that are independent of ZFC.

---

[2] We assume that theory contains only sentences, since it is more difficult to accept a formula with some free variables as an axiom. (Intuitively whether it is true or false can depend on the value of the free variables.) However, logically any formula φ is equivalent to a sentence $\forall x_1 \cdots \forall x_n \varphi$, where $\{x_1, \ldots, x_n\}$ is a list of all free variables in φ. So our restriction is not essential.

The forcing method will be a tool to show that a given sentence ψ is consistent with ZFC. So, in order to prove that a sentence ψ is independent of ZFC, we will be proving that the theories ZFC+"ψ" and ZFC+"$\neg\psi$" are both consistent.

EXERCISES

1 Complete the details of the proof of Lemma 9.1.1 by showing the minimality of $\mathrm{trcl}(x)$.

2 Prove (9.1).

3 Prove (9.2). *Hint:* Use the fact that $(T \cup \{\varphi\}) \vdash \psi$ if and only if $T \vdash (\varphi \rightarrow \psi)$.

9.2   Forcing method and a model for ¬CH

Consider a transitive set M. For every formula ψ of the language of set theory (without any shortcuts) we define a formula ψ^M, called its *relativization to M*. It is obtained by replacing in ψ each unbounded quantifier $\forall x$ or $\exists x$ with its bounded counterpart $\forall x \in M$ or $\exists x \in M$. For example, if ψ_0 is a sentence from the axiom of extensionality

$$\forall x\, \forall y\, [\forall z\, (z \in x \leftrightarrow z \in y) \rightarrow x = y]$$

then ψ_0^M stands for

$$\forall x \in M\, \forall y \in M\, [\forall z \in M\, (z \in x \leftrightarrow z \in y) \rightarrow x = y].$$

In particular, if $\psi(x_1, \dots, x_n)$ is a formula with free variables $x_1, \dots, x_n$ and $t_1, \dots, t_n \in M$ then $\psi^M(t_1, \dots, t_n)$ says that $\psi(t_1, \dots, t_n)$ is true under the interpretation that all variables under quantifiers are bound to M. In other words, $\psi^M(t_1, \dots, t_n)$ represents the formula $\psi(t_1, \dots, t_n)$ as seen by a "person living inside M," that is, thinking that M represents the entire class of all sets. (This is the best way to think of ψ^M.)

For a transitive set M and a formula ψ (with possible parameters from M) we say that "ψ *is true in M*" and write $M \models \psi$ if ψ^M is true. For a theory T we say that "T *is true in M*" or that "M *is a model for T*" if ψ^M holds for every ψ from T.

Note that a model for a theory T is just a set satisfying some properties.[3] Thus, if M is the empty set, then M is a model for the theory T consisting

---

[3] For those readers who have been exposed previously to any kind of model theory we give here a bit of an explanation. We do not need to interpret any constants of set theory in M, since the language of set theory does not contain any constants. And the only relations that we have to take care of are the relation symbols "$\in$" and "$=$," which are interpreted as the real relations of "being an element of" and "being equal to."

of all the axioms of ZFC except the set existence and the infinity axioms. This is so since any other axiom starts with a general quantifier $\forall x$, and its relativization to $M = \emptyset$ starts with $\forall x \in \emptyset$, that is, is satisfied vacuously. Similarly, if M is any transitive set, then the axiom of extensionality is satisfied in M, since for any set $x \in M$

$$z \in x \text{ if and only if } z \in M \ \& \ z \in x.$$

It is also relatively easy to see that $M = R(\omega)$ is a model for ZFC minus the infinity axiom (see Exercise 1).

Forcing consistency proofs will be based on the following fundamental principle.

Forcing principle In order to prove that the consistency of ZFC implies the consistency of ZFC+"ψ" it is enough to show (in ZFC) that

(F)   every countable transitive model M of ZFC can be extended to a countable transitive model N of ZFC+"ψ."

We use here the word "extend" in the sense of inclusion, that is, N extends M means that $M \subset N$.

In what follows we will use the letters CTM as an abbreviation for "countable transitive model." Thus the forcing principle asserts that if we can prove in ZFC the implication

$$[\exists M(M \text{ is a CTM for ZFC})] \Rightarrow [\exists N(M \subset N \ \& \ N \text{ is a CTM for ZFC+"}\psi")]$$

then we can conclude from it that $\text{Con}(ZFC)$ implies $\text{Con}(ZFC + "\psi")$.

We will not prove the forcing principle here, since its proof is not important for the applications of the forcing method. However, a sketch of its proof can be found in Appendix B.

To use the forcing principle we will assume that we have a countable transitive model M of ZFC and then will extend it to an appropriate model N. To describe the extension method we need a few more definitions. Let M be an arbitrary family of sets and let $\mathbb{P}$ be a partially ordered set. A filter G in $\mathbb{P}$ is M-generic if $D \cap G \neq \emptyset$ for every dense subset D of $\mathbb{P}$ that belongs to M. Notice that the Rasiowa–Sikorski lemma (Theorem 8.1.2) immediately implies the following lemma since the family $\{D \in M : D \subset \mathbb{P} \ \& \ D \text{ is dense in } \mathbb{P}\}$ is countable for any countable set M.

Lemma 9.2.1 *For every partially ordered set $\mathbb{P}$, $p \in \mathbb{P}$, and countable set M there exists an M-generic filter G in $\mathbb{P}$ such that $p \in G$.*

The model N from (F) will be constructed by using the following theorem.

Theorem 9.2.2 *For any countable transitive model M of ZFC, partially ordered set $\mathbb{P} = \langle \mathbb{P}_0, \leq \rangle \in M$, and M-generic filter G in $\mathbb{P}$ there exists a smallest countable transitive model N of ZFC such that $M \subset N$ and $G \in N$.*

The model N from Theorem 9.2.2 is usually denoted by $M[G]$ and is called a *generic extension of M*. We will also refer to M as the *ground model* of $M[G]$.

Before we present a construction of $M[G]$ we will have a closer look at Theorem 9.2.2. To make good use of it we must expect that a countable transitive model M for some property ψ (such as CH) can be extended to a model $M[G]$ of its negation $\neg\psi$. For this, $M[G]$ must be a proper extension of M. However, Theorem 9.2.2 does not rule out that $M[G] = M$. In fact, if $G \in M$ then clearly $M[G] = M$. Fortunately, the extension will be proper for a large class of partially ordered sets described in Proposition 9.2.3.

Proposition 9.2.3 *Let M be a transitive model of ZFC and let $\mathbb{P} \in M$ be a partially ordered set such that*

$(\star)$ *for every $p \in \mathbb{P}$ there are two incompatible $q, r \in \mathbb{P}$ below p.*

If G is an M-generic filter in $\mathbb{P}$ then $G \notin M$.

Proof Notice that $(\star)$ implies that $D = \mathbb{P} \setminus G$ is dense in $\mathbb{P}$. Now if $D \in M$ then, by the definition of an M-generic filter, we would have $G \cap D \neq \emptyset$, which is impossible, since $G \cap D = G \cap (\mathbb{P} \setminus G) = \emptyset$. So $D \notin M$. But $\mathbb{P} \in M$ and M is closed under taking the difference of two sets, since it is a model of ZFC. Therefore $G \notin M$, since otherwise $D = \mathbb{P} \setminus G \in M$. $\qquad\square$

Without any doubt the reader should notice a similarity between Martin's axiom and the statement of Theorem 9.2.2. In fact, the generic filter G from the theorem will be used in $M[G]$ in a way similar to the way we used generic filters in the proofs where Martin's axiom was used. To see the similarities as well as differences between such uses in both cases consider the following example.

Example 9.1 Let M be a countable transitive model of ZFC. For sets X and Y from M consider the forcing $\mathbb{P}(X,Y) = \langle \mathrm{Func}_\omega(X,Y), \supset \rangle$. Then $\mathbb{P}(X,Y)$ belongs to M, being defined from $X, Y \in M$. It is easy to see that if X is infinite then the sets

$$D_x = \{p \in \mathbb{P}(X,Y) : x \in \mathrm{dom}(p)\} \text{ for } x \in X$$

and

$$R_y = \{p \in \mathbb{P}(X,Y) : y \in \mathrm{range}(p)\} \text{ for } y \in Y$$

are dense in $\mathbb{P}(X,Y)$. They belong to M, since each of them is defined from $\mathbb{P}(X,Y)$ and either $x \in X \subset M$ or $y \in Y \subset M$. In particular, if G is an M-generic filter in $\mathbb{P}(X,Y)$ then, by Proposition 8.1.1, $g = \bigcup G$ is a function from X onto Y.

In the proof that ¬CH is consistent with ZFC we will use the forcing $\mathbb{P}_0 = \mathbb{P}(\omega_2 \times \omega, 2)$. Then, for an M-generic filter G_0 in $\mathbb{P}_0$, we see that $g = \bigcup G_0 \colon \omega_2 \times \omega \to 2$ belongs to $M[G_0]$. Moreover, notice that for every $\zeta < \xi < \omega_2$ the sets

$$E_\zeta^\xi = \{ p \in \mathbb{P}_0 \colon \exists n < \omega \ [\langle \zeta, n \rangle, \langle \xi, n \rangle \in \mathrm{dom}(p) \ \& \ p(\zeta, n) \neq p(\xi, n)] \}$$

are dense in $\mathbb{P}_0$ and belong to M. Thus an M-generic filter intersects every such set, that is,

for every $\zeta < \xi < \omega_2$ there exists an $n < \omega$ such that $g(\zeta, n) \neq g(\xi, n)$. (9.3)

Now for every $\xi < \omega_2$ we can define a function $g_\xi \colon \omega \to 2$ by $g_\xi(n) = g(\xi, n)$. The set $\{ g_\xi \colon \xi < \omega_2 \}$ belongs to $M[G]$, since it has been defined only from $g \in M[G]$. Moreover, by (9.3), all functions g_ξ are distinct and $\{ g_\xi \colon \xi < \omega_2 \} \subset 2^\omega$. So we have proved that (in $M[G_0]$) the family 2^ω contains a subset $\{ g_\xi \colon \xi < \omega_2 \}$ of cardinality ω_2, so $\mathfrak{c} = |2^\omega| \geq \omega_2 > \omega_1$. That is, in $M[G_0]$ the continuum hypothesis is false.

Is this really all we need to prove? Is it that simple? Unfortunately the preceding argument contains several imprecisions and gaps. To see this let us consider another forcing $\mathbb{P}^\star = \mathbb{P}(\omega, \omega_1) = \langle \mathrm{Func}_\omega(\omega, \omega_1), \supset \rangle$ and let $G^\star$ be an M-generic filter in $\mathbb{P}^\star$. Then, as we noticed before, $g = \bigcup G^\star$ is a function from ω *onto* ω_1! This looks like a clear contradiction. What is wrong? Is Theorem 9.2.2 false?

The answer is that there is nothing wrong with Theorem 9.2.2. What is wrong with our "contradiction" is that there is a misconception of what the theorem really says. It tells us that $M[G]$ is a model for ZFC as long as M is such a model. In particular, φ^M and $\varphi^{M[G]}$ are true for every sentence φ that is a consequence of ZFC axioms. However, the constants are not a part of the formal language of set theory, and we use them only as shortcuts for the formulas representing them. Thus we write "x is equal to ω_1" to express the fact that x is the unique set satisfying the formula $\varphi(x)$: "x is the smallest uncountable ordinal number." Theorem 9.2.2 tells us that the sentence ψ defined by $\exists! x \ \varphi(x)$ is true both in M and in $M[G]$, since the existence of a unique ω_1 can be proved in ZFC. Thus ψ^N holds for $N = M$ or $N = M[G]$, and there exists a unique $\omega_1^N \in N$ such that $\varphi^N(\omega_1^N)$ is true, that is, that ω_1^N is an "ω_1 with respect to N." However, Theorem 9.2.2 does not tell us that $\omega_1^M = \omega_1^{M[G]}$! In fact, since M is countable and transitive, every element of M is countable. Thus by

manipulating M from outside we may be able to take any set $x \in M$ that
is uncountable in M and find a generic extension $M[G]$ of M in which x
is countable. And this is precisely what happened in $M[G^\star]$! The ordinal
number ω_1^M became a countable object in $M[G^\star]$, and ω_1^M is not equal to
$\omega_1^{M[G^\star]}$. (The argument that indeed $\omega_1^M \neq \omega_1^{M[G^\star]}$ will be completed after
Lemma 9.2.5.)

In fact, so far we are not even sure whether what is an ordinal number
in M remains so in $M[G]$, or even whether what is 2 in M remains 2 in
$M[G]$. To address this issue we have to examine our set-theoretic vocab-
ulary more carefully. In particular, the general question we would like to
examine is, Which properties of elements of M are preserved in its extension $M[G]$? (This can be viewed as an analog of the preservation problems
for transfinite-induction constructions discussed in Section 6.1.)

To address this issue recall once more that every set-theoretic term we
use can be expressed by a formula of set theory. So the question may be
rephrased as follows:

> For which properties P does there exist a formula $\psi(x_1, \dots, x_n)$
> describing P such that if $\psi(t_1, \dots, t_n)$ is true in M for some
> $t_1, \dots, t_n \in M$ then $\psi(t_1, \dots, t_n)$ is also true in $M[G]$?

The properties that are always preserved in such extensions will be called
absolute properties. More precisely, a property is absolute provided it can
be expressed by a formula $\psi(x_1, \dots, x_n)$ such that for every transitive
models M and N of ZFC with $M \subset N$ and for every $t_1, \dots, t_n \in M$
formula $\psi^M(t_1, \dots, t_n)$ holds if and only if $\psi^N(t_1, \dots, t_n)$ holds.

The absoluteness of most properties we are interested in follows just
from the fact that the models we consider are transitive. To identify a
large class of such properties we need the following notion. A formula ψ is
a Δ_0-*formula* if it can be written such that the only quantifiers it contains
are bounded quantifiers, that is, quantifiers of the form $\exists x \in y$ or $\forall x \in y$.

Lemma 9.2.4 *Let M be a transitive set and let $\psi(x_1, \dots, x_n)$ be a Δ_0-
formula. If $t_1, \dots, t_n \in M$ then $\psi^M(t_1, \dots, t_n)$ is equivalent to $\psi(t_1, \dots, t_n)$
in the sense that the sentence*

$$\forall M[M \text{ is transitive} \rightarrow (\forall t_1, \dots, t_n \in M)(\psi^M(t_1, \dots, t_n) \leftrightarrow \psi(t_1, \dots, t_n))]$$

is provable in ZFC. In particular, the property expressed by ψ is absolute.

Proof The proof goes by induction on the length of formula ψ.

Clearly it is true for any basic formula $x \in y$ or $x = y$, since for any
quantifier-free formula ψ its relativization ψ^M is identical to ψ.

So assume that formula ψ is built from the less complicated formulas
φ_0 and φ_1, that is, ψ is in one of the following forms: $\varphi_0 \& \varphi_1$, $\varphi_0 \vee \varphi_1$,

$\varphi_0 \rightarrow \varphi_1$, $\varphi_0 \leftrightarrow \varphi_1$, $\neg\varphi_0$, $\exists x \in t \; \varphi_0$, or $\forall x \in t \; \varphi_0$. For example, if $\psi = \varphi_1 \vee \varphi_2$ and the lemma is true for all formulas of length less than the length of ψ, then it holds for φ_0 and φ_1. In particular,

$$\psi^M = (\varphi_0 \vee \varphi_1)^M = \varphi_0^M \vee \varphi_1^M = \varphi_0 \vee \varphi_1 = \psi.$$

We argue similarly if ψ is the negation of a formula, or is obtained from two formulas by using the logical operations &, $\rightarrow$, or $\leftrightarrow$.

If ψ is of the form $\exists x \in t \; \varphi_0(x)$ for some $t \in M$ then its formal representation is $\exists x \; [x \in t \; \& \; \varphi_0(x)]$. Then ψ^M stands for $\exists x \in M \; [x \in t \; \& \; \varphi_0^M(x)]$, which is equivalent to $(\exists x \in t \cap M)\varphi_0(x)$ since $\varphi_0(x)$ is equivalent to $\varphi_0^M(x)$ by the inductive assumption. But $t \in M$ and M is transitive. So $t \subset M$ and $t \cap M = t$. Therefore ψ^M is equivalent to $\exists x \in t \; \varphi_0(x)$, that is, to ψ.

The case when ψ is of the form $\forall x \in t \; \varphi_0(x)$ is similar. $\qquad\square$

For example, the property "y is a union of x," $y = \bigcup x$, is absolute, since it can be written as the following Δ_0-formula $\psi_1(x, y)$:

$$\forall w \in y \exists z \in x(w \in z) \; \& \; \forall z \in x \forall w \in z(w \in y).$$

Lemma 9.2.5 *The following properties can be expressed by a Δ_0-formula. In particular, they are absolute.*

(0) $x \subset y$.

(1) $y = \bigcup x$.

(2) $y = \bigcap x$.

(3) $z = x \cup y$.

(4) $z = x \cap y$.

(5) $z = x \setminus y$.

(6) $z = \{x, y\}$.

(7) z *is an unordered pair.*

(8) $z = \langle x, y \rangle$ *(i.e., $z = \{\{x\}, \{x, y\}\}$).*

(9) z *is an ordered pair.*

(10) $z = x \times y$.

(11) r *is a binary relation.*

(12) d *is the domain of binary relation r.*

(13) R is the range of binary relation r.

(14) f is a function.

(15)   Function f is injective.

(16) $\leq$ is a partial-order relation on $\mathbb{P}$ (i.e., $\langle \mathbb{P}, \leq \rangle$ is a partially ordered set).

(17) D is a dense subset of the partially ordered set $\langle \mathbb{P}, \leq \rangle$.

(18) A is an antichain in the partially ordered set $\langle \mathbb{P}, \leq \rangle$.

(19)   Set x is transitive.

(20) α is an ordinal number.

(21) α is a limit ordinal number.

(22) α is the first nonzero limit ordinal number (i.e., $\alpha = \omega$).

(23) α is a finite ordinal number.

(24) α is a successor ordinal number.

Proof The Δ_0-formulas for these properties can be found in Appendix B.
$\square$

Now, for a property P expressed by a formula $\psi(x)$, a transitive model M of ZFC, and $t \in M$ the statement "t has the property P" may be interpreted in two different ways: as either that t satisfies the formula $\psi(x)$, that is, that $\psi(t)$ holds, or that $\psi(x)$ is satisfied by t in M, that is, that $\psi^M(t)$ is true. In the latter case we will say that "t has the property P in M," reserving the statement "t has the property P" for the former case. However, for the absolute properties, such as "α is an ordinal number," such a distinction is redundant. Thus, we will say "α is an ordinal number" independently of whether we think of α as an ordinal number in M or outside it.

A similar distinction has to be made when we talk about constants. When we write ω we think of it as "ω is the first infinite ordinal number." In principle, an object ω^M that satisfies this property in M might be different from the "real" ω. However, for objects represented by absolute formulas this cannot happen. So in all these cases we will drop the superscript ^{M} from the symbol representing an absolute constant, even when we think of it as being "in M." In particular, we will use the same symbol ω regardless of the context, since $\omega^M = \omega$. For the cases of the constants that are not absolute, such as ω_1 or $\mathfrak{c}$, we will keep using the superscript ^{M} to denote their counterparts in M. For example, we will write ω_1^M and $\mathfrak{c}^M$ to express ω_1 and $\mathfrak{c}$ in M.

In fact, we have already used this convention when arguing that $\omega_1^M \neq \omega_1^{M[G^\star]}$, with $G^\star$ being an M-generic filter in $\mathbb{P}^\star = \langle \mathrm{Func}_\omega(\omega, \omega_1), \supset \rangle$. And now we can indeed see that $\omega_1^M \neq \omega_1^{M[G^\star]}$. This is so since ω_1^M and $\omega_1^{M[G^\star]}$ are (real) ordinal numbers, and in $M[G^\star]$ there is a (real) function f from (true) ω onto ω_1^M. So ω_1^M is a countable ordinal in $M[G^\star]$, whereas $\omega_1^{M[G^\star]}$ is an uncountable ordinal in $M[G^\star]$ by its definition. Thus they cannot be equal.

Proof of Theorem 9.2.2 To construct the set $M[G]$ fix a countable transitive model M of ZFC, a partially ordered set $\mathbb{P} \in M$, and an M-generic filter G in $\mathbb{P}$.

We say that τ is a $\mathbb{P}$-*name* provided τ is a binary relation and

$$\forall \langle \sigma, p \rangle \in \tau \ (\sigma \text{ is a } \mathbb{P}\text{-name} \ \& \ p \in \mathbb{P}).$$

Formally, we define the property "τ is a $\mathbb{P}$-name" recursively by induction on $\mathrm{rank}(\tau)$. Thus τ is a $\mathbb{P}$-name if it satisfies the following recursive formula:

$\psi(\tau)$: there exists an ordinal number α and a function f such that $\mathrm{dom}(f) = \alpha + 1$, $f(\alpha) = \{\tau\}$, and for every $\beta \in \alpha + 1$, $t \in f(\beta)$, and $z \in t$ there exist $\gamma \in \beta$, $\sigma \in f(\gamma)$, and $p \in \mathbb{P}$ such that $z = \langle \sigma, p \rangle$.

It is not difficult to see that ψ is absolute (see Exercise 2). So the property "τ is a $\mathbb{P}$-name" is absolute too.

The set of all $\mathbb{P}$-names that belong to M will be denoted by $M^{\mathbb{P}}$, that is,

$$M^{\mathbb{P}} = \{\tau \in M : \tau \text{ is a } \mathbb{P}\text{-name}\}.$$

It can be proved that the class of all $\mathbb{P}$-names is a proper class. Thus $M^{\mathbb{P}}$ is a proper class in M. It particular, it does not belong to M.

For a $\mathbb{P}$-name τ we define the *valuation* of τ by

$$\mathrm{val}_G(\tau) = \{\mathrm{val}_G(\sigma) : \exists p \in G \ (\langle \sigma, p \rangle \in \tau)\}.$$

Once again, the formal definition of $\mathrm{val}_G(\tau)$ should be given by a recursive formula. Now we define $M[G]$ by

$$M[G] = \{\mathrm{val}_G(\tau) : \tau \in M^{\mathbb{P}}\}.$$

This is the model from the statement of Theorem 9.2.2.

We will not check here that all ZFC axioms are true in $M[G]$, since it is tedious work and has little to do with the forcing applications. However, we will argue here for its other properties listed in Theorem 9.2.2.

Transitivity of $M[G]$ follows immediately from the definition of $M[G]$.
Take $y \in M[G]$. Thus $y = \text{val}_G(\tau) = \{\text{val}_G(\sigma) \colon \exists p \in G \ (\langle \sigma, p \rangle \in \tau)\}$ for
some $\tau \in M^{\mathbb{P}}$. So any $x \in y$ is in the form $\text{val}_G(\sigma)$ for some $\langle \sigma, p \rangle \in \tau \in M$.
Thus σ is a $\mathbb{P}$-name that belongs to M, and $x = \text{val}_G(\sigma) \in M[G]$.

The minimality of $M[G]$ is clear, since every element of $M[G]$ is of
the form $\text{val}_G(\tau)$ and so must belong to any transitive model of ZFC that
contains G and τ.

To see that $M[G]$ is countable notice first that it is a set by the axiom
of replacement since val_G is a function defined by a formula and $M[G] =$
$\text{val}_G[M^{\mathbb{P}}]$. It is countable, since $M^{\mathbb{P}} \subset M$ is countable.

To argue for $M \subset M[G]$ and $G \in M[G]$ we need a few more definitions.
For a set x define, by recursion, its $\mathbb{P}$-name $\hat{x}$ by

$$\hat{x} = \{\langle \hat{y}, p \rangle \colon y \in x \ \& \ p \in \mathbb{P}\}.$$

(Note that the definition of $\hat{x}$ depends on $\mathbb{P}$.)

Lemma 9.2.6 *If M is a countable transitive model of ZFC, $\mathbb{P} \in M$ is a
partially ordered set, and G is an M-generic filter over $\mathbb{P}$ then for every
$x \in M$*

$$\hat{x} \in M^{\mathbb{P}} \quad and \quad \text{val}_G(\hat{x}) = x.$$

In particular, $M \subset M[G]$.

Proof It is an easy inductive argument. First prove by induction on
$\text{rank}(x)$ that $\hat{x}$ is a $\mathbb{P}$-name. Then notice that the definition of $\hat{x}$ is absolute
and $x \in M$. So $\hat{x} \in M$. Finally, another induction on $\text{rank}(x)$ shows that
$\text{val}_G(\hat{x}) = x$. The details are left as an exercise. $\qquad\square$

To see that $G \in M[G]$ let $\Gamma = \{\langle \hat{p}, p \rangle \colon p \in \mathbb{P}\}$. It is clearly a $\mathbb{P}$-name
defined in M from $\mathbb{P}$. So $\Gamma \in M^{\mathbb{P}}$. Moreover,

$$\text{val}_G(\Gamma) = \{\text{val}_G(\hat{p}) \colon \exists p \in G \ (\langle \hat{p}, p \rangle \in \Gamma)\} = \{p \in \mathbb{P} \colon p \in G\} = G.$$

So $G \in M[G]$.

This finishes the proof of Theorem 9.2.2. $\qquad\square$

The last important basic problem we have to face is how to check that a
given property φ is true in $M[G]$. For this, we will introduce the following
definition. For a formula $\varphi(x_1, \ldots, x_n)$, a countable transitive model M
of ZFC, a partial order $\mathbb{P} \in M$, and $\mathbb{P}$-names $\tau_1, \ldots, \tau_n \in M$ we say that
$p \in \mathbb{P}$ *forces* $\varphi(\tau_1, \ldots, \tau_n)$ and write

$$p \Vdash_{\mathbb{P}, M} \varphi(\tau_1, \ldots, \tau_n)$$

provided $\varphi(\text{val}_G(\tau_1), \ldots, \text{val}_G(\tau_n))$ is true in $M[G]$ for every M-generic
filter G in $\mathbb{P}$ that contains p. We will usually just write $\Vdash$ in place of

$\Vdash_{\mathbb{P},M}$ when $\mathbb{P}$ and M are fixed. We will also write $\mathbb{P} \Vdash \varphi(\tau_1, \dots, \tau_n)$ if $p \Vdash \varphi(\tau_1, \dots, \tau_n)$ for every $p \in \mathbb{P}$. Notice that

$$\text{if } p \Vdash \varphi(\tau_1, \dots, \tau_n) \text{ and } q \leq p \text{ then } q \Vdash \varphi(\tau_1, \dots, \tau_n), \qquad (9.4)$$

since every M-generic filter G that contains q contains also p.

Clearly, the definition of the relation $\Vdash$ depends on a model M and a set $\mathbb{P}$ and involves the knowledge of all M-generic filters in $\mathbb{P}$. Thus, it is clearly defined outside M. One of the most important facts concerning forcing is that an equivalent form of $\Vdash$ can also be found inside M, in a sense described in the following theorem.

Theorem 9.2.7 *For every formula $\varphi(x_1, \dots, x_n)$ of set theory there exists another formula $\psi(\pi, P, x_1, \dots, x_n)$, denoted by $\pi \Vdash_P^\star \varphi(x_1, \dots, x_n)$, such that for every countable transitive model M of ZFC, partial order $\mathbb{P} \in M$, and $\mathbb{P}$-names $\tau_1, \dots, \tau_n \in M$*

$$p \Vdash_{\mathbb{P},M} \varphi(\tau_1, \dots, \tau_n) \Leftrightarrow M \models (p \Vdash_P^\star \varphi(\tau_1, \dots, \tau_n)) \qquad (9.5)$$

for every $p \in \mathbb{P}$. Moreover,

$$M[G] \models \varphi(\mathrm{val}_G(\tau_1), \dots, \mathrm{val}_G(\tau_n)) \Leftrightarrow \exists p \in G(p \Vdash \varphi(\tau_1, \dots, \tau_n)) \qquad (9.6)$$

for every M-generic filter G in $\mathbb{P}$.

This theorem will be left without proof.

The relation $\Vdash^\star$ from Theorem 9.2.7 will justify the use of all forcing arguments within a model M and will let us conclude all essential properties of $M[G]$ without knowledge of G. We will often write $\Vdash$ where formally the relation $\Vdash^\star$ should be used.

This completes all general details concerning the forcing method. However, before we give a proof of $\mathrm{Con}(ZFC + \neg CH)$ we still need some technical lemmas. The first of them tells us that a $\mathbb{P}$-name representing a function with domain and range belonging to a ground model can be chosen to have a particularly nice form.

Lemma 9.2.8 *Let M be a countable transitive model of ZFC and $\mathbb{P} \in M$ be a partially ordered set. If $X, Y \in M$, $p_0 \in \mathbb{P}$, and $\sigma \in M^{\mathbb{P}}$ are such that $p_0 \Vdash$ "σ is a function from $\hat{X}$ into $\hat{Y}$" then there exists a $\tau \in M^{\mathbb{P}}$ such that*

(a) $\tau \subset \{\langle \widehat{\langle x, y \rangle}, p \rangle : x \in X \ \& \ y \in Y \ \& \ p \in \mathbb{P} \ \& \ p \leq p_0\}$;

(b) $p_0 \Vdash \sigma = \tau$; *and*

(c) $A_x = \{p \in \mathbb{P}: \exists y \in Y \ (\langle \widehat{\langle x, y \rangle}, p \rangle \in \tau)\}$ *is an antichain for every $x \in X$.*

Proof For $x \in X$ let

$$D_x = \{p \in \mathbb{P} : p \leq p_0 \ \& \ \exists y \in Y \ (p \Vdash \widehat{\langle x, y \rangle} \in \sigma)\}.$$

Notice that $D_x \in M$ since, by condition (9.5) of Theorem 9.2.7,

$$D_x = \{p \in \mathbb{P} : p \leq p_0 \ \& \ \exists y \in Y \ (p \Vdash^* \widehat{\langle x, y \rangle} \in \sigma)^M\}.$$

Let $\bar{A}_x$ be a maximal subset of D_x of pairwise-incompatible elements. Thus every $\bar{A}_x$ is an antichain. Define

$$\tau = \{\langle \widehat{\langle x, y \rangle}, p \rangle : x \in X \ \& \ y \in Y \ \& \ p \in \bar{A}_x \ \& \ p \Vdash \widehat{\langle x, y \rangle} \in \sigma\}.$$

Once again $\tau \in M$ follows from (9.5), since

$$\tau = \{\langle \widehat{\langle x, y \rangle}, p \rangle : x \in X \ \& \ y \in Y \ \& \ p \in \bar{A}_x \ \& \ (p \Vdash^* \widehat{\langle x, y \rangle} \in \sigma)^M\}$$

and the family $\{\bar{A}_x : x \in X\}$ can be chosen from M.

Clearly τ is a $\mathbb{P}$-name satisfying (a).

To see (c) it is enough to notice that $A_x = \bar{A}_x$ for every $x \in X$. This is the case since

$$
\begin{aligned}
p \in A_x \quad &\Leftrightarrow \quad \exists y \in Y \ \langle \widehat{\langle x, y \rangle}, p \rangle \in \tau \\
&\Leftrightarrow \quad \exists y \in Y \ \left(p \in \bar{A}_x \ \& \ p \Vdash \widehat{\langle x, y \rangle} \in \sigma \right) \\
&\Leftrightarrow \quad p \in \bar{A}_x.
\end{aligned}
$$

To prove that $p_0 \Vdash \sigma = \tau$, let G be an M-generic filter in $\mathbb{P}$ such that $p_0 \in G$. Then $g = \mathrm{val}_G(\sigma)$ is a function from X into Y, since p_0 forces it. Also, clearly $\mathrm{val}_G(\tau) \subset X \times Y$. So let $x \in X$ and $y \in Y$. It is enough to show that $\langle x, y \rangle \in \mathrm{val}_G(\sigma)$ if and only if $\langle x, y \rangle \in \mathrm{val}_G(\tau)$.

But if $\langle x, y \rangle \in \mathrm{val}_G(\tau)$ then there exists a $p \in G$ such that $\langle \widehat{\langle x, y \rangle}, p \rangle \in \tau$. In particular, $p \Vdash \widehat{\langle x, y \rangle} \in \sigma$. But $p \in G$. Therefore $\langle x, y \rangle \in \mathrm{val}_G(\sigma)$.

To prove the other implication first note that the set

$$E_x = \{q \leq p_0 : \exists p \in \bar{A}_x \ (q \leq p)\} \cup \{q \in \mathbb{P} : q \text{ is incompatible with } p_0\}$$

is dense in $\mathbb{P}$. To see it, let $s \in \mathbb{P}$. We have to find $q \in E_x$ with $q \leq s$.

If s is incompatible with p_0 then $q = s \in E_x$. So assume that s is compatible with p_0 and choose $p_1 \in \mathbb{P}$ extending s and p_0. Let H be an M-generic filter in $\mathbb{P}$ containing p_1. Then $h = \mathrm{val}_H(\sigma)$ is a function from X into Y since the condition $p_0 \in H$ forces it. Let $y = h(x)$. Then, by (9.6), there exists an $r \in H$ forcing it, that is, such that $r \Vdash \widehat{\langle x, y \rangle} \in \sigma$. Choose $p_2 \in H$ extending $r \in H$ and $p_1 \in H$. Then, by (9.4), $p_2 \Vdash \widehat{\langle x, y \rangle} \in \sigma$. In particular, $p_2 \in D_x$. But $\bar{A}_x$ is a maximal antichain in D_x. So there exists a $q \le p_2$ extending some p from $\bar{A}_x$. Hence $q \le p_2 \le p_0$, so $q \in E_x$, and $q \le p_2 \le p_1 \le s$. The density of E_x has been proved.

Now let $\langle x, y \rangle \in \mathrm{val}_G(\sigma)$. Then, by (9.6), there exists an $r \in G$ that forces it, that is, such that $r \Vdash \widehat{\langle x, y \rangle} \in \sigma$. Moreover, there exists a $q \in E_x \cap G$, since G is M-generic and $E_x \in M$ is dense. But $q \in G$ is compatible with $p_0 \in G$. So there exists a $p \in \bar{A}_x \subset D_x$ with $q \le p$. Thus, by the definition of D_x, there exists a $y_1 \in Y$ such that $p \Vdash \widehat{\langle x, y_1 \rangle} \in \sigma$. In particular, $\langle \widehat{\langle x, y_1 \rangle}, p \rangle \in \tau$ and $\langle x, y_1 \rangle \in \mathrm{val}_G(\tau)$, since $p \in G$. To finish the proof it is enough to notice that $y_1 = y$. So let $s \in G$ be a common extension of r, p, and p_0. Then, by (9.4),

$$ s \Vdash \left(\widehat{\langle x, y \rangle} \in \sigma \ \& \ \widehat{\langle x, y_1 \rangle} \in \sigma \ \& \ \sigma \text{ is a function} \right). $$

Therefore, if H is an M-generic filter in $\mathbb{P}$ containing s then $h = \mathrm{val}_H(\sigma)$ is a function and $y = h(x) = y_1$. □

Corollary 9.2.9 *Let M be a countable transitive model of ZFC and $\mathbb{P} \in M$ be a partially ordered set that is ccc in M. Let G be an M-generic filter in $\mathbb{P}$, and let $g \in M[G]$ be such that $g \colon X \to Y$ and $X, Y \in M$. Then there exists a function $F \in M$ such that*

$$ M \models F \colon X \to [Y]^{\le \omega} $$

and $g(x) \in F(x)$ for all $x \in X$.

Proof Let $\sigma \in M$ be a $\mathbb{P}$-name such that $\mathrm{val}_G(\sigma) = g$. Since

$$ M[G] \models g \text{ is a function from } X \text{ into } Y, $$

by Theorem 9.2.7(9.6) there exists a $p_0 \in G$ such that

$$ p_0 \Vdash \sigma \text{ is a function from } \hat{X} \text{ into } \hat{Y}. $$

Let $\tau \in M$ be a $\mathbb{P}$-name from Lemma 9.2.8 and define

$$ F(x) = \{ y \in Y \colon \exists p^y \in \mathbb{P} \, (\langle \widehat{\langle x, y \rangle}, p^y \rangle \in \tau) \}. $$

Then clearly $F \in M$ and $M \models F \colon X \to \mathcal{P}(Y)$. To see that for every $x \in X$ the set $F(x)$ is countable in M note that the map $g \colon F(x) \to A_x$, $g(y) = p^y$, is one-to-one.

Indeed, if $y, z \in F(x)$ and $p^y = p^z$ then $p = p^y \le p_0$ and so

$$ p \Vdash \left(\widehat{\langle x, y \rangle} \in \tau \ \& \ \widehat{\langle x, z \rangle} \in \tau \ \& \ \tau \text{ is a function} \right). $$

Thus $y = z$. In particular,

$$M \models |F(x)| \le |A_x| \le \omega,$$

as $\mathbb{P}$ is ccc in M. Finally, since $g = \mathrm{val}_G(\sigma) = \mathrm{val}_G(\tau)$

$$y = g(x) \Leftrightarrow \exists p \in G \left(\langle \widehat{\langle x, y \rangle}, p \rangle \in \tau \right) \Rightarrow y \in F(x).$$

Corollary 9.2.9 has been proved.   □

Next we will show that ccc forcings are very nice with respect to the cardinal numbers. To formulate this more precisely we need a few definitions.

Let M be a countable transitive model of ZFC, $\mathbb{P} \in M$ be a partial order, and $\kappa \in M$ be an ordinal number such that $M \models$ "κ is a cardinal number." If $M[G] \models$ "κ is a cardinal number" for every M-generic filter G in $\mathbb{P}$ then we say that *forcing $\mathbb{P}$ preserves the cardinal κ*. On the other hand, if $M[G] \models$ "κ is not a cardinal number" for every M-generic filter G in $\mathbb{P}$ then we say that *forcing $\mathbb{P}$ collapses the cardinal κ*. If $\mathbb{P}$ preserves all cardinal numbers from M then we simply say that $\mathbb{P}$ *preserves cardinals*.

Lemma 9.2.10 *Let M be a countable transitive model of ZFC and $\mathbb{P} \in M$ be a partial order that preserves cardinals. If G is an M-generic in $\mathbb{P}$ then $\omega_\alpha^M = \omega_\alpha^{M[G]}$ for every ordinal number $\alpha \in M$.*

Proof This follows by an easy induction; it is left as an exercise.   □

Theorem 9.2.11 *Let M be a countable transitive model of ZFC. If $\mathbb{P} \in M$ is a partial order such that $M \models$ "$\mathbb{P}$ is ccc" then $\mathbb{P}$ preserves cardinals.*

Proof Let M and $\mathbb{P}$ be as in the theorem and assume, to obtain a contradiction, that there is an ordinal number κ and an M-generic filter G in $\mathbb{P}$ such that κ is a cardinal in M but is not in $M[G]$. By the absoluteness of finite cardinals and ω it must be that $\kappa > \omega$.

Since κ is still an ordinal in $M[G]$ it means that it is not initial, that is, there exist an ordinal $\alpha < \kappa$ and a function $g \in M[G]$ such that g maps α onto κ. So, by Corollary 9.2.9, there exists a function $F \in M$ such that $M \models$ "$F: \alpha \to [\kappa]^{\le \omega}$" and $g(\xi) \in F(\xi)$ for all $\xi \in \alpha$. Hence $\kappa = g[\alpha] \subset \bigcup_{\xi \in \alpha} F(\xi)$.

But $\bigcup_{\xi \in \alpha} F(\xi) \in M$, since $F \in M$. So the following holds in M:

$$\kappa \le \left| \bigcup_{\xi \in \alpha} F(\xi) \right| \le |\alpha| \otimes \omega < \kappa,$$

since all sets $F(\xi)$ are countable in M. This contradiction finishes the proof.   □

To apply this theorem to the proof of the consistency of ZFC+¬CH we need one more fact (compare Exercise 1 from Section 8.2).

Lemma 9.2.12 *Forcing* $\mathbb{P} = \langle \mathrm{Func}_\omega(X,Y), \supset \rangle$ *is ccc for* $|Y| \leq \omega$.

Proof Let $\langle p_\xi \colon \xi < \omega_1 \rangle$ be a one-to-one sequence of elements of $\mathbb{P}$. We have to find $\zeta < \xi < \omega_1$ such that p_ζ and p_ξ are compatible.

Consider the family $\mathcal{A} = \{\mathrm{dom}(p_\xi) \colon \xi < \omega_1\}$ of finite sets. Notice that $\mathcal{A}$ must be uncountable, since for every $A \in \mathcal{A}$ there is at most $|Y^A| \leq \omega$ p_ξs with $A = \mathrm{dom}(p_\xi)$. So, by the Δ-system lemma (Lemma 8.2.11), there exist an uncountable set $S \subset \omega_1$ and a finite A with the property that $\mathrm{dom}(p_\zeta) \cap \mathrm{dom}(p_\xi) = A$ for any distinct $\zeta, \xi \in S$. But the set Y^A is at most countable. So there exist distinct ζ and ξ from S such that $p_\zeta|_A = p_\xi|_A$. This implies that $p = p_\zeta \cup p_\xi$ is a function, that is, $p \in \mathbb{P}$. Since p clearly extends p_ζ and p_ξ we conclude that p_ζ and p_ξ are compatible.   □

Theorem 9.2.13 *Theory ZFC+¬CH is consistent.*

Proof We will follow the path described in Example 9.1. Let M be a countable transitive model of ZFC. Define (in M)

$$\mathbb{P} = \langle \mathrm{Func}_\omega(\omega_2^M \times \omega, 2), \supset \rangle$$

and let G be an M-generic filter in $\mathbb{P}$. We will show that $\mathfrak{c} > \omega_1$ in $M[G]$.
   For $x \in \omega_2^M \times \omega$ let

$$D_x = \{p \in \mathbb{P} \colon x \in \mathrm{dom}(p)\}.$$

It is easy to see that all sets D_x belong to M since they are defined using $x, \mathbb{P} \in M$. They are dense, since for every $p \in \mathbb{P}$ either $x \in \mathrm{dom}(p)$ and $p \in D_x$, or $x \notin \mathrm{dom}(p)$ and $p \cup \{\langle x, 0 \rangle\} \in D_x$ extends p. Thus, by Proposition 8.1.1, $g = \bigcup G$ is a function from $\omega_2^M \times \omega$ to 2. Clearly $g \in M[G]$, since it is constructed from $G \in M[G]$.
   Similarly as for D_x we argue that for every $\zeta < \xi < \omega_2^M$ the sets

$$E_\zeta^\xi = \{p \in \mathbb{P} \colon \exists n < \omega\ [\langle \zeta, n \rangle, \langle \xi, n \rangle \in \mathrm{dom}(p)\ \&\ p(\zeta, n) \neq p(\xi, n)]\}$$

are dense in $\mathbb{P}$ and belong to M. Thus an M-generic filter intersects every such set, that is, for every $\zeta < \xi < \omega_2^M$ there exists an $n < \omega$ such that $g(\zeta, n) \neq g(\xi, n)$.
   Now for every $\xi < \omega_2^M$ we can define (in $M[G]$) a function $g_\xi \colon \omega \to 2$ by $g_\xi(n) = g(\xi, n)$. The set $\{g_\xi \colon \xi < \omega_2^M\}$ belongs to $M[G]$, since it has

been constructed only from $g \in M[G]$. Moreover, $\{g_\xi : \xi < \omega_2^M\} \subset 2^\omega$ and all the functions g_ξ are different. So

$$M[G] \models \mathfrak{c} = |2^\omega| \geq \omega_2^M.$$

But, by Lemma 9.2.12, forcing $\mathbb{P}$ is ccc in M and so, by Theorem 9.2.7, it preserves cardinals. In particular, by Lemma 9.2.10, $\omega_2^M = \omega_2^{M[G]}$, that is,

$$M[G] \models \mathfrak{c} \geq \omega_2^{M[G]} > \omega_1.$$

This finishes the proof.   □

A model $M[G]$ obtained as a generic extension of a ground model M via the forcing $\mathbb{P} = \langle \mathrm{Func}_\omega(X, 2), \supset \rangle$ (with $X \in M$) is called a *Cohen model*. The numbers g_ξ from Theorem 9.2.13 are called *Cohen (real) numbers*.

EXERCISES

1 Prove that $x \in R(\omega)$ if and only if $|\operatorname{trcl}(x)| < \omega$. Use this to show that $M = R(\omega)$ is a model for ZFC minus the infinity axiom.

2 Let $\psi(\tau)$ be a formula expressing the property that "τ is a $\mathbb{P}$-name" and let M be a CTM of ZFC with $\mathbb{P} \in M$. Show, by induction on $\operatorname{rank}(\tau)$, that for every $\tau \in M$

$$\psi(\tau) \quad \text{if and only if} \quad \psi^M(\tau).$$

Thus the property "τ is a $\mathbb{P}$-name" is absolute.

3 Complete the details of the proof of Lemma 9.2.6.

4 Let M be a countable transitive model of ZFC, $\mathbb{P} \in M$ be a partial order, and G be an M-generic filter in $\mathbb{P}$. Show that $\omega_\alpha^M \leq \omega_\alpha^{M[G]}$ for every ordinal number $\alpha \in M$.

5 Prove Lemma 9.2.10.

9.3 Model for CH and $\diamondsuit$

The main goal of this section is to prove the following theorem.

Theorem 9.3.1 *Let M be a countable transitive model of ZFC, and let $\mathbb{P} = (\mathrm{Func}_{\omega_1}(\omega_1, 2))^M$. If G is an M-generic filter in $\mathbb{P}$ then $\omega_1^M = \omega_1^{M[G]}$, $\mathcal{P}(\omega) \cap M = \mathcal{P}(\omega) \cap M[G]$, and $\diamondsuit$ holds in $M[G]$.*

Since $\diamondsuit$ implies CH the theorem immediately implies the following corollary.

Corollary 9.3.2 *Theories ZFC+CH and ZFC+$\Diamond$ are consistent.*

Combining this with Theorem 9.2.13 we immediately obtain the following.

Corollary 9.3.3 *The continuum hypothesis is independent of ZFC.*

Note that $\mathbb{P} = \mathrm{Func}_{\omega_1}(\omega_1, 2)$ stands for the set of all functions $p\colon A \to 2$, where A is an at most countable subset of ω_1. Since the terms "ω_1" and "being countable" are not absolute, the superscript ^{M} in the definition of $\mathbb{P}$ in Theorem 9.3.1 is essential.

We will prove Theorem 9.3.1 with the help of several general facts that are, in most cases, as important as the theorem itself.

Let $\mathbb{P}$ be a partially ordered set and let $p \in \mathbb{P}$. A set $D \subset \mathbb{P}$ is *dense below p* if for every $q \le p$ there exists a $d \in D$ such that $d \le q$.

Proposition 9.3.4 *Let M be a countable transitive model of ZFC, $\mathbb{P} \in M$ be a partially ordered set, and $p \in \mathbb{P}$. If $D \in M$ is a subset of $\mathbb{P}$ that is dense below p then $G \cap D \ne \emptyset$ for every M-generic filter G in $\mathbb{P}$ such that $p \in G$.*

Proof Let $R = \{r \in \mathbb{P}\colon r$ is incompatible with $p\}$. Then $D_0 = D \cup R \in M$ and D_0 is dense in $\mathbb{P}$. So if G is an M-generic filter in $\mathbb{P}$ then $G \cap D_0 \ne \emptyset$. But if $p \in G$ then $G \cap R = \emptyset$, so $G \cap D \ne \emptyset$. $\qquad\qquad\square$

A partially ordered set $\mathbb{P}$ is *countably closed* if for every decreasing sequence $\langle p_n \in \mathbb{P}\colon n < \omega \rangle$ there is a $p_\omega \in \mathbb{P}$ such that $p_\omega \le p_n$ for every $n < \omega$. Notice that $\mathbb{P}^\star = \mathrm{Func}_{\omega_1}(\omega_1, 2)$ is countably closed since $p_\omega = \bigcup_{n<\omega} p_n \in \mathbb{P}$ extends every p_n. So $\mathbb{P} = (\mathrm{Func}_{\omega_1}(\omega_1, 2))^M$ is countably closed in M.

Theorem 9.3.5 *Let M be a countable transitive model of ZFC and $\mathbb{P} \in M$ be a partially ordered set that is countably closed in M. Let $A, B \in M$ be such that A is countable in M and let G be an M-generic filter in $\mathbb{P}$. If $f \in M[G]$ is a function from A into B then $f \in M$.*

Proof Let $\tau \in M$ be a $\mathbb{P}$-name such that $\mathrm{val}_G(\tau) = f$. Then, by Theorem 9.2.7, there exists a $p \in G$ such that

$$p \Vdash \tau \text{ is a function from } \hat{A} \text{ into } \hat{B}.$$

Define in M

$$D = \{q \in \mathbb{P}\colon \exists g \in B^A \ (q \Vdash \tau = \hat{g})\}.$$

(D belongs to M since the relation $\Vdash$ can be replaced by the *formula* $\Vdash^\star$.) By Proposition 9.3.4 it is enough to prove that D is dense below p since then there exists a $q \in G \cap D$, that is, there is a $g \in M$ such that g maps A to B and

$$f = \mathrm{val}_G(\tau) = \mathrm{val}_G(\hat{g}) = g \in M.$$

To prove that D is dense below p, fix $r \in \mathbb{P}$ such that $r \leq p$. We have to find a $q \in D$ with $q \leq r$. Let $\{a_n : n < \omega\}$ be an enumeration of A in M. Define, in M, sequences $\langle p_n \in \mathbb{P} : n < \omega \rangle$ and $\langle b_n \in B : n < \omega \rangle$ such that for all $n < \omega$

(1) $p_0 = r$,

(2) $p_{n+1} \leq p_n$,

(3) $p_{n+1} \Vdash \tau(\hat{a}_n) = \hat{b}_n$.

To see that such a construction can be made assume that p_i for $i \leq n$ and b_j for $j < n$ have already been constructed. It is enough to show that there exist p_{n+1} and b_n satisfying (2) and (3).

The proof of the existence of such p_{n+1} and b_n will be done outside M. For this, let H be an M-generic filter in $\mathbb{P}$ with $p_n \in H$. Then $h = \mathrm{val}_H(\tau)$ is a function in $M[H]$ from A into B, since $p_n \leq p$ forces it. Thus there is a $b \in B$ such that $h(a_n) = b$ and we can find an $r \in H$ which forces it, that is, such that $r \Vdash \tau(\hat{a}_n) = \hat{b}$. Refining r, if necessary, we can also assume that $r \leq p_n$. Then $p_{n+1} = r$ and $b_n = b$ satisfy (2) and (3).

Now let $p_\omega \in \mathbb{P}$ be such that $p_\omega \leq p_n$ for every $n < \omega$ and let $g : A \to B$ be such that $g(a_n) = b_n$. Then $g \in M$ and

$$p_\omega \Vdash \tau(\hat{a}_n) = \hat{b}_n$$

for every $n < \omega$. To finish the proof it is enough to show that

$$p_\omega \Vdash \tau = \hat{g}$$

since then $q = p_\omega \leq r$ and belongs to D.

So let H be an arbitrary M-generic filter in $\mathbb{P}$ with $p_\omega \in H$. Then $h = \mathrm{val}_H(\tau)$ is a function in $M[H]$ from A into B. Moreover, $h(a_n) = b_n$ for every $n < \omega$ since $p_\omega \leq p_n$ forces it. In particular,

$$M[H] \models h = \mathrm{val}_H(\tau) = g$$

that is, $p_\omega \Vdash \tau = \hat{g}$.   □

Theorem 9.3.5 tells us that in a generic extension of a model M obtained from a countably closed forcing the extension will have the same countable sequences with elements from a fixed set from M. This implies the following corollary.

Corollary 9.3.6 *Let M be a countable transitive model of ZFC and $\mathbb{P} \in M$ be a partially ordered set that is countably closed in M. If G is an M-generic filter in $\mathbb{P}$ then $\mathcal{P}(\omega) \cap M = \mathcal{P}(\omega) \cap M[G]$ and $\omega_1^M = \omega_1^{M[G]}$.*

Proof Let $\beta = \omega_1^M$. If $\alpha \le \beta$ then α is an ordinal number in M and in $M[G]$. Moreover, if $\alpha < \beta$ then there exists an $f \in M \subset M[G]$ such that f maps ω onto α. Thus every $\alpha < \beta$ is at most countable (in M and in $M[G]$). In particular, it is enough to show that β is uncountable in $M[G]$, since then it is the first uncountable ordinal in $M[G]$, that is, $\omega_1^M = \beta = \omega_1^{M[G]}$.

To see that β is uncountable in $M[G]$, take $f \colon \omega \to \beta$ from $M[G]$ and notice that, by Theorem 9.3.5, $f \in M$. So f cannot be onto $\beta = \omega_1^M$. Thus β is uncountable in $M[G]$.

To see that $\mathcal{P}(\omega) \cap M = \mathcal{P}(\omega) \cap M[G]$ it is enough to prove that $\mathcal{P}(\omega) \cap M[G] \subset \mathcal{P}(\omega) \cap M$. So let $A \in \mathcal{P}(\omega) \cap M[G]$ and let $f = \chi_A \colon \omega \to 2$. Then $f \in M[G]$ and, by Theorem 9.3.5, $f \in M$. Therefore $A = f^{-1}(1) \in \mathcal{P}(\omega) \cap M$. $\square$

Notice that Corollary 9.3.6 implies the statements $\omega_1^M = \omega_1^{M[G]}$ and $\mathcal{P}(\omega) \cap M = \mathcal{P}(\omega) \cap M[G]$ of Theorem 9.3.1 since the forcing $\mathbb{P}$ from the theorem is countably closed in M. Notice also that Corollary 9.3.6 does not imply that countably closed forcings preserve all cardinal numbers. In fact, by proving Theorem 9.3.1 we will show that in $M[G]$ there exists a bijection between $\omega_1^M = \omega_1^{M[G]}$ and $\mathcal{P}^M(\omega) = \mathcal{P}^{M[G]}(\omega)$, that is, there is a bijection between ω_1^M and $\mathfrak{c}^M$. Thus, if $\omega_1^M < \mathfrak{c}^M$ (which can happen by a result from the previous section) then the cardinal $\mathfrak{c}^M$ is collapsed in $M[G]$.

To prove that ◇ holds in the model $M[G]$ from Theorem 9.3.1 the following lemma will be useful.

Lemma 9.3.7 *Let M be a countable transitive model of ZFC and let $\mathbb{P}, \mathbb{P}_1 \in M$ be partially ordered sets such that there is an order isomorphism $f \in M$ between $\mathbb{P}$ and $\mathbb{P}_1$. If G is an M-generic filter in $\mathbb{P}$ then $G_1 = f[G]$ is an M-generic filter in $\mathbb{P}_1$ and $M[G] = M[G_1]$.*

Proof It is easy to see that $G_1 = f[G]$ is a filter in $\mathbb{P}_1$. To see that G_1 is M-generic let $D_1 \in M$ be a dense subset of $\mathbb{P}_1$. Then $D = f^{-1}(D_1)$ belongs to M and is dense in $\mathbb{P}$. So $D \cap G \ne \emptyset$ and

$$D_1 \cap G_1 = f[D] \cap f[G] = f[D \cap G] \ne \emptyset.$$

To see that $M[G] = M[G_1]$ notice that $M \subset M[G]$ and $M \subset M[G_1]$. Moreover, $G_1 = f[G] \in M[G]$; thus, by the minimality of $M[G_1]$, we conclude $M[G_1] \subset M[G]$. Similarly, $G = f^{-1}(G_1) \in M[G_1]$. Thus $M[G] \subset M[G_1]$. $\square$

Lemma 9.3.7 tells us that if we replace a forcing by its isomorphic copy then the appropriate generic extensions will be identical. In particular, if $X \in M$ is such that $M \models |X| = \omega_1$ then forcings $\mathbb{P} = (\text{Func}_{\omega_1}(\omega_1, 2))^M$ and $\mathbb{P}_1 = (\text{Func}_{\omega_1}(X, 2))^M$ are isomorphic in M (by an isomorphism $F \colon \mathbb{P} \to \mathbb{P}_1$ with $F(p)(x) = p(f(x))$, where $f \colon X \to \omega_1^M$ is a bijection with $f \in M$).

Before we prove that $\diamondsuit$ holds in the model $M[G]$ from Theorem 9.3.1 we will show that $M[G] \models CH$. This is not used in the proof of $M[G] \models \diamondsuit$. However, the proof of CH in $M[G]$ is simpler than the one for $\diamondsuit$, and it will prepare the reader for the later one. Moreover, the reader interested only in the consistency of ZFC+CH can skip the latter part of this section.

By Lemma 9.3.7 we can replace $\mathbb{P}$ with $\mathbb{P}_1 = (\text{Func}_{\omega_1}(\omega_1 \times \omega, 2))^M$. Let G_1 be an M-generic filter in $\mathbb{P}_1$. Thus $M[G] = M[G_1]$. Clearly $g = \bigcup G_1$ belongs to $M[G]$ and, by Proposition 8.1.1, it is a function from $\omega_1^M \times \omega$ to 2. Notice that for every $f \in 2^\omega \cap M$ the set

$$D_f = \{p \in \mathbb{P}_1 \colon \exists \alpha < \omega_1^M \, [\{\alpha\} \times \omega \subset \text{dom}(p) \ \& \ \forall n < \omega \, (p(\alpha, n) = f(n))]\}$$

belongs to M and is dense in $\mathbb{P}_1$. (To see its denseness, choose $p \in \mathbb{P}_1$, find $\alpha < \omega_1^M$ with $(\{\alpha\} \times \omega) \cap \text{dom}(p) = \emptyset$, and notice that the condition $p \cup \{\langle\langle\alpha, n\rangle, f(n)\rangle \colon n < \omega\} \in D_f$ extends p.) So if $f_\alpha \colon \omega \to 2$ for $\alpha < \omega_1^M$ is defined by $f_\alpha(n) = g(\alpha, n)$ then, by Theorem 9.3.5,

$$2^\omega \cap M \subset \{f_\alpha \colon \alpha < \omega_1^M\} \subset 2^\omega \cap M[G] = 2^\omega \cap M.$$

Thus, in $M[G]$,

$$\omega_1^{M[G]} = \omega_1^M = |\omega_1^M| = |\{f_\alpha \colon \alpha < \omega_1^M\}| = |2^\omega \cap M[G]| = \mathfrak{c}^{M[G]}.$$

So CH holds in $M[G]$.

Proof of Theorem 9.3.1 The proof is similar to that of Theorem 9.3.5.

Let $X = \{\langle\zeta, \xi\rangle \colon \zeta < \eta < \omega_1^M\}$ and replace $\mathbb{P} = (\text{Func}_{\omega_1}(\omega_1, 2))^M$ with $\mathbb{P}_2 = (\text{Func}_{\omega_1}(X, 2))^M$. We can do this by Lemma 9.3.7 since these forcings are isomorphic in M. Let G be an M-generic filter in $\mathbb{P}_2$ and let $g = \bigcup G \in M[G]$. Then g is a function from X to 2. For $\alpha < \omega_1^M$ let $g_\alpha \colon \alpha \to 2$ be defined by $g_\alpha(\zeta) = g(\zeta, \alpha)$ and put $A_\alpha = g_\alpha^{-1}(1)$. It is easy to see that the sequence $\langle A_\alpha \colon \alpha < \omega_1^M\rangle$ belongs to $M[G]$. We will show that it is a $\diamondsuit$-sequence in $M[G]$.

So let $A \in M[G]$ be such that $A \subset \omega_1^{M[G]} = \omega_1^M$ and let $f = \chi_A$. We have to show that the set

$$S = \{\alpha < \omega_1^M \colon A \cap \alpha = A_\alpha\} = \{\alpha < \omega_1^M \colon f|_\alpha = g_\alpha\}$$

is stationary in $M[G]$ (remember that $\omega_1^{M[G]} = \omega_1^M$). For this, let $C \subset \omega_1^M$, $C \in M[G]$, be such that

$$M[G] \models C \text{ is a closed unbounded subset of } \omega_1.$$

We will show that $S \cap C \neq \emptyset$, that is, that there exists an $\alpha \in C$ such that $f|_\alpha = g_\alpha$.

Let $\tau, \sigma \in M$ be $\mathbb{P}_2$-names such that $\mathrm{val}_G(\tau) = f$ and $\mathrm{val}_G(\sigma) = C$. Then there exists a $p \in G$ such that

$$p \Vdash (\sigma \text{ is a closed unbounded subset of } \hat{\omega}_1 \ \& \ \tau \colon \hat{\omega}_1 \to \hat{2}).$$

Define in M the set D by

$$\left\{ q \in \mathbb{P}_2 \colon \exists \alpha < \omega_1^M \left(q \Vdash \left(\hat{\alpha} \in \sigma \ \& \ \forall \zeta < \hat{\alpha} \left(\tau(\zeta) = \left(\bigcup \Gamma \right) (\zeta, \hat{\alpha}) \right) \right) \right) \right\},$$

where Γ is the standard $\mathbb{P}_2$-name for a generic filter. It is enough to prove that D is dense below p since then there exists a $q \in G \cap D$ and so there is an $\alpha \in \mathrm{val}_G(\sigma) = C$ such that

$$f(\zeta) = g(\zeta, \alpha) \quad \text{for all } \zeta < \alpha,$$

that is, $f|_\alpha = g_\alpha$.

To prove that D is dense below p fix an $r \in \mathbb{P}$ such that $r \leq p$. We have to find a $q \in D$ with $q \leq r$.

Define in M the sequences $\langle p_n \in \mathbb{P}_2 \colon n < \omega \rangle$, $\langle \beta_n < \omega_1^M \colon n < \omega \rangle$, $\langle \delta_n < \omega_1^M \colon n < \omega \rangle$, and $\langle b_n \in 2^{\beta_n} \colon n < \omega \rangle$ such that for every $n < \omega$

(1) $p_0 = r$ and $\delta_0 = 0$;

(2) $\mathrm{dom}(p_n) \subset \{ \langle \zeta, \xi \rangle \colon \zeta < \xi < \beta_n \}$;

(3) $p_n \leq p_{n-1}$ if $n > 0$;

(4) $\beta_{n-1} < \delta_n < \beta_n$ if $n > 0$;

(5) $p_n \Vdash \left(\hat{\delta}_n \in \sigma \ \& \ \tau|_{\hat{\beta}_{n-1}} = \hat{b}_n \right)$ if $n > 0$.

To see that such a construction can be made assume that for some $n < \omega$ the sequences are defined for all $i < n$. If $n = 0$ then p_0 and δ_0 are already constructed and β_0 and b_0 can be easily chosen, since the conditions (3), (4), and (5) do not concern this case. So assume that $n > 0$. We have to show that the inductive step can be made, that is, that appropriate p_n, β_n, δ_n, and b_n exist.

So let G_0 be an M-generic filter in $\mathbb{P}_2$ with $p_{n-1} \in G_0$. Then $f_0 = \mathrm{val}_{G_0}(\tau) \in M[G_0]$ is a function from ω_1^M to 2 and $C_0 = \mathrm{val}_{G_0}(\sigma)$ is a closed and unbounded subset of ω_1 in $M[G_0]$, since $p_{n-1} \leq p$ forces it. In particular, there exists a $d \in C_0$ with $d > \beta_{n-1}$. Also, $b = h|_{\beta_{n-1}} \in M$, since $\mathbb{P}_2$ is countably closed. So there exists a $q \in G_0$, with $q \leq p_{n-1}$, that forces it. Then $p_n = q$, $b_n = b$, and $\delta_n = d$ satisfy (3), (5), and the first part of (4). The choice of β_n satisfying (2) and (4) finishes the construction.

Now let $p_\omega = \bigcup_{n<\omega} p_n \in \mathbb{P}_2$. Then for all $n < \omega$

$$p_\omega \Vdash \left(\hat{\delta}_{n+1} \in \sigma \ \& \ \tau|_{\hat{\beta}_n} = \hat{b}_{n+1} \right).$$

Thus the functions b_{n+1} must be compatible, and $b = \bigcup_{n<\omega} b_{n+1} \in M$ is a function from $\alpha = \bigcup_{n<\omega} \beta_n = \bigcup_{n<\omega} \delta_n$ into 2.

Put $q = p_\omega \cup \{\langle \langle \zeta, \alpha \rangle, b(\zeta) \rangle \colon \zeta < \alpha\}$. Notice that q is a function since, by (2), $\mathrm{dom}(p_\omega)$ is disjoint from $\{\langle \zeta, \alpha \rangle \colon \zeta < \alpha\}$. Since $q \leq r$ it is enough to show that q belongs to D.

So let G_1 be an M-generic filter in $\mathbb{P}_2$ containing q and let $C_1 = \mathrm{val}_{G_1}(\sigma)$ and $f_1 = \mathrm{val}_{G_1}(\tau)$. It is enough to show that $\alpha \in C_1$ and $f_1(\zeta) = (\bigcup \Gamma)(\zeta, \alpha)$ for every $\zeta < \alpha$. But C_1 is a closed unbounded subset of $\omega_1^{M[G_1]}$ and $\delta_{n+1} \in C_1$ for every $n < \omega$, since q forces it. Thus $\alpha = \bigcup_{n<\omega} \delta_n \in C_1$. Also, for every $\zeta < \alpha$ there exists an $n < \omega$ such that $\zeta < \beta_n$. So

$$f_1(\zeta) = \mathrm{val}_{G_1}(\tau)|_{\beta_n}(\zeta) = b_{n+1}(\zeta) = b(\zeta) = q(\zeta, \alpha) = \left(\bigcup \Gamma \right)(\zeta, \alpha).$$

This finishes the proof. $\square$

The proof just presented can be easily generalized to the following theorem. Its proof is left as an exercise.

Theorem 9.3.8 *Let M be a countable transitive model of ZFC, κ be a regular cardinal in M, and $\mathbb{P} = (\mathrm{Func}_{\kappa^+}(\kappa^+ \times \kappa, 2))^M$. Then forcing $\mathbb{P}$ preserves all cardinals $\leq \kappa^+$ and $2^\kappa = \kappa^+$ holds in $M[G]$, where G is an M-generic filter in $\mathbb{P}$.*

Theorem 9.3.8 implies immediately the following corollary, which will be used in Section 9.5.

Corollary 9.3.9 *It is relatively consistent with ZFC that $2^{\omega_1} = \omega_2$.*

EXERCISES

1 Complete the details of Lemma 9.3.7 by showing that for every isomorphism f between partially ordered sets $\mathbb{P}$ and $\mathbb{P}_1$ and every $G, D \subset \mathbb{P}$

(1) G is a filter in $\mathbb{P}$ if and only if $f[G]$ is a filter in $\mathbb{P}_1$;

(2) D is dense in $\mathbb{P}$ if and only if $f[D]$ is dense in $\mathbb{P}_1$.

2 Show that the functions b_{n+1} from the proof of Theorem 9.3.1 are indeed compatible and that $b = \bigcup_{n<\omega} b_{n+1}$ is a function from α to 2.

3 Prove Theorem 9.3.8 in the following steps.

(a)   Generalize the proof of Theorem 9.3.5 to show that if $f \in M[G]$ is a function from κ into $A \in M$ then $f \in M$.

(b)   Use (a) and the ideas from the proof of Corollary 9.3.6 to show that $\mathbb{P}$ preserves cardinals $\leq \kappa^+$ and $\mathcal{P}(\kappa) \cap M = \mathcal{P}(\kappa) \cap M[G]$.

(c)   Show that the function $f \colon \kappa^+ \to 2^\kappa$, $f(\alpha)(\beta) = (\bigcup G)(\alpha, \beta)$, is onto 2^κ.

9.4   Product lemma and Cohen model

In the next section we will prove the consistency of MA+¬CH. The method used in its proof is called *iterated forcing*. The idea behind this method is to repeat the forcing extension process recursively, that is, to construct by transfinite induction of some length α a sequence

$$M = M_0 \subset M_1 \subset \cdots \subset M_\xi \subset \cdots \subset M_\alpha$$

of countable transitive models of ZFC such that $M_{\xi+1} = M_\xi[G_\xi]$ for every $\xi < \alpha$, where G_ξ is an M_ξ-generic filter in some forcing $\mathbb{P}_\xi \in M_\xi$.

   This description is very specific and easy to handle at successor stages. The limit stage, however, presents a problem. For a limit ordinal $\lambda \leq \alpha$ we cannot simply take $M_\lambda = \bigcup_{\xi < \lambda} M_\xi$, since such an M_λ does not have to be a model of ZFC. Intuitively, we can avoid this problem by defining M_λ as a generic extension of M by a sequence $\langle G_\xi \colon \xi < \lambda \rangle$. But in what forcing should the sequence $\langle G_\xi \colon \xi < \lambda \rangle$ be a generic filter?

   To solve this problem, we will find one forcing in the ground model that will encode all forcings used in our intuitive inductive construction. In this project, however, we will find that the successor step creates more trouble than the limit one. In fact, at the moment we don't even know whether the model $M_2 = M[G_0][G_1]$ obtained by two consecutive generic extensions can also be obtained by a single generic extension, that is, whether $M_2 = M[G]$ for some $\mathbb{P} \in M$ and an M-generic filter G in $\mathbb{P}$.

   In this section we will concentrate on this problem only in the easier case where both partially ordered sets $\mathbb{P}_0$ and $\mathbb{P}_1$ leading to the extensions $M_1 = M[G_1]$ and $M_2 = M_1[G_1]$, respectively, belong to the ground model M. The general iteration, including that of two-stage iteration, will be described in the next section.

   In the simple case when $\mathbb{P}_0, \mathbb{P}_1 \in M$ the forcing $\mathbb{P}$ is just a product of forcings $\langle \mathbb{P}_0, \leq_0 \rangle$ and $\langle \mathbb{P}_1, \leq_1 \rangle$, that is, $\mathbb{P} = \mathbb{P}_0 \times \mathbb{P}_1$ and

$$\langle p_0, p_1 \rangle \leq \langle q_0, q_1 \rangle \Leftrightarrow p_0 \leq_0 q_0 \ \& \ p_1 \leq_1 q_1.$$

The relation between the forcing extensions obtained by these partially ordered sets is described in the next theorem. We will start, however, with the following simple fact.

Proposition 9.4.1 *Let $\mathbb{P}$ be a product of the forcings $\mathbb{P}_0$ and $\mathbb{P}_1$. Then G is a filter in $\mathbb{P}$ if and only if $G = G_0 \times G_1$ for some filters G_0 and G_1 in $\mathbb{P}_0$ and $\mathbb{P}_1$, respectively.*

Proof First, assume that G is a filter in $\mathbb{P}$. Let G_0 and G_1 be the projections of G onto the first and the second coordinates, respectively. Then $G \subset G_0 \times G_1$.

To see the other inclusion take $p_0 \in G_0$ and $q_1 \in G_1$. Then there exist $q_0 \in G_0$ and $p_1 \in G_1$ such that $\langle p_0, p_1 \rangle, \langle q_0, q_1 \rangle \in G$. Let $\langle r_0, r_1 \rangle \in G$ be an extension of $\langle p_0, p_1 \rangle$ and $\langle q_0, q_1 \rangle$. Then $\langle r_0, r_1 \rangle \leq \langle p_0, q_1 \rangle$, so $\langle p_0, q_1 \rangle \in G$.

The proof that G_0 and G_1 are filters is left as an exercise. It is also easy to see that $G = G_0 \times G_1$ is a filter, provided G_0 and G_1 are. $\square$

Proposition 9.4.1 tells us, in particular, that any filter in $\mathbb{P}_0 \times \mathbb{P}_1$ is a product of filters in $\mathbb{P}_0$ and in $\mathbb{P}_1$. This gives a better perspective on the next theorem.

Theorem 9.4.2 (Product lemma) *Let M be a countable transitive model of ZFC, let $\mathbb{P}_0, \mathbb{P}_1 \in M$ be partially ordered sets, and let $G_0 \subset \mathbb{P}_0$ and $G_1 \subset \mathbb{P}_1$ be filters. Then the following conditions are equivalent.*

(i) $G_0 \times G_1$ *is an M-generic filter in $\mathbb{P}_0 \times \mathbb{P}_1$.*

(ii) G_0 *is an M-generic filter in $\mathbb{P}_0$ and G_1 is an $M[G_0]$-generic filter in $\mathbb{P}_1$.*

(iii) G_1 *is an M-generic filter in $\mathbb{P}_1$ and G_0 is an $M[G_1]$-generic filter in $\mathbb{P}_0$.*

Moreover, if any of these conditions hold then $M[G_0 \times G_1] = M[G_0][G_1] = M[G_1][G_0]$.

Proof Note first that the natural isomorphism between $\mathbb{P}_0 \times \mathbb{P}_1$ and $\mathbb{P}_1 \times \mathbb{P}_0$ maps $G_0 \times G_1$ onto $G_1 \times G_0$. Thus, by Lemma 9.3.7, condition (i) is equivalent to the condition

(i′) $G_1 \times G_0$ is an M-generic filter in $\mathbb{P}_1 \times \mathbb{P}_0$.

But the equivalence (i′)⇔(iii) is obtained from (i)⇔(ii) by exchanging indices. So all equivalences follow from (i)⇔(ii).

To prove (i)$\Rightarrow$(ii), assume (i). To see that G_0 is M-generic in $\mathbb{P}_0$ take a dense subset $D_0 \in M$ of $\mathbb{P}_0$. Then $D_0 \times \mathbb{P}_1 \in M$ is dense in $\mathbb{P}_0 \times \mathbb{P}_1$. Thus, by (i), $(D_0 \times \mathbb{P}_1) \cap (G_0 \times G_1) \neq \emptyset$, and we conclude that $D_0 \cap G_0 \neq \emptyset$.

To see that G_1 is an $M[G_0]$-generic filter in $\mathbb{P}_1$ take a dense subset $D_1 \in M[G_0]$ of $\mathbb{P}_1$. Let $\tau \in M$ be a $\mathbb{P}_0$-name such that $D_1 = \mathrm{val}_{G_0}(\tau)$ and let $p_0 \in G_0$ be such that

$$p_0 \Vdash \tau \text{ is dense in } \hat{\mathbb{P}}_1.$$

Choose $p_1 \in G_1$, define

$$D = \{\langle q_0, q_1 \rangle \colon q_0 \leq p_0 \ \& \ q_0 \Vdash \hat{q}_1 \in \tau\} \in M,$$

and note that D is dense below $\langle p_0, p_1 \rangle$.

Indeed, let $\langle r_0, r_1 \rangle \leq \langle p_0, p_1 \rangle$ and let G_0' be an M-generic filter in $\mathbb{P}_0$ containing r_0. Then $D_1' = \mathrm{val}_{G_0'}(\tau)$ is dense in $\mathbb{P}_1$ since $r_0 \leq p_0$, and p_0 forces it. Take $q_1 \leq r_1$ with $q_1 \in D_1'$ and find a $q_0 \in G_0'$ that forces it:

$$q_0 \Vdash \hat{q}_1 \in \tau.$$

Since q_0 and r_0 belong to the same filter G_0' they are compatible, and taking their common extension, if necessary, we can assume that $q_0 \leq r_0$. But then $\langle q_0, q_1 \rangle \leq \langle r_0, r_1 \rangle$ and $\langle q_0, q_1 \rangle \in D$. Thus D is dense below $\langle p_0, p_1 \rangle \in G_0 \times G_1$.

So we can find $\langle q_0, q_1 \rangle \in D \cap (G_0 \times G_1)$. But $q_0 \in G_0$ and $q_0 \Vdash \hat{q}_1 \in \tau$. Thus $q_1 \in \mathrm{val}_{G_0}(\tau) = D_1$, so we have found $q_1 \in D_1 \cap G_1$.

To prove that (ii) implies (i) take a dense subset D of $\mathbb{P}_0 \times \mathbb{P}_1$ that belongs to M. We will show that $D \cap (G_0 \times G_1) \neq \emptyset$.

For this, define

$$D_1 = \{p_1 \in \mathbb{P}_1 \colon \exists p_0 \in G_0 \ (\langle p_0, p_1 \rangle \in D)\} \in M[G_0]$$

and note that D_1 is dense in $\mathbb{P}_1$. Indeed, if $r_1 \in \mathbb{P}_1$ then the set

$$D_0 = \{p_0 \in \mathbb{P}_0 \colon \exists p_1 \leq r_1 \ (\langle p_0, p_1 \rangle \in D)\} \in M$$

is dense in $\mathbb{P}_0$ and so there exists a $p_0 \in D_0 \cap G_0$. Thus there exists a $p_1 \leq r_1$ with $\langle p_0, p_1 \rangle \in D$ and $p_1 \in D_1$.

Now, by genericity, there exists a $p_1 \in G_1 \cap D_1$. So, by the definition of D_1, we can find a $p_0 \in G_0$ with $\langle p_0, p_1 \rangle \in D$. But then $\langle p_0, p_1 \rangle \in D \cap (G_0 \times G_1)$, that is, $D \cap (G_0 \times G_1) \neq \emptyset$.

Finally, if conditions (i)–(iii) hold then the equation $M[G_0 \times G_1] = M[G_0][G_1]$ holds by the minimality of different generic extensions.

To see that $M[G_0 \times G_1] \subset M[G_0][G_1]$ note that $G_0 \in M[G_0] \subset M[G_0][G_1]$ and $G_1 \in M[G_0][G_1]$. So $G_0 \times G_1 \in M[G_0][G_1]$. Moreover,

$M \subset M[G_1] \subset M[G_0][G_1]$. Thus, by the minimality of $M[G_0 \times G_1]$, the inclusion holds.

To see the reverse inclusion note that $M \subset M[G_0 \times G_1]$ and $G_0 \in M[G_0 \times G_1]$. So, by the minimality of $M[G_0]$, we have $M[G_0] \subset M[G_0 \times G_1]$. But we have also $G_1 \in M[G_0 \times G_1]$. Thus $M[G_0][G_1] \subset M[G_0 \times G_1]$ by the minimality of $M[G_0][G_1]$.

The equation $M[G_0 \times G_1] = M[G_1][G_0]$ is proved similarly. $\square$

Next, we will see an application of Theorem 9.4.2 to the Cohen model, that is, the model from Section 9.2.

First note that if $\mathbb{P}(Z) = \langle \mathrm{Func}_\omega(Z,2), \supset \rangle$ and $\{A, B\}$ is a partition of Z then $\mathbb{P}(Z)$ is isomorphic to $\mathbb{P}(A) \times \mathbb{P}(B)$ via the isomorphism $i\colon \mathbb{P}(Z) \to \mathbb{P}(A) \times \mathbb{P}(B)$, $i(s) = \langle s|_A, s|_B \rangle$. Also, if G is a filter in $\mathbb{P}(Z)$ and $C \subset Z$ then

$$G_C = \{s|_C \colon s \in G\} = G \cap \mathbb{P}(C)$$

and $i[G] = G_A \times G_B = (G \cap \mathbb{P}(A)) \times (G \cap \mathbb{P}(B))$.

Now, if M is a countable transitive model of ZFC, $Z, A, B \in M$ are such that $\{A, B\}$ is a partition of Z, and G is an M-generic filter in $\mathbb{P}(Z)$ then, by Theorem 9.4.2, $G_A = G \cap \mathbb{P}(A)$ is an M-generic filter in $\mathbb{P}(A)$, $G_B = G \cap \mathbb{P}(B)$ is an $M[G_A]$-generic filter in $\mathbb{P}(B)$, and

$$M[G] = M[G_A \times G_B] = M[G_A][G_B].$$

This fact and the next lemma are powerful tools for proving different facts about the Cohen model.

Lemma 9.4.3 *Let M be a countable transitive model of ZFC, $X, Y, Z \in M$, and G an M-generic filter in $\mathbb{P}(Z)$. If $f\colon X \to Y$ is in $M[G]$ then there exists an $A \in \mathcal{P}^M(Z)$ such that*

$$M \models |A| \le |X| + \omega$$

and $f \in M[G \cap \mathbb{P}(A)]$.

Proof By Lemma 9.2.8 we can find a $\mathbb{P}(Z)$-name $\tau \in M$ such that $f = \mathrm{val}_G(\tau)$, $\tau \subset \{\langle \widehat{\langle x, y \rangle}, p \rangle \colon x \in X \ \& \ y \in Y \ \& \ p \in \mathbb{P}(Z)\}$, and

$(\star)$ $A_x = \{p \in \mathbb{P}(Z) \colon \exists y \in Y \ (\langle \widehat{\langle x, y \rangle}, p \rangle \in \tau)\}$ is an antichain for every $x \in X$.

So every A_x is countable in M, since $\mathbb{P}(Z)$ is ccc in M.

Let

$$A = \bigcup_{x \in X} \bigcup_{p \in A_x} \mathrm{dom}(p).$$

Then $A \in M$, $A \subset Z$, and $|A| \leq |X| + \omega$ in M, since every set $\bigcup_{p \in A_x} \operatorname{dom}(p)$ is countable in M. To finish the proof it is enough to show that $f \in M[G \cap \mathbb{P}(A)]$.

But for every $s = \langle \langle x, y \rangle, p \rangle \in \tau$ we have $p \in A_x \subset \mathbb{P}(A)$. Thus

$$\tau^\star = \{\langle \widehat{\langle x, y \rangle}^\star, p \rangle \colon \langle \widehat{\langle x, y \rangle}, p \rangle \in \tau\} \in M$$

is a $\mathbb{P}(A)$-name, where $\widehat{\langle x, y \rangle}^\star$ is a standard $\mathbb{P}(A)$-name for $\langle x, y \rangle \in M$. So

$$\begin{aligned} f = \operatorname{val}_G(\tau) &= \{\langle x, y \rangle \colon \exists p \in G \, (\langle \widehat{\langle x, y \rangle}, p \rangle \in \tau)\} \\ &= \{\langle x, y \rangle \colon (\exists p \in G \cap \mathbb{P}(A))(\langle \widehat{\langle x, y \rangle}^\star, p \rangle \in \tau^\star)\} \\ &= \operatorname{val}_{G \cap \mathbb{P}(A)}(\tau^\star) \in M[G \cap \mathbb{P}(A)]. \end{aligned}$$

This finishes the proof. $\qquad\square$

Since every real number is identified with a function $r \colon \omega \to 2$, we obtain immediately the following corollary.

Corollary 9.4.4 *Let M be a countable transitive model of ZFC, $Z \in M$, and G be an M-generic filter in $\mathbb{P}(Z)$. If $r \in \mathbb{R}^{M[G]}$ then, in M, there exists a countable subset A of Z such that $r \in M[G \cap \mathbb{P}(A)]$.*

Our next goal is to prove that Martin's axiom is false in the Cohen model obtained by the forcing $\mathbb{P}(Z)$ with $|Z| > \omega_1$. For this, first note that the formula "r is a real number," which is identified with "r is a function from ω into 2," is absolute. So for every countable transitive model M of ZFC we have

$$\mathbb{R}^M = \mathbb{R} \cap M \quad \text{and} \quad [0, 1]^M = [0, 1] \cap M.$$

In particular, if $M[G]$ is any generic extension of M then

$$[0, 1] \cap M, \mathbb{R} \cap M \in M[G],$$

since $\mathbb{R} \cap M = \mathbb{R}^M \in M \subset M[G]$ and $[0, 1] \cap M = [0, 1]^M \in M \subset M[G]$.

Lemma 9.4.5 *Let M be a countable transitive model of ZFC, let $Z \in M$ be countable in M, and let G be an M-generic filter in $\mathbb{P}(Z)$. Then $M \cap [0, 1] \in M[G]$ has Lebesgue measure zero in $M[G]$.*

Proof Since the forcings $\mathbb{P}(Z)$ and $\mathbb{P}(\omega)$ are isomorphic in M we can assume, by Lemma 9.3.7, that $Z = \omega$. Also, we will identify numbers from $[0, 1]$ with their binary expansions, that is, functions from ω into 2.

For $s \in \mathbb{P}(\omega)$ let $[s] = \{t \in [0,1]: s \subset t\}$. Notice that the Lebesgue measure of $[s]$ is equal to $l([s]) = 2^{-n}$, where $n = |s|$. Now let $r = \bigcup G \in 2^\omega$ and for $k < \omega$ put

$$S_k = \bigcup_{m > k} [r|_{[2^m, 2^{m+1})}].$$

Notice that

$$l(S_k) \leq \sum_{m > k} l([r|_{[2^m, 2^{m+1})}]) = \sum_{m > k} 2^{-2^m} \leq \sum_{m > k} 2^{-m} = 2^{-k}.$$

Thus the set $S = \bigcap_{k < \omega} S_k$ has measure zero. It is enough to prove that $M \cap [0,1] \subset S_k$ for every $k < \omega$.

So fix $k < \omega$ and $t \in M \cap [0,1]$ and define

$$D = \left\{ s \in \mathbb{P}(\omega): \exists m > k \ \left(s|_{[2^m, 2^{m+1})} = t|_{[2^m, 2^{m+1})} \right) \right\} \in M.$$

It is easy to see that D is dense in $\mathbb{P}(\omega)$. Thus, $G \cap D \neq \emptyset$, that is, there is an $m > k$ such that $r|_{[2^m, 2^{m+1})} = t|_{[2^m, 2^{m+1})}$. So $t \in [r|_{[2^m, 2^{m+1})}] \subset S_k$. □

Theorem 9.4.6 *Let M be a countable transitive model of ZFC, $Z \in M$ uncountable in M, and G an M-generic filter in $\mathbb{P}(Z)$. Then, in $M[G]$, the interval $[0,1]$ is a union of ω_1 sets of Lebesgue measure zero.*

Proof First assume that Z has cardinality ω_1 in M. Then the forcings $\mathbb{P}(Z)$ and $\mathbb{P}(\omega_1 \times \omega)$ are isomorphic. So, by Lemma 9.3.7, we can assume that $Z = \omega_1 \times \omega$.

Now, for every $\xi < \omega_1$,

$$M[G] = M[G \cap \mathbb{P}((\omega_1 \setminus \{\xi\}) \times \omega)][G \cap \mathbb{P}(\{\xi\} \times \omega)],$$

so, by Lemma 9.4.5 used with $M[G \cap \mathbb{P}((\omega_1 \setminus \{\xi\}) \times \omega)]$ as a ground model, there exists a set $S_\xi \in M[G]$ of Lebesgue measure zero such that

$$M[G \cap \mathbb{P}((\omega_1 \setminus \{\xi\}) \times \omega)] \cap [0,1] \subset S_\xi.$$

But, by Corollary 9.4.4, for every $r \in M[G] \cap [0,1]$ there exists in M a countable subset A of Z with $r \in M[G \cap \mathbb{P}(A)]$. Hence there exists a $\xi < \omega_1$ such that $A \subset (\omega_1 \setminus \{\xi\}) \times \omega$ and

$$r \in M[G \cap \mathbb{P}(A)] \subset M[G \cap \mathbb{P}((\omega_1 \setminus \{\xi\}) \times \omega)].$$

Therefore $M[G] \cap [0,1] \subset \bigcup_{\xi < \omega_1} S_\xi$. This finishes the proof of the case when $M \models |Z| = \omega_1$.

For the general case, take $Z_0 \in \mathcal{P}^M(Z)$ of cardinality ω_1 in M. Then $M[G] = M_1[G_1]$, where $M_1 = M[G \cap \mathbb{P}(Z \setminus Z_0)]$ and $G_1 = G \cap \mathbb{P}(Z_0)$ is an M_1-generic filter in $\mathbb{P}(Z_0)$. The conclusion of the theorem follows from the first case. □

Corollary 9.4.7 *Let M be a countable transitive model of ZFC and assume that $Z \in M$ has cardinality at least ω_2 in M. If G is an M-generic filter in $\mathbb{P}(Z)$ then, in $M[G]$, the interval $[0,1]$ is a union of less than continuum many sets of Lebesgue measure zero.*

In particular, Martin's axiom is false in $M[G]$.

Proof Theorem 9.2.13 implies that $\mathfrak{c} > \omega_1$ in $M[G]$. Hence, by Theorem 9.4.6, we can conclude the first part of the corollary. The second part follows from the first part and Theorem 8.2.7. $\qquad\square$

We will finish this section with an estimation of the size of the continuum in the models obtained by ccc forcing extensions.

Theorem 9.4.8 *Let M be a countable transitive model of ZFC, $\mathbb{P} \in M$ be an infinite ccc forcing in M, and G be an M-generic filter in $\mathbb{P}$. If $\kappa = |\mathbb{P}^\omega|^M$ then*

$$M[G] \models \mathfrak{c} \leq \kappa.$$

Proof Since, by Theorem 9.2.11, the models M and $M[G]$ have the same cardinals, we will not distinguish between them. Let $\mathcal{F}$ be the family of all $\mathbb{P}$-names $\tau \subset \{\langle \widehat{\langle n, i \rangle}, p \rangle \colon n \in \omega \ \& \ i \in 2 \ \& \ p \in \mathbb{P}\}$ such that

$(\star)$ $\quad A_n = \{p \in \mathbb{P} \colon \exists i \in 2 \ (\langle \widehat{\langle n, i \rangle}, p \rangle \in \tau)\} \neq \emptyset$ is an antichain for every $n \in \omega$.

Then $\mathcal{F} \in M$ and every A_n is at most countable.

By Lemma 9.2.8 for every $f \in 2^\omega \cap M[G]$ we can find a $\tau \in \mathcal{F}$ such that $f = \mathrm{val}_G(\tau)$. In particular, the image $\mathrm{val}_G[\mathcal{F}]$ of $\mathcal{F}$ under val_G contains $2^\omega \cap M[G]$. So $\mathfrak{c} \leq |\mathcal{F}|$ in $M[G]$, and it is enough to prove that $|\mathcal{F}| \leq \kappa$. We will show this in M.

To see it, define in M a map $h \colon (2 \times \mathbb{P})^{\omega \times \omega} \to \mathcal{F}$ such that for every $t \in (2 \times \mathbb{P})^{\omega \times \omega}$ if $t(n, j) = \langle i_{n,j}, p_{n,j} \rangle$ then

$$h(t) = \{\langle \widehat{\langle n, i_{n,j} \rangle}, p_{n,j} \rangle \colon n, j \in \omega\}.$$

Note that h is onto $\mathcal{F}$, since every set A_n is at most countable. Thus

$$|\mathcal{F}| \leq \left|(2 \times \mathbb{P})^{\omega \times \omega}\right| = \kappa.$$

This finishes the proof. $\qquad\square$

Corollary 9.4.9 *Let M be a countable transitive model of ZFC, κ an infinite cardinal in M, and G an M-generic filter in $\mathbb{P}(\kappa)$. If $\kappa = |\kappa^\omega|^M$ then*

$$M[G] \models \mathfrak{c} = \kappa.$$

In particular, if CH holds in M and $\kappa = \omega_2^M$ then $\mathfrak{c}^{M[G]} = \omega_2^{M[G]}$.

Proof First note that $|\mathbb{P}(\kappa)| = \kappa$, and so $\kappa = |\kappa^\omega|^M = |\mathbb{P}(\kappa)^\omega|^M$. Thus, by Theorem 9.4.8, $\mathfrak{c}^{M[G]} \leq \kappa$. The other inequality for $\kappa = \omega_2$ follows from Theorem 9.2.13. The general case is left as an exercise.

To see the additional part note that under CH the regularity of ω_2 implies

$$\omega_2 \leq \omega_2^\omega = \left| \bigcup_{\xi < \omega_2} \xi^\omega \right| = \omega_2 \otimes \omega_1^\omega = \omega_2 \otimes \mathfrak{c}^\omega = \omega_2 \otimes \mathfrak{c} = \omega_2 \otimes \omega_1 = \omega_2.$$

So, under CH, $\kappa = \omega_2$ satisfies the assumptions of the main part of the corollary. $\square$

EXERCISES

1 Complete the proof of Proposition 9.4.1.

2 Show that the set D from the proof of Lemma 9.4.5 is indeed dense in $\mathbb{P}(\omega)$.

3 Complete the proof of Corollary 9.4.9 for arbitrary κ by showing that $\mathfrak{c}^{M[G]} \geq \kappa$.

4 Let M be a countable transitive model of ZFC, $\mathbb{P} \in M$ a partially ordered set, and G an M-generic filter in $\mathbb{P}$. Prove that $M[G]$ and M have the same ordinal numbers.

9.5 Model for MA+¬CH

The goal of this section is to prove the consistency of MA+"$\mathfrak{c} = \omega_2$." For this, we will follow the idea mentioned at the beginning of the previous section. We will start with a countable transitive model M of ZFC+"$2^{\omega_1}=\omega_2$," which exists by Corollary 9.3.9, and find its generic extension N via a ccc forcing $\mathbb{P}_\mu \in M$ for which there exists a sequence

$$M = M_0 \subset M_1 \subset \cdots \subset M_\xi \subset \cdots \subset M_\mu = N$$

with $\mu = \omega_2^M = \omega_2^N$ and the following properties:

P1 $M_{\xi+1} = M_\xi[G^\xi]$ for every $\xi < \mu$, where G^ξ is an M_ξ-generic filter in some forcing $\mathbb{P}^\xi \in M_\xi$.

P2 For every $Y \in M$ and $S \subset Y$ with $N \models |S| < \omega_2$ there exists an $\alpha < \mu$ such that $S \in M_\alpha$.

P3 $|(\mathbb{P}_\mu)^\omega|^M \leq \omega_2^M$.

P4 For every forcing $\langle \lambda, \ll \rangle \in N$ such that

$$N \models (\lambda \text{ is an ordinal number, } \lambda \leq \omega_1, \text{ and } \langle \lambda, \ll \rangle \text{ is ccc})$$

and every $\alpha < \mu$ there is a $\xi < \mu$, with $\xi > \alpha$, such that $\mathbb{P}^\xi = \langle \lambda, \ll \rangle$.

To argue that P1–P4 imply $N \models MA+\text{“}\mathfrak{c} = \omega_2\text{”}$ we also need the following definitions and lemma. For an uncountable cardinal number κ and a partially ordered set $\mathbb{P}$ we will write $MA_\kappa(\mathbb{P})$ to denote the statement

$MA_\kappa(\mathbb{P})$ For every family $\mathcal{D}$ of dense subsets of $\mathbb{P}$ such that $|\mathcal{D}| < \kappa$ there exists a $\mathcal{D}$-generic filter F in $\mathbb{P}$.

We will also use the symbol MA_κ for the statement that $MA_\kappa(\mathbb{P})$ holds for every ccc forcing $\mathbb{P}$. Thus MA is equivalent to $MA_\mathfrak{c}$.

Lemma 9.5.1 *For an uncountable cardinal number κ the following statements are equivalent:*

(i) MA_κ;

(ii) $MA_\kappa(\mathbb{P})$ *holds for every ccc forcing* $\mathbb{P} = \langle P, \leq \rangle$ *with* $|P| < \kappa$;

(iii) $MA_\kappa(\mathbb{P})$ *holds for every ccc forcing* $\mathbb{P} = \langle \lambda, \ll \rangle$*, where λ is a cardinal number less than κ.*

Before we prove Lemma 9.5.1 let us see how it can be used to conclude $N \models MA+\text{“}\mathfrak{c} = \omega_2\text{”}$ from P1–P4.

First, we will prove that

$$N \models MA_{\omega_2}. \tag{9.7}$$

By Lemma 9.5.1 it is enough to show in N that $MA_{\omega_2}(\mathbb{P})$ holds for every ccc forcing $\mathbb{P} = \langle \lambda, \ll \rangle$ with λ being an ordinal number less than or equal to ω_1^N. For this, let $\langle \lambda, \ll \rangle \in N$ be ccc in N with $\lambda \leq \omega_1^N$, and let $\mathcal{D} = \{D_\xi : \xi < \omega_1^N\} \in N$ be a family of dense subsets of $\langle \lambda, \ll \rangle$. Since $\mathbb{P}_\mu$ is ccc we have $\lambda \leq \omega_1^N = \omega_1^M$, and $\ll$ as well as $\bigcup_{\xi < \omega_1^M} \{\xi\} \times D_\xi$ are subsets of $\omega_1^M \times \omega_1^M \in M$. Hence, by P2, there exists an $\alpha < \mu$ such that $\langle \lambda, \leq \rangle, \mathcal{D} \in M_\alpha$. Moreover, by P4 there exists $\alpha < \xi < \mu$ with $\mathbb{P}^\xi = \langle \lambda, \ll \rangle$. In particular, $G^\xi \in M_{\xi+1}$ is an M_ξ-generic filter in $\mathbb{P}^\xi = \langle \lambda, \ll \rangle$. But $\mathcal{D} \subset M_\alpha \subset M_\xi$. So G^ξ is also $\mathcal{D}$-generic in $\langle \lambda, \ll \rangle$. Condition (9.7) has been proved.

Now MA_{ω_2} and $\mathfrak{c} = \omega_2$ clearly imply MA. Thus, it is enough to argue that $N \models \mathfrak{c} = \omega_2$. But the inequality $\mathfrak{c}^N \leq \omega_2^N$ follows from P3 and Theorem 9.4.8. The other inequality follows from the fact that

$$\text{MA}_{\omega_2} \ \Rightarrow \ \forall G \in [2^\omega]^{\leq \omega_1} \ \exists f \in 2^\omega \setminus G. \tag{9.8}$$

To see it, put $\mathbb{P}(\omega) = \langle \text{Func}_\omega(\omega, 2), \supset \rangle$ and for every $g \in 2^\omega$ define $D_g = \{s \in \mathbb{P}(\omega) : s \not\subset g\}$. Then the sets D_g are dense in $\mathbb{P}(\omega)$. Let F be a $\{D_g : g \in G\}$-generic filter in $\mathbb{P}(\omega)$. Then any extension of $\bigcup F$ to a function $f : \omega \to 2$ will have the desired property. The details are left as an exercise.

Before we move to the technical aspects of the construction, it is worthwhile to reflect for a moment on the idea behind the foregoing argument. It is clearly of the transfinite-induction nature and the argument is of a diagonal character. Condition P1 represents an inductive step in which we take care of "one problem at a time." Condition P2 represents a kind of closure argument. It tells us that the "small objects" from N can already be found in the earlier steps of our construction and thus we will have a chance to "take care of them" in the later part of the induction.

The construction of the forcing $\mathbb{P}_\mu$ leading to a model satisfying P1 and P2 is of a very general nature and can be compared to the recursion theorem. The specific aspects of our model are addressed mainly by condition P4, with Lemma 9.5.1 allowing us to reduce the induction to one of reasonably short length.

This construction closes, in a way, a full circle that we have made in this course. We started with simple recursion proofs. Next, when the difficulties with the length of induction mounted, we started to use refined recursive arguments by introducing additional axioms such as CH, $\diamondsuit$, and MA. This led us all the way to the forcing arguments of the previous sections, and finally to finish here by coming back to a relatively simple transfinite-induction argument.

The preceding argument shows that in order to prove the consistency of MA+"$\mathfrak{c} = \omega_2$" it is enough to prove Lemma 9.5.1 and find a generic extension N of M satisfying P1–P4. We will start with Lemma 9.5.1.

Proof of Lemma 9.5.1 Clearly (i)$\Rightarrow$(iii). Thus it is enough to show (iii)$\Rightarrow$(ii) and (ii)$\Rightarrow$(i).

(ii)$\Rightarrow$(i): Let $\mathbb{P} = \langle P, \leq \rangle$ be a ccc partially ordered set and let $\mathcal{D}$ be a family of dense subsets of $\mathbb{P}$ such that $|\mathcal{D}| < \kappa$. Choose a function $f : P \times P \to P$ such that $f(p, q)$ extends p and q provided p and q are compatible in $\mathbb{P}$, and for every $D \in \mathcal{D}$ pick $f_D : P \to D$ such that $f_D(p) \leq p$ for every $p \in P$. Let $p_0 \in \mathbb{P}$ and let Q be the smallest subset of P containing p_0 and being closed under the operations f and f_D for every $D \in \mathcal{D}$. Thus

$$Q = \bigcup_{n < \omega} Q_n,$$

where $Q_0 = \{p_0\}$ and $Q_{n+1} = Q_n \cup f[Q_n \times Q_n] \cup \bigcup_{D \in \mathcal{D}} f_D[Q_n]$ (compare with Lemma 6.1.6). Then $|Q| \leq |\mathcal{D}| + \omega < \kappa$.

Note that any $p, q \in Q$ that are compatible in $\mathbb{P}$ are also compatible in $\mathbb{P}_0 = \langle Q, \leq \rangle$, since Q is closed under the action of f. Thus $\mathbb{P}_0$ is ccc and, by (ii), $\mathrm{MA}_\kappa(\mathbb{P}_0)$ holds. But, by the closure of Q under the action of f_D, for every $D \in \mathcal{D}$ the set $D \cap Q$ is dense in $\mathbb{P}_0$. So there exists a $\{D \cap Q \colon D \in \mathcal{D}\}$-generic filter F_0 in $\mathbb{P}_0$. To finish the proof it is enough to notice that $F = \{p \in P \colon \exists q \in F_0 \ (q \leq p)\}$ is a $\mathcal{D}$-generic filter in $\mathbb{P}$. The details are left as an exercise.

(iii)$\Rightarrow$(ii): The proof is very similar to that of Lemma 9.3.7. To see it, let $\mathbb{P} = \langle P, \leq \rangle$ be a ccc partially ordered set such that $\lambda = |P| < \kappa$ and let $\mathcal{D}$ be a family of dense subsets of $\mathbb{P}$ such that $|\mathcal{D}| < \kappa$. Choose a bijection f between λ and P and define a partial-order relation $\ll$ on λ by putting

$$\alpha \ll \beta \quad \Leftrightarrow \quad f(\alpha) \leq f(\beta).$$

Notice that f is an order isomorphism between $\mathbb{P}_0 = \langle \lambda, \ll \rangle$ and $\mathbb{P} = \langle P, \leq \rangle$. In particular, $f^{-1}(D)$ is dense in $\mathbb{P}_0$ for every $D \in \mathcal{D}$. So $\mathcal{D}_0 = \{f^{-1}(D) \colon D \in \mathcal{D}\}$ is a family of dense subsets of $\mathbb{P}_0$ and $|\mathcal{D}_0| = |\mathcal{D}| < \kappa$. Hence, by (iii), there exists a $\mathcal{D}_0$-generic filter F_0 in $\mathbb{P}_0$. To finish the proof it is enough to notice that $f[F_0]$ is a $\mathcal{D}$-generic filter in $\mathbb{P}$. The details are left as an exercise. $\qquad\square$

To find a model N satisfying P1–P4 we first need to come back to the problem of expressing a model obtained by two consecutive generic extensions as a single generic extension. More precisely, if $M_1 = M[G_0]$, where G_0 is an M-generic filter in $\mathbb{P}_0 \in M$, and $M_2 = M[G_0][G_1]$, where G_1 is an $M[G_0]$-generic filter in $\mathbb{P}_1 \in M[G_0]$, we would like to find a $\mathbb{P} \in M$ and an M-generic filter G in $\mathbb{P}$ such that $M_2 = M[G]$. The product lemma (Theorem 9.4.2) gives a solution to this problem when $\mathbb{P}_1 \in M$. Thus, we will concentrate here on the case when $\mathbb{P}_1 \in M[G_0] \setminus M$. To define such $\mathbb{P}$ we need the following definition.

A pair $\langle \pi, \sigma \rangle \in M$ of $\mathbb{P}$-names satisfying the condition

$$\mathbb{P} \Vdash \sigma \text{ is a partial-order relation on } \pi \tag{9.9}$$

will be called a *good* $\mathbb{P}$-*name* for a partially ordered set. (Note that, formally speaking, a good $\mathbb{P}$-name is not a $\mathbb{P}$-name.) If $\langle \pi, \sigma \rangle$ is a good $\mathbb{P}$-name for a partially ordered set we will often write π for $\langle \pi, \sigma \rangle$.

The following lemma tells us that any forcing $\mathbb{P}_1$ in $M[G_0]$ has a good $\mathbb{P}_0$-name representing it. So it explains the restriction of our attention to good names only. However, the lemma will not be used in what follows, so we will leave it without a proof.

Lemma 9.5.2 *Let M be a countable transitive model of ZFC, $\mathbb{P} \in M$ a partially ordered set, and G_0 an M-generic filter in $\mathbb{P}$. If $\mathbb{P}_1 = \langle P, \leq \rangle \in M[G_0]$ is a partially ordered set in $M[G_0]$ then there are $\mathbb{P}$-names $\pi, \sigma \in M$ satisfying (9.9) and such that $\langle P, \leq \rangle = \langle \mathrm{val}_{G_0}(\pi), \mathrm{val}_{G_0}(\sigma) \rangle$.*

Now let $\mathbb{P}$ be a partially ordered set and $\langle \pi, \leq_\pi \rangle$ be a good $\mathbb{P}$-name for a partially ordered set. Then their *iteration* $\mathbb{P} \star \pi$ is defined by

$$\mathbb{P} \star \pi = \{\langle p, \tau \rangle \colon p \in \mathbb{P} \ \& \ \tau \in \mathrm{dom}(\pi) \ \& \ p \Vdash \tau \in \pi\}.$$

Moreover, on $\mathbb{P} \star \pi$ we define a binary relation $\preceq$ by

$$\langle p, \tau \rangle \preceq \langle q, \sigma \rangle \quad \Leftrightarrow \quad p \leq q \ \& \ p \Vdash \tau \leq_\pi \sigma.$$

Notice that the relation $\preceq$ is reflexive and transitive on $\mathbb{P} \star \pi$. It is reflexive since for any $\langle p, \tau \rangle \in \mathbb{P} \star \pi$ condition p forces $\tau \in \pi$ and $\leq_\pi$ to be a partial-order relation on π (since $\langle \pi, \leq_\pi \rangle$ is good). The proof that $\preceq$ is transitive is left as an exercise.

Unfortunately, $\preceq$ does not have to be antisymmetric on $\mathbb{P} \star \pi$. To solve this problem, we notice that the relation $\sim$ on $\mathbb{P} \star \pi$ defined by

$$\langle p, \tau \rangle \sim \langle q, \sigma \rangle \Leftrightarrow \langle p, \tau \rangle \preceq \langle q, \sigma \rangle \ \& \ \langle q, \sigma \rangle \preceq \langle p, \tau \rangle$$

is an equivalence relation on $\mathbb{P} \star \pi$ (see Exercise 4 from Section 2.4). Thus, in order to consider $\langle \mathbb{P} \star \pi, \preceq \rangle$ as a partially ordered set, we will identify the elements of $\mathbb{P} \star \pi$ that are $\sim$-equivalent. (Formally, we should replace $\mathbb{P} \star \pi$ by the quotient class $\mathbb{P} \star \pi / _\sim$ and the relation $\preceq$ by the quotient relation $\leq$ defined as in part (b) of Exercise 4 in Section 2.4. However, this would obscure the clarity of the "simple-identification" approach.) Upon such an identification we will consider $\mathbb{P} \star \pi$ as a partially ordered set.

Let us also notice that if the partially ordered sets $\mathbb{P}_0$ and $\mathbb{P}_1$ are in the ground model M and π and $\leq_\pi$ are the standard $\mathbb{P}_0$-names for $\mathbb{P}_1$ and its order relation then $\langle \pi, \leq_\pi \rangle$ is a good name for a partially ordered set and

$$\mathbb{P}_0 \star \pi \ \text{is isomorphic to} \ \mathbb{P}_0 \times \mathbb{P}_1. \tag{9.10}$$

Thus the process of iteration is truly a generalization of the product of forcings as described in the previous section. To formulate an iteration analog of the product lemma, we need one more definition.

Let M be a countable transitive model of ZFC, $\mathbb{P}_0 \in M$ be a partially ordered set, and G_0 be an M-generic filter in $\mathbb{P}_0$. If π is a good $\mathbb{P}_0$-name for a partially ordered set and $G_1 \subset \mathrm{val}_{G_0}(\pi)$ then we define

$$G_0 \star G_1 = \{\langle p, \tau \rangle \in \mathbb{P}_0 \star \pi \colon p \in G_0 \ \& \ \mathrm{val}_{G_0}(\tau) \in G_1\}.$$

Theorem 9.5.3 *Let M be a countable transitive model of ZFC and $\mathbb{P}_0 \star \pi$ be an iteration of a partially ordered set $\mathbb{P}_0 \in M$ and a good $\mathbb{P}_0$-name $\pi \in M$ for a partially ordered set. If G is an M-generic filter in $\mathbb{P}_0 \star \pi$ and*

$$G_0 = \{p \in \mathbb{P}_0 \colon \exists \langle q, \tau \rangle \in G \ (q \le p)\}$$

then G_0 is an M-generic filter in $\mathbb{P}_0$. Moreover, if

$$G_1 = \{\mathrm{val}_{G_0}(\tau) \colon \langle q, \tau \rangle \in G\}$$

then G_1 is an $M[G_0]$-generic filter in $\mathbb{P}_1 = \mathrm{val}_{G_0}(\pi)$, $G = G_0 \star G_1$, and $M[G] = M[G_0][G_1]$.

Proof It is easy to see that G_0 is a filter in $\mathbb{P}_0$. To show that G_0 is M-generic take a $D_0 \in M$ that is dense in $\mathbb{P}_0$. Then $D = \{\langle p, \tau \rangle \in \mathbb{P}_0 \star \pi \colon p \in D_0\}$ is dense in $\mathbb{P}_0 \star \pi$, since $\langle p, \tau \rangle \preceq \langle q, \tau \rangle$ and $\langle p, \tau \rangle \in D$ for every $\langle q, \tau \rangle \in \mathbb{P}_0 \star \pi$ and $p \in D_0$ with $p \le q$. Take $\langle p, \tau \rangle \in G \cap D$. Then $p \in G_0 \cap D_0$. Thus G_0 is an M-generic filter in $\mathbb{P}_0$.

Next, we will show that G_1 is a filter in $\mathbb{P}_1$. So take $\mathrm{val}_{G_0}(\tau) \in G_1$ with $\langle q, \tau \rangle \in G$ witnessing it and let $\sigma \in \mathrm{dom}(\pi)$ be such that $\mathrm{val}_{G_0}(\sigma) \in \mathbb{P}_1$ and $\mathrm{val}_{G_0}(\tau) \le \mathrm{val}_{G_0}(\sigma)$. We will show that $\mathrm{val}_{G_0}(\sigma) \in G_1$. For this, take $p_0 \in G_0$ with $p_0 \Vdash \tau \le \sigma$ and $\langle p, \rho \rangle \in G$ such that $p \le p_0$. Pick $\langle r, \eta \rangle \in G$ with $\langle r, \eta \rangle \preceq \langle q, \tau \rangle$ and $\langle r, \eta \rangle \preceq \langle p, \rho \rangle$. Then $r \Vdash \eta \le \tau$ and $r \Vdash \tau \le \sigma$, since $r \le p_0$. So $r \Vdash \eta \le \sigma$, and $\langle r, \eta \rangle \preceq \langle r, \sigma \rangle$. Hence $\langle r, \sigma \rangle \in G$ and $\mathrm{val}_{G_0}(\sigma) \in G_1$.

To see that any two elements of G_1 have a common extension in G_1 take $\mathrm{val}_{G_0}(\tau) \in G_1$ with $\langle q, \tau \rangle \in G$ and $\mathrm{val}_{G_0}(\sigma) \in G_1$ with $\langle p, \sigma \rangle \in G$. Then there exists $\langle r, \eta \rangle \in G$ such that $\langle r, \eta \rangle \preceq \langle q, \tau \rangle$ and $\langle r, \eta \rangle \preceq \langle p, \sigma \rangle$. So $r \in G_0$ and $r \Vdash (\eta \le \tau \ \& \ \eta \le \sigma)$. Thus $\mathrm{val}_{G_0}(\eta) \in G_1$ extends $\mathrm{val}_{G_0}(\tau)$ and $\mathrm{val}_{G_0}(\sigma)$.

For the proof that G_1 is $M[G_0]$-generic in $\mathbb{P}_1$ take a $D_1 \in M[G_0]$ that is dense in $\mathbb{P}_1$ and let δ be a $\mathbb{P}_0$-name such that $D_1 = \mathrm{val}_{G_0}(\delta)$. Pick $p \in G_0$ that forces it, that is,

$$p \Vdash \delta \text{ is a dense subset of } \pi,$$

and define

$$D = \{\langle q, \tau \rangle \in \mathbb{P}_0 \star \pi \colon q \Vdash \tau \in \delta\}. \tag{9.11}$$

Now, if $\langle q, \eta \rangle \in G$ is such that $q \le p$ then it is easy to see that D is dense below $\langle q, \eta \rangle$ (the proof is left as an exercise). So there exists $\langle r, \tau \rangle \in D \cap G$, and $\mathrm{val}_{G_0}(\tau) \in D_1 \cap G_1$.

To argue that $G = G_0 \star G_1$ take $\langle p, \tau \rangle \in G$. Then $p \in G_0$ and $\mathrm{val}_{G_0}(\tau) \in G_1$ by the definitions of G_0 and G_1. So $G \subset G_0 \star G_1$. To see the other inclusion, take $\langle p, \tau \rangle \in G_0 \star G_1 \subset \mathbb{P}_0 \star \pi$. Then $p \in G_0$ and $\mathrm{val}_{G_0}(\tau) \in G_1$. So there exist $\langle q, \sigma \rangle \in G$, with $q \leq p$, and $\langle r, \tau \rangle \in G$. Let $\langle s, \eta \rangle \in G$ be a common extension of $\langle q, \sigma \rangle$ and $\langle r, \tau \rangle$. Then $s \leq q \leq p$ and $s \Vdash \eta \leq \tau$. So $\langle s, \eta \rangle \preceq \langle p, \tau \rangle$ and $\langle p, \tau \rangle \in G$.

Finally, $G_0, G_1 \in M[G]$, as they are defined using G. So $M[G_0] \subset M[G]$ by the minimality of $M[G_0]$, and $M[G_0][G_1] \subset M[G]$ by the minimality of $M[G_0][G_1]$. Also, $G = G_0 \star G_1 \in M[G_0][G_1]$. So $M[G] \subset M[G_0][G_1]$. $\square$

Theorem 9.5.3 is an analog of the implication (i)$\Rightarrow$(ii) from the product lemma. The iteration analog of the reverse implication, stated next, is also true. However, since we will not use this fact, we will leave its proof as an exercise.

Theorem 9.5.4 *Let M be a countable transitive model of ZFC, G_0 an M-generic filter in $\mathbb{P}_0 \in M$, and G_1 an $M[G_0]$-generic filter in $\mathbb{P}_1 \in M[G_0]$. If $\pi \in M$ is a good $\mathbb{P}_0$-name representing $\mathbb{P}_1$ then $G_0 \star G_1$ is an M-generic filter in $\mathbb{P}_0 \star \pi$ and $M[G_0][G_1] = M[G_0 \star G_1]$.*

In the proof of the consistency of MA+$\neg$CH we will be interested only in the ccc forcings. The next lemma tells us that we will remain within this class if the iteration concerns two ccc forcings.

Lemma 9.5.5 *Let M be a countable transitive model of ZFC, $\mathbb{P}_0 \in M$ be a ccc forcing in M, and $\langle \pi, \leq_\pi \rangle$ be a good $\mathbb{P}_0$-name for a forcing such that*

$$\mathbb{P}_0 \Vdash \langle \pi, \leq_\pi \rangle \text{ is ccc.} \tag{9.12}$$

Then $\mathbb{P}_0 \star \pi$ is ccc in M.

Proof To obtain a contradiction assume that there exists an uncountable antichain $A = \{\langle p_\xi, \tau_\xi \rangle \colon \xi < \omega_1\} \in M$ in $\mathbb{P}_0 \star \pi$. Let

$$\sigma = \{\langle \hat{\xi}, p_\xi \rangle \colon \xi < \omega_1\}.$$

Then $\sigma \in M$ is a $\mathbb{P}_0$-name. Define, in M,

$$D = \{p \in \mathbb{P}_0 \colon \exists \beta < \omega_1 \, (p \Vdash \sigma \subset \hat{\beta})\}$$

and notice that D is dense in $\mathbb{P}_0$.

To see this, let $q \in \mathbb{P}_0$ and let H be an M-generic filter containing q. Then

$$S = \mathrm{val}_H(\sigma) = \{\xi < \omega_1 \colon p_\xi \in H\}.$$

Notice that for distinct $\xi, \eta \in S$ the conditions $\mathrm{val}_H(\tau_\xi)$ and $\mathrm{val}_H(\tau_\eta)$ are incompatible in $\mathrm{val}_H(\pi)$.

Indeed, if $\xi, \eta \in S$, $\xi \neq \eta$, and $\mathrm{val}_H(\tau_\xi)$ and $\mathrm{val}_H(\tau_\eta)$ are compatible in $\mathrm{val}_H(\pi)$ then there are $r \in H$ extending p_ξ and p_η and $\rho \in \mathrm{dom}(\pi)$ such that $r \Vdash \rho \leq_\pi \tau_\xi$ and $r \Vdash \rho \leq_\pi \tau_\eta$. But then $\langle r, \rho \rangle$ would be a common extension of $\langle p_\xi, \tau_\xi \rangle$ and $\langle p_\eta, \tau_\eta \rangle$, contradicting our assumption that A is an antichain.

Thus S is countable in $M[H]$ since, by (9.12), $\mathrm{val}_H(\pi)$ is ccc in $M[H]$. So there exists a $\beta < \omega_1$ containing S. Now, if $p \in H$ is such that $p \leq q$ and $p \Vdash S \subset \hat{\beta}$ then $p \in D$. So D is dense.

Next consider the set

$$T = \{\beta < \omega_1 \colon \exists p_\beta \in \mathbb{P}_0 \ (p_\beta \Vdash \sup \sigma = \hat{\beta})\}$$

and notice that $A_1 = \{p_\beta \colon \beta \in T\}$ is an antichain in $\mathbb{P}_0$. Thus A_1 and T are countable in M. Let $\gamma < \omega_1$ be such that $T \subset \gamma$. Then the set

$$D' = \{p \in \mathbb{P}_0 \colon p \Vdash \sigma \subset \hat{\gamma}\}$$

is dense in $\mathbb{P}_0$, since it contains D.

Now, to obtain a contradiction, take an M-generic filter H in $\mathbb{P}_0$ containing p_γ. Then, by the denseness of D', $\mathrm{val}_H(\sigma) \subset \gamma$. But, straight from the definition of σ, $\gamma \in \mathrm{val}_H(\sigma)$! This contradiction finishes the proof. □

Now we turn our attention to the α-stage iteration for an arbitrary ordinal number α. More precisely, we will describe a general method of defining inductively a sequence $P = \langle\langle \mathbb{P}_\xi, \leq_\xi \rangle \colon \xi \leq \alpha\rangle$ of forcings such that $\mathbb{P}_{\xi+1}$ can be identified with $\mathbb{P}_\xi \star \pi_\xi$, where π_ξ is a good $\mathbb{P}_\xi$-name for a partially ordered set. This, with help of Theorem 9.5.3, will take care of the condition P1. The technical difficulty in this construction is that we have to know $\mathbb{P}_\xi$ in order to talk about a good $\mathbb{P}_\xi$-name π_ξ. So, along with the sequence P, we have to construct a sequence $\Pi = \langle\langle \pi_\xi, \leq_{\pi_\xi} \rangle \colon \xi < \alpha\rangle$ of appropriate good $\mathbb{P}_\xi$-names. This obscures a simple idea that stands behind the construction. Thus we will first describe this construction with the sequence Π replaced by a sequence $\langle\langle Q_\xi, \leq_{Q_\xi} \rangle \colon \xi < \alpha\rangle$ of forcings. In this case for every $\beta \leq \alpha$ we define

$$\mathbb{P}_\beta = \bigcup \left\{ \prod_{\eta \in S} Q_\eta \colon S \in [\beta]^{<\omega} \right\} = \left\{ f|_S \colon f \in \prod_{\eta < \beta} Q_\eta \ \& \ S \in [\beta]^{<\omega} \right\} \quad (9.13)$$

and order it by

$$p \leq_\beta q \quad \Leftrightarrow \quad \mathrm{dom}(q) \subset \mathrm{dom}(p) \ \& \ \big(\forall \eta \in \mathrm{dom}(q)\big)\big(p(\eta) \leq_{Q_\eta} q(\eta)\big).$$

In particular,

$$\mathbb{P}_0 = \{\emptyset\}, \quad \mathbb{P}_\lambda = \bigcup_{\beta < \lambda} \mathbb{P}_\beta \ \text{ for every limit ordinal } 0 < \lambda \leq \alpha, \quad (9.14)$$

and

$$\mathbb{P}_\gamma \subset \mathbb{P}_\beta \quad \text{and} \quad \leq_\gamma = \leq_\beta \cap (\mathbb{P}_\gamma \times \mathbb{P}_\gamma) \quad \text{for every} \quad \gamma < \beta \leq \alpha. \qquad (9.15)$$

Notice also that $\emptyset$ is a maximal element of every $\mathbb{P}_\beta$ with $\beta \leq \alpha$ and that for every $\xi < \beta \leq \alpha$

$$\text{if } r \in \mathbb{P}_\beta \text{ extends } p \in \mathbb{P}_\xi \text{ then } r|_\xi \in \mathbb{P}_\xi \text{ also extends } p \qquad (9.16)$$

and

$$\text{if } q \in \mathbb{P}_\beta \text{ and } r \in \mathbb{P}_\xi \text{ extends } q|_\xi \text{ then } r \cup q|_{(\beta \setminus \xi)} \in \mathbb{P}_\beta \text{ extends } q. \qquad (9.17)$$

Moreover, if $\mathbb{P}_{\xi+1}^\star = \mathbb{P}_{\xi+1} \setminus \mathbb{P}_\xi = \{p \in \mathbb{P}_{\xi+1} : \xi \in \mathrm{dom}(p)\}$ for $\xi < \alpha$ then

$$\langle \mathbb{P}_{\xi+1}^\star, \leq_{\xi+1} \rangle \text{ is isomorphic to } \langle \mathbb{P}_\xi \times Q_\xi, \preceq \rangle \qquad (9.18)$$

via the isomorphism $p \mapsto \langle p|_\xi, p(\xi) \rangle$. The easy proofs of (9.14)–(9.18) are left as an exercise.

For the general case, when $\mathbb{P}_\xi$-names π_ξ do not necessary represent the forcings from the ground model, the sequence $\langle \langle \mathbb{P}_\xi, \leq_\xi \rangle : \xi \leq \alpha \rangle$ will be defined by simultaneous induction with a sequence $\langle \langle \pi_\xi, \leq_{\pi_\xi} \rangle : \xi < \alpha \rangle$. In particular, we say that a sequence $\langle \langle \mathbb{P}_\xi, \leq_\xi \rangle : \xi \leq \alpha \rangle$ of partially ordered sets is an *α-stage forcing iteration (with finite support)* if there exists a sequence $\langle \langle \pi_\xi, \leq_{\pi_\xi} \rangle : \xi < \alpha \rangle$ such that every $\langle \pi_\xi, \leq_{\pi_\xi} \rangle$ is a good $\mathbb{P}_\xi$-name for a partially ordered set, and if for every $\beta \leq \alpha$

$$\mathbb{P}_\beta = \bigcup_{S \in [\beta]^{<\omega}} \left\{ p \in \prod_{\eta \in S} \mathrm{dom}(\pi_\eta) : \forall \eta \in S \ (p|_\eta \Vdash p(\eta) \in \pi_\eta) \right\} \qquad (9.19)$$

is ordered by

$$p \leq_\beta q \quad \Leftrightarrow \quad \mathrm{dom}(q) \subset \mathrm{dom}(p) \ \& \ \big(\forall \eta \in \mathrm{dom}(q)\big)\big(p|_\eta \Vdash p(\eta) \leq_{\pi_\eta} q(\eta)\big).$$

It is not difficult to see that the definition (9.19) gives the same notion of forcing as (9.13) if every π_ξ is the standard $\mathbb{P}_\xi$-name $\hat{Q}_\xi$ of Q_ξ. Moreover, it is also easy to notice that the sequence defined by (9.19) has the properties (9.14)–(9.17) as well (the inductive proof of (9.17) is left as an exercise). Moreover, for $\mathbb{P}_{\xi+1}^\star = \{p \in \mathbb{P}_{\xi+1} : \xi \in \mathrm{dom}(p)\}$ a counterpart of (9.18) says that for every $\xi < \alpha$

$$\langle \mathbb{P}_{\xi+1}^\star, \leq_{\xi+1} \rangle \text{ is isomorphic to } \langle \mathbb{P}_\xi \star \pi_\xi, \preceq \rangle, \qquad (9.20)$$

where an isomorphism is given by the map $p \mapsto \langle p|_\xi, p(\xi) \rangle$. To see this, it is enough to notice that for every $p, q \in \mathbb{P}^\star_{\xi+1}$

$$
\begin{aligned}
p \leq_{\xi+1} q \;\;\Leftrightarrow\;\; & \operatorname{dom}(q) \subset \operatorname{dom}(p) \;\&\; \big(\forall \eta \in \operatorname{dom}(q) \cap \xi\big)\big(p|_\eta \Vdash p(\eta) \leq_{\pi_\eta} q(\eta)\big) \\
& \&\; p|_\xi \Vdash p(\xi) \leq_{\pi_\xi} q(\xi) \\
\Leftrightarrow\;\; & p|_\xi \leq_\xi q|_\xi \;\&\; p|_\xi \Vdash p(\xi) \leq_{\pi_\xi} q(\xi) \\
\Leftrightarrow\;\; & \langle p|_\xi, p(\xi) \rangle \preceq \langle q|_\xi, q(\xi) \rangle.
\end{aligned}
$$

The next theorem is an α-stage-iteration version of Theorem 9.5.3. It implies the condition P1, where M_ξ is defined as $M[G_\xi]$ from the theorem.

Theorem 9.5.6 *Let M be a countable transitive model of ZFC, let $\alpha \in M$ be an ordinal number, and let $\langle \langle \mathbb{P}_\xi, \leq_\xi \rangle : \xi \leq \alpha \rangle \in M$ be an α-stage forcing iteration based on the sequence $\langle \langle \pi_\xi, \leq_{\pi_\xi} \rangle : \xi < \alpha \rangle$ of appropriate $\mathbb{P}_\xi$-names. If G is an M-generic filter in $\mathbb{P}_\alpha$ then $G_\xi = G \cap \mathbb{P}_\xi$ is an M-generic filter in $\mathbb{P}_\xi$ for every $\xi < \alpha$. Moreover, for every $\xi < \alpha$, if*

$$
G^\xi = \{ \operatorname{val}_{G_\xi}(p(\xi)) \colon p \in G_{\xi+1} \;\&\; \xi \in \operatorname{dom}(p) \}
$$

then G^ξ is an $M[G_\xi]$-generic filter in a forcing $\mathbb{P}^\xi = \langle \operatorname{val}_{G_\xi}(\pi_\xi), \operatorname{val}_{G_\xi}(\leq_{\pi_\xi}) \rangle$ and $M[G_{\xi+1}] = M[G_\xi][G^\xi]$.

Proof To see that G_ξ is a filter in $\mathbb{P}_\xi$ take $p, q \in G_\xi \subset G$. Then there exists an $r \in G$ extending p and q. But $r|_\xi \in \mathbb{P}_\xi$ belongs to G, since $r \leq r|_\xi$. So $r|_\xi \in G_\xi$ and, by (9.16), $r|_\xi$ extends p and q. Also, if $q \in G_\xi$, $p \in \mathbb{P}_\alpha$, and $q \leq p$ then $p \in \mathbb{P}_\xi$, and so $p \in G_\xi$. Thus G_ξ is a filter in $\mathbb{P}_\xi$.

To prove that G_ξ is M-generic in $\mathbb{P}_\xi$ take a $D \in M$ that is a dense subset of $\mathbb{P}_\xi$ and notice that

$$
D^\star = \{ p \in \mathbb{P}_\alpha \colon p|_\xi \in D \} \in M
$$

is dense in $\mathbb{P}_\alpha$. Indeed, if $q \in \mathbb{P}_\alpha$ then $q|_\xi \in \mathbb{P}_\xi$ and there exists an $r \in D$ with $r \leq_\xi q|_\xi$. But, by (9.17), $p = r \cup q|_{(\alpha \setminus \xi)} \in \mathbb{P}_\alpha$ extends q and it is easy to see that $p \in D^\star$. So $D^\star$ is dense and there exists an $r \in G \cap D^\star$. Hence $r|_\xi \in D \cap G_\xi$.

The proof of the additional part will be done in several steps. First notice that

$$
\mathbb{P}^\star_{\xi+1} \text{ is dense in } \mathbb{P}_{\xi+1}, \tag{9.21}
$$

since every $p \in \mathbb{P}_{\xi+1} \setminus \mathbb{P}^\star_{\xi+1} = \mathbb{P}_\xi$ forces $\langle \pi_\xi, \leq_{\pi_\xi} \rangle$ to be a partially ordered set, so there exists a $\mathbb{P}_\xi$-name τ such that $p \Vdash \tau \in \pi_\xi$ and $p \cup \{\langle \xi, \tau \rangle\} \in \mathbb{P}^\star_{\xi+1}$ extends p.

Next notice that

$$G^\star_{\xi+1} = G \cap \mathbb{P}^\star_{\xi+1} = G_{\xi+1} \setminus \mathbb{P}_\xi$$

is also an M-generic filter in $\mathbb{P}^\star_{\xi+1}$, since, by (9.21), every dense $D \in M$ subset of $\mathbb{P}^\star_{\xi+1}$ is also dense in $\mathbb{P}_{\xi+1}$ and $D \cap G^\star_{\xi+1} = D \cap G_{\xi+1} \neq \emptyset$. Also, $G_{\xi+1} = G^\star_{\xi+1} \cup \{p|_\xi : p \in G^\star_{\xi+1}\} \in M[G^\star_{\xi+1}]$. So $M[G_{\xi+1}] = M[G^\star_{\xi+1}]$.

Now, if i is an isomorphism from (9.20) then

$$i[G^\star_{\xi+1}] = \{\langle p|_\xi, p(\xi) \rangle : p \in G^\star_{\xi+1}\}$$

is an M-generic filter in $\mathbb{P}_\xi \star \pi_\xi$ and $M[G_{\xi+1}] = M[G^\star_{\xi+1}] = M[i[G^\star_{\xi+1}]]$. So, by Theorem 9.5.3, $i[G^\star_{\xi+1}] = G_\xi \star G^\xi$, G^ξ is an $M[G_\xi]$-generic filter in $\mathbb{P}^\xi$, and $M[G_{\xi+1}] = M[i[G^\star_{\xi+1}]] = M[G_\xi][G^\xi]$. $\square$

The next theorem says that an α-stage iteration (with finite support) of ccc forcings is ccc. This will guarantee that the forcing $\mathbb{P}_\mu$ from the beginning of the section will be ccc. The theorem is a generalization of Lemma 9.5.5.

Theorem 9.5.7 *Let M be a countable transitive model of ZFC, let $\alpha \in M$ be an ordinal number, and let $\langle\langle \mathbb{P}_\xi, \leq_\xi \rangle : \xi \leq \alpha \rangle \in M$ be an α-stage forcing iteration based on the sequence $\langle\langle \pi_\xi, \leq_{\pi_\xi} \rangle : \xi < \alpha \rangle$ of appropriate $\mathbb{P}_\xi$-names. If*

$$\mathbb{P}_\xi \Vdash \pi_\xi \ \text{is ccc} \tag{9.22}$$

for every $\xi < \alpha$, then $\mathbb{P}_\alpha$ is ccc in M.

Proof We will prove, by induction on $\beta \leq \alpha$, that $\mathbb{P}_\beta$ is ccc in M. The proof will be done in M.

Clearly $\mathbb{P}_0 = \{\emptyset\}$ is ccc. So let $0 < \beta \leq \alpha$ be such that $\mathbb{P}_\eta$ is ccc in M for every $\eta < \beta$ and let $A = \{p_\gamma : \gamma < \omega_1\} \subset \mathbb{P}_\beta$. We will find $\delta < \gamma < \omega_1$ such that p_δ and p_γ are compatible.

If β is a successor ordinal, say $\beta = \xi + 1$, take $C_0 = \{\gamma < \omega_1 : p_\gamma \in \mathbb{P}_\xi\}$. If $|C_0| = \omega_1$ then $A_0 = \{p_\gamma : \gamma \in C_0\} \subset \mathbb{P}_\xi$, and $\mathbb{P}_\xi$ is ccc by the inductive assumption. Hence $A_0 \subset A$ contains two compatible elements with different indices. So assume that $|C_0| < \omega_1$. Then $C_1 = \omega_1 \setminus C_0$ is uncountable and $A_1 = \{p_\gamma : \gamma \in C_1\} \subset \mathbb{P}^\star_{\xi+1}$. But $\mathbb{P}^\star_{\xi+1}$ is isomorphic to $\mathbb{P}_\xi \star \pi_\xi$, which, by Lemma 9.5.5, is ccc. Thus, once more, $A_1 \subset A$ contains two compatible elements with different indices.

If β is a limit ordinal, apply to the family $\{\text{dom}(p_\gamma) : \gamma < \omega_1\}$ the Δ-system lemma to find an uncountable set $C \subset \omega_1$ and a finite set $D \subset \beta$ such that $\text{dom}(p_\gamma) \cap \text{dom}(p_\delta) = D$ for all distinct $\gamma, \delta \in C$. Let $\xi < \beta$ be such that $D \subset \xi$ and consider $\{p_\gamma|_\xi : \gamma \in C\} \subset \mathbb{P}_\xi$. Since, by the inductive assumption, $\mathbb{P}_\xi$ is ccc, we can find $p \in \mathbb{P}_\xi$ and distinct $\gamma, \delta \in C$ such that $p \leq_\xi p_\gamma|_\xi$ and $p \leq_\xi p_\delta|_\xi$. To finish the proof it is enough to notice that

$$q = p \cup p_\gamma|_{(\beta \setminus \xi)} \cup p_\delta|_{(\beta \setminus \xi)} \tag{9.23}$$

belongs to $\mathbb{P}_\beta$ and extends p_γ and p_δ; this follows from (9.17). The details are left as an exercise. $\square$

We say that an α-stage forcing iteration $\langle\langle \mathbb{P}_\xi, \leq_\xi\rangle\colon \xi \leq \alpha\rangle$ is a *ccc forcing iteration* if it satisfies condition (9.22).

The next lemma is the last general fact concerning the α-stage iteration and it will imply property P2. It is an analog of Lemma 9.4.3.

Lemma 9.5.8 *Let M be a countable transitive model of ZFC, let $\kappa \in M$ be an infinite regular cardinal number, and let $\langle\langle \mathbb{P}_\xi, \leq_\xi\rangle\colon \xi \leq \kappa\rangle \in M$ be a κ-stage ccc forcing iteration. If $Y \in M$, G is an M-generic filter in $\mathbb{P}_\kappa$, and $S \subset Y$ is such that $M[G] \models |S| < \kappa$ then there exists a $\zeta < \kappa$ such that $S \in M[G_\zeta]$, where $G_\zeta = G \cap \mathbb{P}_\zeta$.*

Proof By Theorem 9.5.7 the forcing $\mathbb{P}_\kappa$ is ccc. So the cardinals in M and $M[G]$ are the same, and we will not distinguish between them.

Let $X = \lambda$, where λ is the cardinality of S in $M[G]$. Then $X \in M$ and $|X| = \lambda < \kappa$. Let $f \in M[G]$ be a map from X onto S. We will find a $\zeta < \kappa$ such that $f \in M[G_\zeta]$.

By Lemma 9.2.8 we can find a $\mathbb{P}_\kappa$-name $\tau \in M$ such that $f = \mathrm{val}_G(\tau)$, $\tau \subset \{\langle \widehat{\langle x, y\rangle}, p\rangle\colon x \in X\ \&\ y \in Y\ \&\ p \in \mathbb{P}_\kappa\}$, and

($\star$) $A_x = \left\{ p \in \mathbb{P}_\kappa\colon \exists y \in Y\ \left(\langle \widehat{\langle x, y\rangle}, p\rangle \in \tau\right)\right\}$ is an antichain for every $x \in X$.

So every A_x is countable in M, since $\mathbb{P}_\kappa$ is ccc in M. Let

$$A = \bigcup_{x \in X} \bigcup_{p \in A_x} \mathrm{dom}(p).$$

Then $A \in M$, $A \subset \kappa$, and $|A| \leq |X| + \omega < \kappa$ in M, since every set $\bigcup_{p \in A_x} \mathrm{dom}(p)$ is countable in M. Thus, by the regularity of κ, there exists a $\zeta < \kappa$ such that $A \subset \zeta$. In particular, $\bigcup_{x \in X} A_x \subset \mathbb{P}_\zeta$. So

$$\tau^\star = \left\{ \langle \widehat{\langle x, y\rangle}^\star, p\rangle\colon \langle \widehat{\langle x, y\rangle}, p\rangle \in \tau\right\} \in M$$

is a $\mathbb{P}_\zeta$-name, where $\widehat{\langle x, y\rangle}^\star$ is a standard $\mathbb{P}_\zeta$-name for $\langle x, y\rangle \in M$, and

$$\begin{aligned}
f = \mathrm{val}_G(\tau) &= \left\{ \langle x, y\rangle\colon \exists p \in G\ \left(\langle \widehat{\langle x, y\rangle}, p\rangle \in \tau\right)\right\} \\
&= \left\{ \langle x, y\rangle\colon (\exists p \in G \cap \mathbb{P}_\zeta)\ \left(\langle \widehat{\langle x, y\rangle}^\star, p\rangle \in \tau^\star\right)\right\} \\
&= \mathrm{val}_{G_\zeta}(\tau) \in M[G_\zeta].
\end{aligned}$$

So $S = f[X] \in M[G_\zeta]$.   □

Now we are ready for the main theorem of this section.

Theorem 9.5.9 *It is consistent with ZFC that* $\mathfrak{c} = \omega_2$ *and MA holds.*

Proof By the foregoing discussion, it is enough to construct a μ-stage ccc finite-support forcing iteration $\langle\langle\mathbb{P}_\xi, \leq_\xi\rangle\colon \xi \leq \mu\rangle$, with $\mu = \omega_2^M$, that satisfies P3 and P4. The construction will be done in a countable transitive model M of ZFC in which $2^{\omega_1} = \omega_2$ (such an M exists by Corollary 9.3.9). In what follows we will use the notation introduced in the first part of this section. We will also use repeatedly the fact, following from (9.15), that for every $\zeta \leq \xi \leq \mu$ any $\mathbb{P}_\zeta$-name is a $\mathbb{P}_\xi$-name too.

Let $\langle\langle\alpha_\xi, \beta_\xi\rangle\colon \xi < \mu\rangle \in M$ be a one-to-one enumeration of $\mu \times \mu$ such that $\beta_\xi \leq \xi$ for every $\xi < \mu$. By induction on $\xi < \mu$ we will construct a finite-support iteration $\langle\langle\mathbb{P}_\xi, \leq_\xi\rangle\colon \xi \leq \mu\rangle$, together with a sequence $\langle\langle\pi_\xi, \leq_{\pi_\xi}\rangle\colon \xi < \mu\rangle$ of appropriate $\mathbb{P}_\xi$-names, and a sequence $\langle S_\xi\colon \xi < \mu\rangle$ such that the following inductive conditions are satisfied for every $\xi < \mu$.

(I_ξ) $|\mathbb{P}_\xi| \leq \omega_1$.

(II_ξ) $S_\xi = \langle\langle\lambda_\alpha^\xi, \sigma_\alpha^\xi\rangle\colon \alpha < \mu\rangle$ contains every pair $\langle\lambda, \sigma\rangle$ with $\lambda \leq \omega_1$ and

$$\sigma \subset \{\langle\langle\widehat{\langle\gamma, \delta\rangle}, p\rangle\colon \gamma, \delta \in \lambda \,\&\, p \in \mathbb{P}_\xi\}.$$

(σ will be used as a $\mathbb{P}_\xi$-name for a partial order on λ.)

(III_ξ) $\pi_\xi = \hat{\lambda}$ is a standard $\mathbb{P}_\xi$-name for $\lambda = \lambda_{\alpha_\xi}^{\beta_\xi}$ and

$$\leq_{\pi_\xi} = \{\langle\langle\widehat{\langle\gamma, \delta\rangle}, p\rangle \in \sigma_{\alpha_\xi}^{\beta_\xi}\colon \exists q \in A \,(q \leq_\xi p)\}$$
$$\cup\{\langle\langle\widehat{\langle\gamma, \delta\rangle}, p\rangle\colon \gamma \leq \delta < \lambda \,\&\, p \in B\},$$

where

$$A = \{q \in \mathbb{P}_\xi\colon q \Vdash \langle\hat{\lambda}, \sigma_{\alpha_\xi}^{\beta_\xi}\rangle \text{ is a ccc forcing}\}$$

and $B = \{p \in \mathbb{P}_\xi\colon p$ is incompatible with every $q \in A\}$. In particular,

$$\mathbb{P}_\xi \Vdash \langle\pi_\xi, \leq_{\pi_\xi}\rangle \text{ is a ccc forcing.} \tag{9.24}$$

To see that this inductive construction preserves (I_ξ) consider two cases. If $0 < \xi < \mu$ is a limit ordinal, then $\mathbb{P}_\xi = \bigcup_{\eta<\xi} \mathbb{P}_\eta$ has cardinality $\leq \omega_1$ by the inductive assumption. If $\xi = \eta + 1$ is a successor ordinal, then $\mathbb{P}_\xi = \mathbb{P}_\eta \cup \mathbb{P}_\xi^\star$, and $\mathbb{P}_\xi^\star$ is isomorphic to $\mathbb{P}_\eta \star \pi_\eta$. But by ($I_\eta$) and ($III_\eta$) we have $|\pi_\eta| \leq \omega_1$. So $|\mathbb{P}_\xi| = |\mathbb{P}_\eta| \oplus (|\mathbb{P}_\eta| \otimes |\pi_\eta|) \leq \omega_1$ as well.

To see that the sequence S_ξ in (II_ξ) can be found note that there are only ω_2 pairs $\langle \lambda, \sigma \rangle$ to list since, by (I_ξ), the set $\{\langle \widehat{\langle \gamma, \delta \rangle}, p \rangle : \gamma, \delta \in \lambda \; \& \; p \in \mathbb{P}_\xi\}$ has cardinality $\leq \omega_1$ and so it has at most ω_2 subsets, as $2^{\omega_1} = \omega_2$ in M.

To finish the construction it is enough to argue for (9.24). For this, notice first that

$$A \cup B \text{ is dense in } \mathbb{P}_\xi \tag{9.25}$$

and

$$p \Vdash (\sigma_{\alpha_\xi}^{\beta_\xi} = \leq_{\pi_\xi}) \qquad \text{for every } p \in A. \tag{9.26}$$

To see (9.26) take $p_0 \in A$ and an M-generic filter H in $\mathbb{P}_\xi$ containing p_0. Then

$$
\begin{aligned}
\mathrm{val}_H(\sigma_{\alpha_\xi}^{\beta_\xi}) &= \{\langle \gamma, \delta \rangle : \langle \widehat{\langle \gamma, \delta \rangle}, p \rangle \in \sigma_{\alpha_\xi}^{\beta_\xi} \; \& \; p \in H\} \\
&= \{\langle \gamma, \delta \rangle : \langle \widehat{\langle \gamma, \delta \rangle}, p \rangle \in \sigma_{\alpha_\xi}^{\beta_\xi} \; \& \; \exists q \in A \; (q \leq_\xi p) \; \& \; p \in H\} \\
&= \{\langle \gamma, \delta \rangle : \langle \widehat{\langle \gamma, \delta \rangle}, p \rangle \in \leq_{\pi_\xi} \; \& \; \exists q \in A \; (q \leq_\xi p) \; \& \; p \in H\} \\
&= \{\langle \gamma, \delta \rangle : \langle \widehat{\langle \gamma, \delta \rangle}, p \rangle \in \leq_{\pi_\xi} \; \& \; p \in H\} \\
&= \mathrm{val}_H(\leq_{\pi_\xi}),
\end{aligned}
$$

proving (9.26).

To argue for (9.24) let H be an M-generic filter in $\mathbb{P}_\xi$. We have to show that $\langle \mathrm{val}_H(\pi_\xi), \mathrm{val}_H(\leq_{\pi_\xi}) \rangle = \langle \lambda, \mathrm{val}_H(\leq_{\pi_\xi}) \rangle$ is a ccc partially ordered set in $M[H]$. But, by (9.25), either $A \cap H \neq \emptyset$, in which case this is true since an element of $A \cap H$ forces $\langle \pi_\xi, \leq_{\pi_\xi} \rangle = \langle \hat{\lambda}, \sigma_{\alpha_\xi}^{\beta_\xi} \rangle$ to be a ccc forcing, or $B \cap H \neq \emptyset$, in which case $\mathrm{val}_H(\leq_{\pi_\xi})$ is just a standard well ordering of λ, so it is obviously ccc. This finishes the inductive construction.

Now, to see that P3 is satisfied note that $\mathbb{P}_\mu = \bigcup_{\xi < \mu} \mathbb{P}_\xi$ has cardinality $\leq \omega_2$. Thus $|(\mathbb{P}_\mu)^\omega| \leq (\omega_2)^{\omega_1} = 2^{\omega_1} = \omega_2$.

To see that P4 holds let G be an M-generic filter in $\mathbb{P}_\mu$, let $N = M[G]$, and let $\langle \lambda, \ll \rangle \in N$ be such that

$$N \models \lambda \leq \omega_1 \text{ and } \langle \lambda, \ll \rangle \text{ is ccc.}$$

Choose an $\alpha_0 < \mu$. By Theorem 9.5.6 it is enough to find $\alpha_0 < \xi < \mu$ such that $\mathbb{P}^\xi = \langle \mathrm{val}_{G_\xi}(\pi_\xi), \mathrm{val}_{G_\xi}(\leq_{\pi_\xi}) \rangle = \langle \lambda, \ll \rangle$. But, by Lemma 9.5.8, there is $\alpha_0 < \zeta < \mu$ such that $\langle \lambda, \ll \rangle \in M[G_\zeta]$. Take a $\mathbb{P}_\zeta$-name $\sigma^\star \in M$ such that $\mathrm{val}_{G_\zeta}(\sigma^\star)$ is equal to $\ll$ and let

$$\sigma = \{\langle \widehat{\langle \gamma, \delta \rangle}, p \rangle : \gamma, \delta \in \lambda \; \& \; p \in \mathbb{P}_\zeta \; \& \; p \Vdash \widehat{\langle \gamma, \delta \rangle} \in \sigma^\star\} \in M.$$

Note that

$$\mathrm{val}_{G_\zeta}(\sigma) = \mathrm{val}_{G_\zeta}(\sigma^\star)$$

since for every $\langle \gamma, \delta \rangle \in \lambda \times \lambda$

$$\langle \gamma, \delta \rangle \in \text{val}_{G_\zeta}(\sigma) \;\Leftrightarrow\; \exists p \in G_\zeta \; (p \Vdash \widehat{\langle \gamma, \delta \rangle} \in \sigma^\star) \;\Leftrightarrow\; \langle \gamma, \delta \rangle \in \text{val}_{G_\zeta}(\sigma^\star).$$

Next, choose $\alpha < \mu$ such that $\langle \lambda_\alpha^\zeta, \sigma_\alpha^\zeta \rangle = \langle \lambda, \sigma \rangle$ and pick $\xi < \mu$ with $\langle \alpha_\xi, \beta_\xi \rangle = \langle \alpha, \zeta \rangle$. Then $\alpha_0 < \zeta = \beta_\xi \leq \xi$ and $\lambda_{\alpha_\xi}^{\beta_\xi} = \lambda_\alpha^\zeta = \lambda$, so $\text{val}_{G_\xi}(\pi_\xi) = \lambda$. Also, $\sigma_{\alpha_\xi}^{\beta_\xi} = \sigma_\alpha^\zeta = \sigma$ and

$$\text{val}_{G_\xi}(\sigma) = \text{val}_{G_\zeta}(\sigma) = \text{val}_{G_\zeta}(\sigma^\star) = (\ll) \in M[G_\zeta] \subset M[G_\xi]$$

since $G_\zeta = G_\xi \cap \mathbb{P}_\zeta$.

Thus it suffices to argue that

$$\text{val}_{G_\xi}(\leq_{\pi_\xi}) = \text{val}_{G_\xi}(\sigma). \tag{9.27}$$

But $\langle \lambda, \ll \rangle$ is a partially ordered set in $M[G_\xi]$, since it is a partially ordered set in N and this is an absolute property. Also, $\langle \lambda, \ll \rangle$ is ccc in $M[G_\xi]$, because every antichain $A \in M[G_\xi]$ is also an antichain in N and

$$\omega = |A|^N = |A|^M = |A|^{M[G_\xi]},$$

since $\langle \lambda, \ll \rangle$ is ccc in N and both extensions of M preserve the cardinals. Thus there is a $p \in G_\xi$ that forces this to be true for $\langle \hat{\lambda}, \sigma \rangle = \langle \hat{\lambda}, \sigma_{\alpha_\xi}^{\beta_\xi} \rangle$. In particular, $G_\xi \cap A \neq \emptyset$ so (9.26) and $\sigma_{\alpha_\xi}^{\beta_\xi} = \sigma$ imply (9.27). This finishes the proof. $\qquad\square$

EXERCISES

1 Complete the details of the proof of (9.8).

2 Complete the details of the proof of Lemma 9.5.1.

3 Prove that the relation $\preceq$ defined on an iteration $\mathbb{P} \star \pi$ is transitive.

4 Prove (9.10).

5 Complete the proof of Theorem 9.5.3 by showing that the set D from (9.11) is indeed dense below $\langle q, \eta \rangle$.

6 Prove Theorem 9.5.4.

7 Prove properties (9.14)–(9.18).

8 Prove (9.17) for the forcing defined by (9.19).

9 Complete the proof of Theorem 9.5.7 by showing that the condition q defined by (9.23) has the desired properties.

Appendix A

Axioms of set theory

Axiom 0 *(Set Existence)* There exists a set:

$$\exists x(x = x).$$

Axiom 1 *(Extensionality)* If x and y have the same elements, then x is equal to y:

$$\forall x \forall y \, [\forall z(z \in x \leftrightarrow z \in y) \rightarrow x = y].$$

Axiom 2 *(Comprehension scheme* or *schema of separation)* For every formula $\varphi(s, t)$ with free variables s and t, for every x, and for every parameter p there exists a set $y = \{u \in x \colon \varphi(u, p)\}$ that contains all those $u \in x$ that have property φ:

$$\forall x \forall p \exists y \, [\forall u(u \in y \leftrightarrow (u \in x \ \& \ \varphi(u, p)))].$$

Axiom 3 *(Pairing)* For any a and b there exists a set x that contains a and b:

$$\forall a \forall b \exists x(a \in x \ \& \ b \in x).$$

Axiom 4 *(Union)* For every family $\mathcal{F}$ there exists a set U containing the *union* $\bigcup \mathcal{F}$ of all elements of $\mathcal{F}$:

$$\forall \mathcal{F} \exists U \forall Y \forall x[(x \in Y \ \& \ Y \in \mathcal{F}) \rightarrow x \in U].$$

Axiom 5 *(Power set)* For every set X there exists a set P containing the set $\mathcal{P}(X)$ (the *power set*) of all subsets of X:

$$\forall X \exists P \forall z \, [z \subset X \rightarrow z \in P].$$

To make the statement of the next axiom more readable we introduce the following abbreviation. We say that y is a successor of x and write $y = S(x)$ if $S(x) = x \cup \{x\}$, that is,

$$\forall z[z \in y \leftrightarrow (z \in x \vee z = x)].$$

Axiom 6 *(Infinity)* (Zermelo 1908) There exists an infinite set (of some special form):

$$\exists x \, [\forall z(z = \emptyset \rightarrow z \in x) \ \& \ \forall y \in x \forall z(z = S(y) \rightarrow z \in x)].$$

Axiom 7 *(Replacement scheme)* (Fraenkel 1922; Skolem 1922) For every formula $\varphi(s, t, U, w)$ with free variables s, t, U, and w, every set A, and every parameter p if $\varphi(s, t, A, p)$ defines a function F on A by $F(x) = y \Leftrightarrow \varphi(x, y, A, p)$, then there exists a set Y containing the range $F[A] = \{F(x) \colon x \in A\}$ of the function F:

$$\forall A \forall p \, [\forall x \in A \exists! y \varphi(x, y, A, p) \rightarrow \exists Y \forall x \in A \exists y \in Y \varphi(x, y, A, p)],$$

where the quantifier $\exists! x \varphi(x)$ is an abbreviation for "there exists precisely one x satisfying φ," that is, is equivalent to the formula

$$\exists x \varphi(x) \ \& \ \forall x \forall y(\varphi(x) \ \& \ \varphi(y) \rightarrow x = y).$$

Axiom 8 *(Foundation* or *regularity)* (Skolem 1922; von Neumann 1925) Every nonempty set has an $\in$-minimal element:

$$\forall x \, [\exists y(y \in x) \rightarrow \exists y(y \in x \ \& \ \neg \exists z(z \in x \ \& \ z \in y))].$$

Axiom 9 *(Choice)* (Levi 1902; Zermelo 1904) For every family $\mathcal{F}$ of disjoint nonempty sets there exists a "selector," that is, a set S that intersects every $x \in \mathcal{F}$ in precisely one point:

$$\forall \mathcal{F} \, [\forall x \in \mathcal{F}(x \neq \emptyset) \ \& \ \forall x \in \mathcal{F} \forall y \in \mathcal{F}(x = y \vee x \cap y = \emptyset)]$$

$$\rightarrow \exists S \forall x \in \mathcal{F} \exists! z(z \in S \ \& \ z \in x),$$

where $x \cap y = \emptyset$ is an abbreviation for

$$\neg \exists z(z \in x \ \& \ z \in y).$$

Using the comprehensive schema for Axioms 0, 3, 4, and 5 we may easily obtain the following strengthening of them, which is often used as the original axioms.

Axiom 0' *(Empty set)* There exists the *empty set* $\emptyset$:

$$\exists x \forall y \neg (y \in x).$$

Axiom 3' *(Pairing)* For any a and b there exists a set x that contains *precisely* a and b.

Axiom 4' *(Union)* (Cantor 1899; Zermelo 1908) For every family $\mathcal{F}$ there exists a set $U = \bigcup \mathcal{F}$, the *union* of all subsets of $\mathcal{F}$.

Axiom 5' *(Power set)* (Zermelo 1908) For every set x there exists a set $Y = \mathcal{P}(x)$, the *power set* of x, that is, the set of all subsets of x.

The system of Axioms 0–8 is usually called Zermelo–Fraenkel set theory and is abbreviated by ZF. The system of Axioms 0–9 is usually denoted by ZFC. Thus, ZFC is the same as ZF+AC, where AC stands for the axiom of choice.

Historical Remark (Levy 1979) The first similar system of axioms was introduced by Zermelo. However, he did not have the axioms of foundation and replacement. Informal versions of the axiom of replacement were suggested by Cantor (1899) and Mirimanoff (1917). Formal versions were introduced by Fraenkel (1922) and Skolem (1922). The axiom of foundation was added by Skolem (1922) and von Neumann (1925).

Notice that Axioms 2 and 7 are in fact the schemas for infinitely many axioms, one for each formula φ. Thus theory ZFC has, in fact, infinitely many axioms. Axiom 1 of extensionality is the most fundamental one. Axioms 0 (or 0') of set existence (of empty set) and 6 of infinity are "existence" axioms that postulate the existence of some sets. It is obvious that Axiom 0 follows from Axiom 6. Axioms 2 of comprehension and 7 of replacement are schemas for infinitely many axioms. Axioms 3 (or 3') of pairing, 4 (or 4') of union, and 5 (or 5') of power set are conditional existence axioms. The existing sets postulated by 3', 4', and 5' are unique. It is not difficult to see that Axiom 3 of pairing follows from the others.

The axiom of choice AC and the axiom of foundation also have the same conditional existence character. However, the sets existing by them do not have to be unique. Moreover, the axiom of foundation has a very set-theoretic meaning and is seldom used outside abstract set theory or logic. It lets us build "hierarchical models" of set theory. During this course we very seldom make use of it. The axiom of choice, on the other hand, is one of the most important tools in this course. It is true that its nonconstructive character caused, in the past, some mathematicians to reject it (for example, Borel and Lebesgue). However, this discussion has been for the most part resolved today in favor of accepting this axiom.

It follows from the foregoing discussion that we can remove Axioms 0 and 3 from 0–8 and still have the same theory ZF.

Appendix B

Comments on the forcing method

This appendix contains some comments and explanations for Section 9.1. It also includes some missing proofs of the results stated there.

We start by explaining a few more terms. In Section 9.1 we said that a formula φ is a consequence of a theory T, $T \vdash \varphi$, if there is a *formal proof* of φ from T. By a formal proof in this statement we mean a formalization of what we really do in proving theorems. Thus a proof of φ from T is a finite sequence $\varphi_0, \ldots, \varphi_n$ of formulas such that $\varphi_n = \varphi$ and a formula φ_k can appear in the sequence only because of one of following two reasons:

- φ_k is an axiom, that is, it belongs to T or is a logic axiom; or

- φ_k is obtained from some φ_i and φ_j $(i, j < k)$ by a *rule of detachment* (also called a *modus ponens rule*), that is, if there exist $i, j < k$ such that φ_j has the form $\varphi_i \rightarrow \varphi_k$.

Since any formal proof is a finite sequence of formulas, it can contain only finitely many sentences from T. Thus, if $T \vdash \varphi$ then there is a finite subtheory $T_0 \subset T$ such that $T_0 \vdash \varphi$. In particular, if theory T is inconsistent, then there is a finite subtheory T_0 of T that is also inconsistent. So theory T is consistent if and only if every one of its finite subtheories is consistent.

Note also that if $\varphi_0, \ldots, \varphi_n$ is a proof of $\varphi = \varphi_n$ in theory $T \cup \{\psi\}$ then $(\psi \rightarrow \varphi_0), \ldots, (\psi \rightarrow \varphi_n)$ is a proof of $\psi \rightarrow \varphi$ in T, where formula $\psi \rightarrow (\varphi_i \rightarrow \varphi_k)$ is identified with $\varphi_i \rightarrow (\psi \rightarrow \varphi_k)$. Thus, if φ is a consequence of a finite theory $T_0 = \{\psi_0, \ldots, \psi_n\}$ then $\emptyset \vdash (\psi_0 \rightarrow (\cdots \rightarrow (\psi_n \rightarrow \varphi) \cdots))$, that is, $(\psi_0 \& \cdots \& \psi_n) \rightarrow \varphi$ is a consequence of the axioms of logic.

To argue for the forcing principle we will need the following theorem.

Theorem B.1 *Let S and T be two theories and assume that for every finite subtheory S_0 of S we can prove in theory T that there exists a nonempty transitive set M that is a model for S_0. Then $\mathrm{Con}(T)$ implies $\mathrm{Con}(S)$.*

Proof Let S and T be as in the theorem and assume that S is inconsistent. It is enough to show that this implies the inconsistency of T.

Since S is inconsistent, there is a finite subtheory $S_0 = \{\psi_1, \ldots, \psi_n\}$ of S such that S_0 is inconsistent, that is, such that $S_0 \vdash \varphi \& \neg\varphi$ for some sentence φ. In particular, $(\psi_1 \& \cdots \& \psi_n) \rightarrow (\varphi \& \neg\varphi)$.

But, by our assumption, in theory T we can prove the existence of a nonempty transitive set M that is a model for S_0, that is, such that ψ^M holds for every ψ from S_0. In particular,

$$T \vdash \exists M \ (\psi_1^M \& \cdots \& \psi_n^M).$$

Now, it is enough to notice that the rules of formal deduction are set up in such a way that if $\varphi_0, \ldots, \varphi_n$ is a formal proof of $\varphi \& \neg\varphi$ from S_0 then $\varphi_0^M, \ldots, \varphi_n^M$ is a formal proof of $\varphi^M \& \neg\varphi^M$ from $S_0^M = \{\psi_1^M, \ldots, \psi_n^M\}$. In particular, the implication

$$(\psi_1^M \& \cdots \& \psi_n^M) \rightarrow (\varphi^M \& \neg\varphi^M)$$

is true for every M. So we have proved in T that there is an M such that $\psi_1^M \& \cdots \& \psi_n^M$, while $(\psi_1^M \& \cdots \& \psi_n^M) \rightarrow (\varphi^M \& \neg\varphi^M)$ is true for every M. So $\exists M \ (\varphi^M \& \neg\varphi^M)$ is a consequence of T, that is, T is inconsistent. □

Now assume that we have proved the following condition (F) from the forcing principle:

(F)   Every CTM M of ZFC can be extended to a CTM N of ZFC+"ψ."

If we could prove in ZFC that

$$\exists M \ (M \text{ is a CTM for ZFC})$$

then we would conclude in ZFC that there exists a model N of ZFC+"ψ," and this, by Theorem B.1, implies that $\mathrm{Con}(ZFC) \Rightarrow \mathrm{Con}(ZFC + "\psi")$. Unfortunately, the existence of a countable transitive model of ZFC cannot be proved from ZFC axioms (this follows from Theorem 1.1.1). Thus we need the following more refined argument.

A closer look at the forcing method shows that it lets us extend *any* countable transitive set M to another countable transitive set $N = M[G]$. Then Theorem 9.2.2 asserts that if M satisfies ZFC then so does $M[G]$. However, an examination of this proof (which is not included in this text) shows that checking whether $M[G]$ has a given property φ is of *finitistic character* in the sense that in the proof of the implication

if M is a model for ZFC then $M[G]$ is a model for "φ"   (B.1) we use the knowledge that M is a model for a given axiom ψ only for finitely many axioms ψ of ZFC. In particular, for any sentence φ for which (B.1) holds there is a finite subtheory T_φ of ZFC such that the implication

$$\text{if } M \text{ is a model for } T_\varphi \text{ then } M[G] \text{ is a model for } \text{``}\varphi\text{''} \qquad \text{(B.2)}$$

has the same proof as (B.1). To show that such a nice proof of (F) implies $\text{Con}(\text{ZFC}) \Rightarrow \text{Con}(\text{ZFC} + \text{``}\psi\text{''})$ we need one more theorem, which will be left without proof.

Theorem B.2 *For every finite subtheory T of ZFC it is provable in ZFC that there exists a countable transitive model M of T.*

Now, to argue for the forcing principle assume, to obtain a contradiction, that (F) holds and ZFC+"ψ" is inconsistent. Then there exists a finite subtheory S of ZFC+"ψ" that is also inconsistent. For every φ from S let T_φ be a finite subset of ZFC for which (B.2) holds and let T be the union of all T_φ for φ from S. By Theorem B.2 there is a CTM M for T, and using (B.2) we conclude that there exists a countable transitive model $N = M[G]$ for S. Since all of this was proved in ZFC, we can use Theorem B.1 to deduce that $\text{Con}(\text{ZFC})$ implies $\text{Con}(S)$. But S is inconsistent, so it can only happen if ZFC is inconsistent. This finishes the argument for the forcing principle.

We will finish this appendix with the list of Δ_0-formulas that constitute the

Proof of Lemma 9.2.5

(0)   Formula $\psi_0(x,y)$ representing $x \subset y$:

$$\forall w \in x (w \in y).$$

(1)   Formula $\psi_1(x,y)$ representing $y = \bigcup x$:

$$\forall w \in y \exists z \in x (w \in z) \ \& \ \forall z \in x \forall w \in z (w \in y).$$

(2)   Formula $\psi_2(x,y)$ representing $y = \bigcap x$:

$$\forall w \in y \forall z \in x (w \in z) \ \& \ \forall z \in x \forall w \in z [\forall u \in x (w \in u) \rightarrow (w \in y)].$$

(3)   Formula $\psi_3(x,y,z)$ representing $z = x \cup y$:

$$\forall w \in z (w \in x \ \vee \ w \in y) \ \& \ \forall w \in x (w \in z) \ \& \ \forall w \in y (w \in z).$$

(4)   Formula $\psi_4(x, y, z)$ representing $z = x \cap y$:

$$\forall w \in z(w \in x \;\&\; w \in y) \;\&\; \forall w \in x[w \in y \rightarrow w \in z].$$

(5)   Formula $\psi_5(x, y, z)$ representing $z = x \setminus y$:

$$\forall w \in z(w \in x \;\&\; w \notin y) \;\&\; \forall w \in x(w \notin y \rightarrow w \in z).$$

(6)   Formula $\psi_6(x, y, z)$ representing $z = \{x, y\}$:

$$x \in z \;\&\; y \in z \;\&\; \forall t \in z(t = x \;\vee\; t = y).$$

(7)   Formula $\psi_7(z)$ representing "z is an unordered pair":

$$\exists x \in z \exists y \in z \; \psi_6(x, y, z).$$

(8)   Formula $\psi_8(x, y, z)$ representing $z = \langle x, y \rangle = \{\{x\}, \{x, y\}\}$:

$$\exists u \in z \exists w \in z[\psi_6(x, x, u) \;\&\; \psi_6(x, y, w) \;\&\; \psi_6(u, w, z)].$$

(9)   Formula $\psi_9(z)$ representing "z is an ordered pair":

$$\exists w \in z \exists x \in w \exists y \in w \psi_8(x, y, z).$$

(10) Formula $\psi_{10}(x, y, z)$ representing $z = x \times y$:

$$\forall w \in z \exists s \in x \exists t \in y \; \psi_8(s, t, w) \;\&\; \forall s \in x \forall t \in y \exists w \in z \; \psi_8(s, t, w).$$

(11) Formula $\psi_{11}(r)$ representing "r is a binary relation":

$$\forall w \in r \; \psi_9(w).$$

(12) Formula $\psi_{12}(d, r)$ representing "d is the domain of a binary relation r":

$$\forall x \in d \exists z \in r \exists w \in z \exists y \in w \; \psi_8(x, y, z)$$

$$\&\; \forall z \in r \forall w \in z \forall x \in w \forall y \in w[\psi_8(x, y, z) \rightarrow x \in d].$$

(13) Formula $\psi_{13}(R, r)$ representing "R is the range of a binary relation r":

$$\forall y \in R \exists z \in r \exists w \in z \exists x \in w \; \psi_8(x, y, z)$$

$$\&\; \forall z \in r \forall w \in z \forall x \in w \forall y \in w[\psi_8(x, y, z) \rightarrow y \in R].$$

(14) Formula $\psi_{14}(f)$ representing "f is a function":

$$\psi_{11}(f) \ \& \ \forall p \in f \forall q \in f \forall u \in p \forall w \in q \forall a \in u \forall b \in u \forall d \in w$$
$$[[\psi_8(a,b,p) \ \& \ \psi_8(a,d,q)] \to b = d].$$

(15) Formula $\psi_{15}(f)$ representing "function f is injective":

$$\psi_{14}(f) \ \& \ \forall p \in f \forall q \in f \forall u \in p \forall w \in q \forall a \in u \forall b \in u \forall c \in w \forall d \in w$$
$$[[(\psi_8(a,b,p) \ \& \ \psi_8(c,d,q) \ \& \ b = d)] \to a = c].$$

(16) Formula $\psi_{16}(\leq, \mathbb{P})$ representing "$\leq$ is a partial-order relation on $\mathbb{P}$":

$$\psi_{11}(\leq) \ \& \ \psi_{12}(\mathbb{P}, \leq) \ \& \ \psi_{13}(\mathbb{P}, \leq) \ \& \ \forall x \in \mathbb{P} \exists w \in \leq \ \psi_8(x, x, w)$$
$$\& \ \forall x \in \mathbb{P} \forall y \in \mathbb{P} \forall v \in \leq \forall w \in \leq [(\psi_8(x, y, v) \ \& \ \psi_8(y, x, w)) \to x = y]$$
$$\& \ \forall x \in \mathbb{P} \forall y \in \mathbb{P} \forall z \in \mathbb{P} \forall v \in \leq \forall w \in \leq$$
$$[(\psi_8(x, y, v) \ \& \ \psi_8(y, z, w)) \to \exists u \in \leq \ \psi_8(x, z, u)].$$

(17) Formula $\psi_{17}(D, \leq, \mathbb{P})$ representing "D is a dense subset of the partially ordered set $\langle \mathbb{P}, \leq \rangle$":

$$\psi_{16}(\leq, \mathbb{P}) \ \& \ \psi_0(D, \mathbb{P}) \ \& \ \forall x \in \mathbb{P} \exists d \in D \exists w \in \leq \ \psi_8(d, x, w).$$

(18) Formula $\psi_{18}(A, \leq, \mathbb{P})$ representing "A is an antichain in the partially ordered set $\langle \mathbb{P}, \leq \rangle$":

$$\psi_{16}(\leq, \mathbb{P}) \ \& \ \psi_0(A, \mathbb{P}) \ \& \ \forall a \in A \forall b \in A \forall x \in \mathbb{P} \forall u \in \leq \forall w \in \leq$$
$$[(\psi_8(x, a, u) \ \& \ \psi_8(x, b, u)) \to a = b].$$

(19) Formula $\psi_{19}(x)$ representing "set x is transitive":

$$\forall y \in x \forall z \in y (z \in x).$$

(20) Formula $\psi_{20}(\alpha)$ representing "α is an ordinal number":

$$\psi_{19}(\alpha) \ \& \ \forall \beta \in \alpha \forall \gamma \in \alpha (\beta = \gamma \ \vee \ \beta \in \gamma \ \vee \ \gamma \in \beta).$$

(21) Formula $\psi_{21}(\alpha)$ representing "α is a limit ordinal number":

$$\psi_{20}(\alpha) \ \& \ \forall \beta \in \alpha \exists \gamma \in \alpha (\beta \in \gamma).$$

(22) Formula $\psi_{22}(\alpha)$ representing "$\alpha = \omega$":

$$\psi_{21}(\alpha) \ \& \ \forall \beta \in \alpha [\psi_{21}(\beta) \to \forall x \in \beta (x \neq x)].$$

(23) Formula $\psi_{23}(\alpha)$ representing "α is a finite ordinal number":

$$\psi_{20}(\alpha) \ \& \ \neg \psi_{22}(\alpha) \ \& \ \forall \beta \in \alpha (\neg \psi_{22}(\beta)).$$

(24) Formula $\psi_{23}(\alpha)$ representing "α is a successor ordinal number":

$$\psi_{20}(\alpha) \ \& \ \neg \psi_{21}(\alpha). \qquad \square$$

Appendix C

Notation

- $x \in y$ – x is an element of y, 6.

- $\neg\varphi$ – the negation of formula φ, 6.

- $\varphi \& \psi$ – the conjunction of formulas φ and ψ, 6.

- $\varphi \vee \psi$ – the disjunction of formulas φ and ψ, 6.

- $\varphi{\rightarrow}\psi$ – the implication, 6.

- $\varphi{\leftrightarrow}\psi$ – the equivalence of formulas φ and ψ, 6.

- $\exists x\varphi$ – the existential quantifier, 6.

- $\forall x\varphi$ – the universal quantifier, 6.

- $\exists x \in A\varphi$ – a bounded existential quantifier, 6.

- $\forall x \in A\varphi$ – a bounded universal quantifier, 6.

- $x \subset y$ – x is a subset of y, 6.

- $\emptyset$ – the empty set, 7.

- $\bigcup \mathcal{F}$ – the union of a family $\mathcal{F}$ of sets, 8.

- $\mathcal{P}(X)$ – the power set of a set X, 8.

- $x \cup y$ – the union of sets x and y, 8.

- $x \setminus y$ – the difference of sets x and y, 8.

- $\bigcap \mathcal{F}$ – the intersection of a family $\mathcal{F}$ of sets, 8.

- $x \cap y$ – the intersection of sets x and y, 9.

- $x \triangle y$ – the symmetric difference of sets x and y, 9.

- $\langle a, b \rangle$ – the ordered pair $\{\{a\}, \{a, b\}\}$, 9.

- $\langle a_1, a_2, \ldots, a_{n-1}, a_n \rangle$ – the ordered n-tuple, 10.

- $X \times Y$ – the Cartesian product of sets X and Y, 10.

- $S(x)$ – the successor of x: $x \cup \{x\}$, 10.

- $\mathrm{dom}(R)$ – the domain of a relation (or function) R, 12.

- $\mathrm{range}(R)$ – the range of a relation (or function) R, 12.

- R^{-1} – the inverse of a relation (or function) R, 13.

- $S \circ R$ – the composition of the relations (or functions) R and S, 13.

- $f \colon X \to Y$ – a function from a set X into a set Y, 16.

- Y^X – the class of all function from a set X into a set Y, 16.

- $f[A]$ – the image of a set A with respect to a function f, 16.

- $f^{-1}(B)$ – the preimage of a set B with respect to a function f, 16.

- $f|_A$ – the restriction of a function f to a set A, 18.

- $\bigcup_{t \in T} F_t$ – the union of an indexed family $\{F_t\}_{t \in T}$, 19.

- $\bigcap_{t \in T} F_t$ – the intersection of an indexed family $\{F_t\}_{t \in T}$, 19.

- $\prod_{t \in T} F_t$ – the Cartesian product of an indexed family $\{F_t\}_{t \in T}$, 20.

- $\mathbb{N}$ – the set of natural numbers, 26.

- ω – the set of natural numbers, 27;
   the order type of an infinite strictly increasing sequence, 39;
   the first infinite ordinal number, 44.

- $\mathbb{Z}$ – the set of integers, 30.

- $\mathbb{Q}$ – the set of rational numbers, 30.

- $\mathbb{R}$ – the set of real numbers, 31.

- $B(p, \varepsilon)$ – the open ball in $\mathbb{R}^n$ with center p and radius ε, 32.

- $\mathrm{int}(S)$ – the interior of a set S in $\mathbb{R}^n$, 33.

- $\mathrm{cl}(S)$ – the closure of a set S in $\mathbb{R}^n$, 33.

- $\omega^\star$ – the order type of an infinite strictly decreasing sequence, 39.

- $O(x_0)$ – the initial segment $\{x \in X : x < x_0\}$, 40.

- $\mathrm{Otp}(W)$ – the order type of a well-ordered set W, 47.

- $|A|$ – the cardinality of a set A, 61.

- 2^κ – the cardinality of the set $\mathcal{P}(\kappa)$, 65.

- χ_A – the characteristic function of a set A, 65.

- κ^+ – the cardinal successor of a cardinal κ, 65.

- ω_α – the αth infinite cardinal number, 66.

- $\aleph_\alpha$ – aleph-alpha: the αth infinite cardinal number, 66.

- $\beth_\alpha$ – bet-alpha, 66.

- $\mathfrak{c}$ – continuum: the cardinality of the set $\mathcal{P}(\omega)$, 66.

- $\kappa \oplus \lambda$ – the cardinal sum of cardinals κ and λ, 68.

- $\kappa \otimes \lambda$ – the cardinal product of cardinals κ and λ, 68.

- κ^λ – cardinal exponentiation: the cardinality of the set κ^λ, 68.

- $A^{<\omega}$ – the set of all finite sequences with values in A: $\bigcup_{n<\omega} A^n$, 71.

- $[X]^{\leq\kappa}$ – the family of all subsets of X of cardinality $\leq \kappa$, 72.

- $[X]^{<\kappa}$ – the family of all subsets of X of cardinality $< \kappa$, 72.

- $[X]^\kappa$ – the family of all subsets of X of cardinality κ, 72.

- $\mathcal{C}(\mathbb{R})$ – the family of all continuous functions $f \colon \mathbb{R} \to \mathbb{R}$, 73.

- $\mathrm{cf}(\alpha)$ – the cofinality of an ordinal number α, 74.

- $\mathrm{cl}_{\mathcal{F}}(Z)$ – closure of a set Z under the action of $\mathcal{F}$, 86.

- F' – limit points of $F \subset \mathbb{R}^n$, 92.

- $\sigma[\mathcal{F}]$ – the smallest σ-algebra containing $\mathcal{F}$, 94.

- $\mathcal{B}or$ – the σ-algebra of Borel sets, 94.

- Σ^0_α – subsets of $\mathcal{B}or$, 94.

- Π^0_α – subsets of $\mathcal{B}or$, 94.

- $\mathcal{ND}$ – the ideal of nowhere-dense sets, 98.

- $\mathcal{M}$ – the σ-ideal of meager (first-category) sets, 98.

- $\mathcal{N}$ – the σ-ideal of Lebesgue measure-zero (null) subsets of $\mathbb{R}^n$, 99.

References

Cantor, G. 1899. Correspondence between Cantor and Dedekind (in German). In Cantor (1932: 443–51).

Cantor, G. 1932. *Gesammelte Abhandlungen: Mathematischen und philosophischen Inhalts.* Berlin: Springer.

Ciesielski, K. 1995–6. Uniformly antisymmetric functions and K_5. *Real Anal. Exchange* **21**: 147–53.

Ciesielski, K. 1996. Sum and difference free partitions of vector spaces. *Colloq. Math.* **71**: 263–71.

Ciesielski, K., and Larson, L. 1993–4. Uniformly antisymmetric functions. *Real Anal. Exchange* **19**: 226–35.

Ciesielski, K., and Miller, A. W. 1994–5. Cardinal invariants concerning functions whose sum is almost continuous. *Real Anal. Exchange* **20**: 657–72.

Erdős, P. 1969. Problems and results in chromatic graph theory. In *Proof techniques in graph theory,* ed. F. Harary, 27–35. New York: Academic Press.

Erdős, P. 1978–9. Measure theoretic, combinatorial and number theoretic problems concerning point sets in Euclidean space. *Real Anal. Exchange* **4**: 113–38.

Erdős, P., and Kakutani, S. 1943. On non-denumerable graphs. *Bull Amer. Math. Soc.* **49**: 457–61.

Fraenkel, A. A. 1922. Zu den Grundlagen der Cantor-Zermeloschen Mengenlehre. *Math. Ann.* **86**: 230–7.

Frege, G. [1893] 1962. *Grundgesetze der Arithmetik.* Vol 1. Reprint. Hildesheim: Olms.

Freiling, C. 1986. Axioms of symmetry: throwing darts at the real number line. *J. Symb. Logic* **51**: 190–200.

Freiling, C. 1989–90. A converse to a theorem of Sierpiński on almost symmetric sets. *Real Anal. Exchange* **15**: 760–7.

Friedman, H. 1980. A consistent Fubini-Tonelli theorem for non-measurable functions. *Illinois J. Math.* **24**: 390–5.

Galvin, F. 1980. Chain conditions and products. *Fund. Math.* **108**: 33–48.

Kirchheim, B., and Natkaniec, T. 1990–1. On universally bad Darboux functions. *Real Anal. Exchange* **16**: 481–6.

Komjáth, P., and Shelah, S. 1993–4. On uniformly antisymmetric functions. *Real Anal. Exchange* **19**: 218–25.

Kunen, K. 1980. *Set Theory.* Amsterdam: North-Holland.

Levi, B. 1902. Intorno alla teoria degli aggregati. *Rendic.*, 2d ser. **35**: 863–8.

Levy, A. 1979. *Basic Set Theory.* New York: Springer-Verlag.

Mazurkiewicz, S. 1914. Sur un ensemble plan (in Polish). *Comptes Rendus Sci. et Lettres de Varsovie* **7**: 382–3. French translation reprinted in *Travaux de Topologie et ses Applications,* 46–7. Warsaw: Polish Scientific Publishers PWN, 1969.

Mirimanoff, D. 1917. Les antinomies de Russell et de Burali-Forti et le problème fondamental de la théorie des ensembles. *L'Enseignement Math.* **19**: 37–52.

Royden, H. L. 1988. *Real Analysis.* 3d ed. New York: Macmillan.

Russell, B. 1903. *The Principles of Mathematics.* Vol. 1. Cambridge: Cambridge University Press.

Sierpiński, W. 1919. Sur un théorèm équivalent à l'hypothèse du continu. *Bull Int. Acad. Sci. Cracovie* **A**: 1–3. Reprinted in *Oeuvres Choisies,* vol. 2, 272–4. Warsaw: Polish Scientific Publishers PWN, 1974.

Sierpiński, W. 1920. Sur les rapports entre l'existence des intégrales $\int_0^1 f(x,y)dx$, $\int_0^1 f(x,y)dy$ et $\int_0^1 dx \int_0^1 f(x,y)dy$. *Fund. Math.* **1**: 142–7. Reprinted in *Oeuvres Choisies,* vol. 2, 341–5. Warsaw: Polish Scientific Publishers PWN, 1974.

Sierpiński, W. 1936. Sur une fonction non measurable partout presque symetrique. *Acta Litt. Scient* (Szeged) **8**: 1–6. Reprinted in *Oeuvres Choisies*, vol. 3, 277–81. Warsaw: Polish Scientific Publishers PWN, 1974.

Skolem, D. 1922. Einige Bemerkungen zur axiomatischen Begründung der Mengenlehre. In *Proceedings of the 5th Scandinavian Mathematicians' Congress in Helsinki, 1922*, 217–32.

Steprāns, J. 1993. A very discontinuous Borel function. *J. Symbolic Logic* **58**: 1268–83.

von Neumann, J. 1925. Zur Einführung der transfiniten Zahlen. *Acta Sci. Math.* (Szeged) **1**: 199–208.

von Neumann, J. 1929. Eine Widerspruchsfreiheitsfrage in der axiomatischen Mengenlehre. *Crelle J.* **160**: 227.

Zermelo, E. 1904. Beweis, daß jede Menge wohlgeordnet werden kann. *Math. Ann.* **59**: 514–16.

Zermelo, E. 1908. Untersuchungen über die Grundlagen der Mengenlehre, I. *Math. Ann.* **65**: 261–81.

Index